Testing
Fluid
Power
Components

Testing Fluid Power Components

Robert A. Nasca

Industrial Press Inc.

For my wife, Chris

Library of Congress Cataloging-in-Publication Data

Nasca, Robert A.
    Testing fluid power components / Robert A. Nasca.—1st ed.
       p. 15.5 × 23.5 cm.
     ISBN 0-8311-3002-4
     1. Fluidic devices—Testing—Handbooks, manuals, etc.
  2. Hydraulic machinery—Testing—Handbooks, manuals, etc.
  3. Pneumatic machinery—Testing—Handbooks, manuals, etc.
  I. Title.
TJ853.N37   1990   90-4284
621.2—dc20   CIP

INDUSTRIAL PRESS INC.
200 Madison Avenue
New York, New York 10016-4078

First Edition
First Printing

TESTING FLUID POWER COMPONENTS

Composition By David E. Seham Associates, Metuchen, N.J.

2   4   6   8   7   5   3

Preface

Testing Fluid Power Components has been designed to help technicians and engineers understand and apply common test procedures for hydraulic and pneumatic components; this book can also serve as a valuable reference for test engineers currently involved in fluid power testing programs. Components covered include pumps and motors (open and closed loop), valves, actuators (linear and rotary), filters, and connectors. Each component's function and operating principle are described to provide an understanding of why certain tests are performed. Common tests for each type of component are listed in the first chapter and are discussed in detail in other chapters.

One of the most important factors to be considered when designing a test circuit is the proper selection of instrumentation. The accuracy of this instrumentation directly affects the test results, so caution must be exercised when deciding which device is to be used. A complete chapter is devoted to various types of instrumentation from pressure gages to highly accurate transducers. The function of each device is described as well as its application in a system. The types of instrumentation covered will allow monitoring of flow, pressure, torque, RPM, force, position, viscosity, temperature, voltage, and current.

Once the operations of both component and instrument are understood, these instruments can be combined to produce a complete test system. Schematics depicting various test systems are explained, in order to present the reader with an idea of the many test methods available that produce the same apparent test results. These procedures are described in detail with an emphasis on the relationship of component and instrumentation interaction. Very often, improper test results are attained when attention is focused on component performance while neglecting the component's interaction with other system components. Sometimes this interation can be deceiving, producing test results that may allow a faulty component to be diagnosed as good or a good component to be diagnosed as faulty.

After the system has been designed, a review of overall accuracy must be conducted to determine whether the system is accurate enough for the test results desired. Many different performance ratings are applied to fluid power components and instruments. The combined system inaccuracy buildup as well as a detailed discussion of the various component ratings are detailed in Chapter 4. In many cases, budget is the prohibiting factor in overall system accuracy. Therefore, there are times when accuracy versus cost must be reviewed to determine what is acceptable within the limitations of capital expenditure.

Finally, the decision to automate must be reviewed. Automation is generally applied to production test equipment, in order to free a tester from manually running tests and recording test results. Of course, automatic test equipment is more expensive than manual, so time studies are usually developed to determine what degree of automation is economically feasible, as well as practical, in the overall scheme of the production schedule. Methods of automatic testing are discussed, and several machines incorporating these principles are presented with a detailed description of their operation.

Contents

viii CONTENTS

Fluid Power Components

There are many fluid power components available that are applied to systems which must perform specific tasks. Manufacturers of these components publish data in their catalogs that detail the minimum and maximum operating characteristics of each component. Without accurate data, a component may be improperly applied, causing the overall system to perform outside its desired parameters.

This chapter details the more common components and ratings applied to these fluid power components. Certain specialized tests are often performed when a component must be applied in a situation that is outside its normal operating range. These tests can be ambient condition tests such as operation in a contaminated environment or in extreme heat or cold, or earthquake endurance. A test can also provide exact system simulation to determine if the component will perform properly to a specific set of parameters.

Many books have been published that describe components, their function, and how they operate. This chapter outlines the more common component ratings and describes what occurs within a component to warrant certain tests. See Chapter 3 for detailed information on exact test procedures and test methods.

1.1 Hydraulic Pump Ratings

There are several common pump designs that are used in today's hydraulic systems. Although the method of producing hydraulic power varies, the results are the same: a given amount of flow and pressure are produced by a given amount of input power.

E F F I C I E N C Y

The ratio of input power to the resultant output power is called the overall efficiency of a pump. Inherent in any pump design is internal leakage that contributes to component inefficiency. This leakage is called volumetric efficiency, which is the ratio of actual pump displacement to theoretical pump displacement. The ratio of overall efficiency to volumetric efficiency is called mechanical efficiency.

These values are determined by the following formulas:

$$\text{Overall efficiency} = \frac{\text{Output horsepower}}{\text{Input horsepower}} \times 100$$

or

$$\text{Overall efficiency} = \text{Volumetric Efficiency}$$

$$\times \text{ Mechanical efficiency} \div 100$$

$$\text{Volumetric efficiency} = \frac{\text{Actual flow rate (gpm)}}{\text{Theoretical flow rate (gpm)}} \times 100$$

$$\text{Mechanical efficiency} = \frac{\text{Theoretical input torque}}{\text{Actual input torque}} \times 100$$

O U T P U T   F L O W   A N D   P R E S S U R E

Output flow and pressure describe the main performance characteristics of a pump. Flow is generally expressed in gpm (gallons per minute) or in.^{3}/sec (cubic inches per second) and pressure is in psi (pounds force per square inch). The total amount of flow produced is based on the continual displacement of the pressurized chambers at the pump outlet. Pressure rating is based on the mechanical design of the flow-producing components and all internal pressurized surfaces. Generally a safety factor of 2–4 is applied to the pump. Specifically, a pump rated at 3000 psi operating pressure usually can withstand 6000–12,000 psi pressure peaks before reaching burst pressure.

Because a pump has several pressurized chambers working together, flow and pressure generally appear quite stable. This is usually true in most systems, but more sophisticated high-response systems are not always tolerant of pressure or flow ripple. The ripple is measured by the ratio of the number of pistons, vanes, or gear teeth pressurized during a specific time period versus pump rotating speed.

INPUT TORQUE OR HORSEPOWER

The amount of input power to a pump determines the amount of flow or pressure a pump can produce. The equation used to calculate the required input horsepower (HP) is

$$HP = \frac{Pressure\ (psi) \times Flow\ (gpm)}{1714 \times (Pump\ overall\ efficiency \div 100)}$$

To convert HP to torque, use

$$Torque\ (ft\text{-}lbf) = \frac{HP \times 5252}{RPM}$$

$$HP = \frac{Torque \times RPM}{5252}$$

As you can see, the exact overall efficiency of a pump is needed to equate the total input power required.

VISCOSITY

One of the key considerations in defining pump characteristics is the evaluation of the pump at the proper oil viscosity. Because viscosity changes the efficiency of a pump, other relationships such as horsepower are affected. When a pump is used with a very low viscosity fluid, efficiency may drop dramatically. This could easily lead to insufficient input horsepower. Pump manufacturers usually publish the minimum and maximum viscosity recommended for each pump series.

INLET PRESSURE

Another key factor in pump performance is the pressure required at the inlet. Insufficient pressure causes cavitation, while too much pressure may exceed the mechanical design of the pump. Open-loop pumps, which usually draw fluid directly from the reservoir, are limited by the suction pull capacity derived from their particular design. Again, viscosity plays an important role, because the more viscous the fluid, the harder it is for the pump to pull suction. Typically, open-loop pumps can pull 5–10 inches of mercury (in. Hg) or require up to 100 psia supercharge pressure. It is also important to note that these inlet pressures are expressed in psia (pounds per square inch absolute). The ambient air pressure plays an important part when a pump is used at higher elevations, especially when the pump is applied to a system running close to its performance limits.

Closed-loop pumps, used in hydrostatic transmissions, are usually supercharged by a make-up pump that replaces fluid lost due to pump inefficiency. Because the return of the system fluid is diverted directly to the pump inlet, a small displacement pump serves this purpose. Again, minimum and maximum inlet pressures are usually well defined.

OPERATING SPEED

Pump design dictates a minimum speed to overcome inefficiency and a maximum speed to prevent cavitation or mechanical fatigue. Again, these values are affected by viscosity and pump mechanical design.

TEMPERATURE

All components are rated at minimum and maximum temperature limits. Minimum temperature usually relates to fluid viscosity, which causes poor suction, while maximum temperature usually relates to pump volumetric efficiency, also due to viscosity. Both values can be overridden by the operating limits of internal seals.

MAXIMUM CASE DRAIN PRESSURE

Some piston-pump designs allow internal leakage to flow into the pump housing. Unless this flow is relieved at low pressure, the housing would be pressurized at the outlet pressure of the pump, which would crack the housing once pressure rises to the housing's mechanical design limit. Tests are performed to determine maximum case pressure of the housing with a generous safety factor.

Monitoring case drain flow of a pump can be useful in determining pump volumetric efficiency. Case drain temperature can be monitored to determine if excessive heat is generated from a faulty rotary group.

Other piston-pump designs allow case drain flow to return to the pump suction port. In this instance, case pressures are no longer a factor to be considered after the pump has been designed.

INTERNAL COMPONENT WEAR

Pump manufacturers run tests to determine pump life or efficiency loss over a period of time. These tests can be performed under ideal operating conditions or under a specific set of conditions such as low viscosity, high temperature, or controlled contamination. This information is valuable to determine how a pump may function under different conditions. During the test, usually at regular intervals, the pump is disassembled and surfaces are inspected for wear. Gear teeth, bearing plates, vanes, cam ring inside diameter (I.D.) piston/bore clearance, swashplates, shaft inside diameter, and bearing outside diameter (O.D.) clearances are all monitored.

BEARING RATING

Bearing rating is a mechanical test that takes drive shaft loading, drive speed, and forces created by pressure into consideration. Bearing life is an important consideration because a large percentage of pump failures are due to bearing wear. Aside from providing near frictionless support for rotating components, the concentricity of bearings helps prevent side loading of gear faces or vanes. Once a bearing begins to wear, efficiency drops and pump failure is not too far away.

One formula used to predict bearing life is:

B_{10} hours of bearing life

$$= \text{Rated life (hr)} \times \frac{\text{Rated speed (RPM)}}{\text{New speed (RPM)}}$$

$$\times \left(\frac{\text{Rated pressure (psi)}}{\text{New pressure (psi)}}\right)^3$$

Pump rotary groups can be designed in either a balanced or an unbalanced configuration. Unbalanced means that the higher the pressure at the outlet, the more side load imposed on the bearing. This type of pump requires heavier bearings and may not provide sufficient bearing life in certain applications.

Fluid lubricity plays a very important role when determining bearing life. Without a good film to protect rotating surfaces, wear increases quickly. In some instances where pumps are running at high temperatures, and where the fluid may not provide sufficient lubrication, a separate source of cooler fluid can be circulated through a pump case to increase bearing life.

1.1.1 Fixed-Displacement Pumps

There are several types of pumps commonly used in hydraulic systems. Fixed-displacement pumps provide flow proportional to drive speed and displacement.

GEAR PUMPS

Figure 1.1 depicts the inner components of a gear pump. As you can see the drive shaft (3) is part of the drive gear, which rotates the driven gear. Pump volumetric efficiency is based on the clearance between the gear faces and bearing blocks. Some pump designs have fixed bearing blocks that decrease pump efficiency due to wear caused by the gears against the bearing surface. Other designs include a hydraulically pressurized bearing plate that maintains a fixed clearance between the gears and bearing surface.

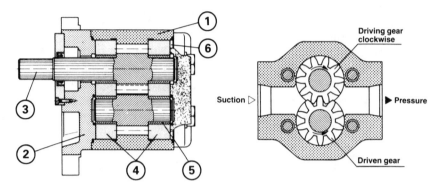

Fig. 1.1. Gear-pump cross section. *(Courtesy Rexroth Corp.)*

You can imagine what happens to efficiency if a pump shaft is side-loaded during use, such as in a belt-driven system. As the bearing bushings (5) wear, the gears are forced into the bearing blocks. The metal-to-metal nonparallel contact eventually scores the components until output flow and pressure decrease or mechanical failure results.

Gear pumps generally have good suction characteristics and can produce maximum pressures in the range of 2500–3500 psi. Gear pumps produce the least efficient output in the common hydraulic-pump category. Typically, overall efficiency runs about 85%, provided the pump is used in a system with the proper viscosity characteristics.

VANE PUMPS

Figure 1.2 depicts the construction of a vane pump. Unlike the gear-pump design, the clearance between the vane and the I.D. of the cam ring is controlled by pressurization of the vanes. Rather

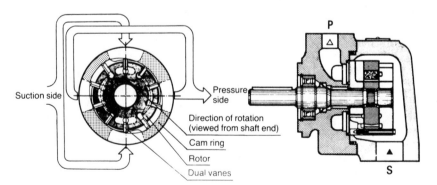

Fig. 1.2. Vane-pump cross section. (The direction of rotation, viewed from the shaft end is counterclockwise.) *(Courtesy Rexroth Corp.)*

than relying on a fixed clearance, the pressurized vanes constantly adjust for wear between the rotating components. Again, the vanes can allow internal leakage across the side bearing surfaces, but because of the controlled O.D. and I.D. clearances, efficiency is greater than gear pumps.

The design shown is more tolerant to side loading of the shaft, because of the main shaft bearing and because only one rotating component is required as opposed to two intermeshing gears.

PISTON PUMPS

There are several variations of piston-pump designs, including axial piston, bent-axis, radial piston, and ball-check piston. Figure 1.3 depicts a bent-axis pump. You can see the larger drive shaft bearings that are required for higher loads owing to increased flow and pressure capabilities. Several pistons are driven back and forth within the cylinder block to pull suction or push fluid out. Since the piston I.D. and bore O.D. clearance is extremely tight and this is the only place for internal leakage to occur, efficiency is very good. Volumetric efficiency of piston pumps run as high as 97–98% under ideal operating conditions. Output pressures can be as high as 10,000 psi.

The only drawback in this design is the increased number of moving components. Tighter tolerances mean that finer filtration is required, and more surfaces can cause failure due to lack of lubricity or viscosity.

Some piston pumps can pull suction, some require a flooded suction, and others require a full flow supercharge. In closed-loop design, all pumps require full-flow-supercharge pressure.

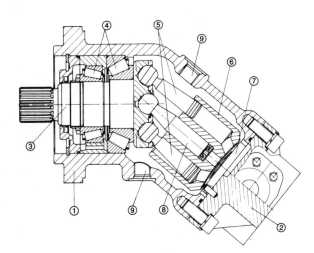

Fig. 1.3. Bent-axis piston-pump cross section. *(Courtesy Rexroth Corp.)*

1.1.2 Variable-Displacement Pumps

We have discussed pump designs with fixed outlet chambers. However, both vane and piston pumps are available with variable outlet chambers to provide controllable flow without creating excessive input power or heat generation during low-demand system cycles. Again, maximum outlet flow is dependent on drive speed, but infinite control below maximum is possible.

VARIABLE-VOLUME VANE PUMPS

In order to vary vane pump displacement the cam ring (4) (Fig. 1.4) moves from an eccentric position at full flow, to a nearly concentric position at compensator setting. The compensator depicted in Fig. 1.4 is a spring (10) that is overcome at compensator pressure. This sliding cam ring creates another set of wear surfaces that add to internal leakage.

VARIABLE-VOLUME PISTON PUMPS

The most efficient pump available with the most control options is the variable-volume piston pump. Figure 1.5 depicts a bent-axis design that is equipped with minimum and maximum volume adjustments. The pump is shown at full stroke, which allows the pistons to reciprocate within the cylinder housing to their full capacity. As the compensator setting is approached, the control piston forces the rotary group toward the minimum volume stop, thus shortening the stroke length of the pistons.

Some pump designs produce full flow at startup and de-stroke upon reaching compensator pressure. Other pump designs start at zero stroke and require a minimum pressure to be seen at the com-

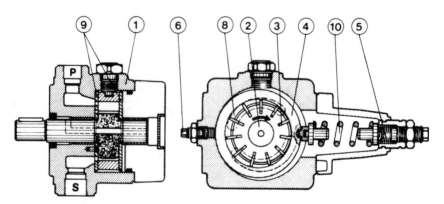

Fig. 1.4. Compensator design cross section. *(Courtesy Rexroth Corp.)*

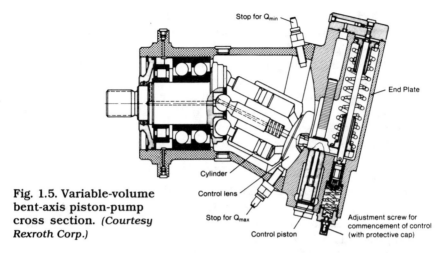

Fig. 1.5. Variable-volume bent-axis piston-pump cross section. *(Courtesy Rexroth Corp.)*

pensator before providing full flow. While this is unimportant in most applications, there are times when this must be considered when applying a pump to a response-time sensitive system.

1.1.3 Pump-Displacement Controllers

Variable-volume pumps have the unique capability of producing the proper amount of either flow or pressure based on system demand. While the methods of controlling displacement may vary, the result is always the same. The pump swashplate or rotating group is adjusted to a position anywhere from full stroke to zero stroke by the control mechanism. In Fig. 1.6, illustrating a vane pump, the

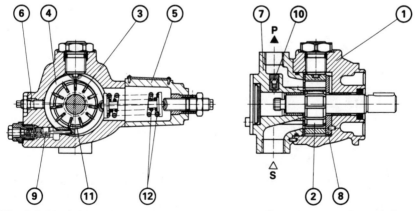

Fig. 1.6. Variable-volume vane-pump cross section. *(Courtesy Rexroth Corp.)*

control mechanism is a spring (12), which limits the flexibility of control options. When a control piston is used as shown in the piston-pump design, many options are available, as the piston responds to external signals generated mechanically, hydraulically, or electronically.

Minimum and maximum control pressure is one common test required. Pumps used in open-loop circuits operating at low pressures in idle may not generate sufficient pressure to control the pump. The outcome is either erratic or total loss of pump control.

PRESSURE COMPENSATION

Pressure compensation is the most common control used in variable-volume hydraulic systems. Figure 1.7 depicts a typical control valve, which transmits a hydraulic signal to the control spool in a piston pump. The spool is basically a directional control valve. Upon a signal from the pressure sensing circuit, the spool shifts, which in turn de-strokes the pump to its displacement setting, which maintains compensator set pressure.

A common test applied to this design is response time from zero stroke to full stroke and full stroke to zero stroke. This response time is usually affected by viscosity owing to control orifices in the valve assembly, so viscosity versus response time is also a consideration.

Valve hysteresis is also tested to determine the actual resultant pressure versus the crack pressure of the pressure-sensing valve. This pressure variation is identical to relief-valve testing, where crack versus full flow setting and reseat pressure are tested for. For more detailed information, see relief valve testing and spool leakage testing (Chapter 3).

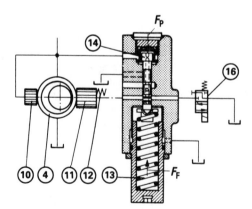

Fig. 1.7. Pump-displacement control valve. (*Courtesy Rexroth Corp.*)

HORSEPOWER LIMITING

Horsepower limiting control senses both system flow and pressure to prevent the pump from trying to produce more output power than input power available. In this situation, the control piston provides an infinitely variable displacement control. This control is used in systems with either limited input power, or with pressure and flow combinations that do not require maximum output of either variable simultaneously.

Again, response time, hysteresis, and spool leakage are tested.

LOAD SENSING

Load sensing control varies pump displacement based on pressure signals between output pressure and load pressure. The pump strokes to maintain a fixed differential between the two.

Tests remain as discussed—response time, hysteresis, and spool leakage.

MANUAL DISPLACEMENT

Manual displacement is commonly used in hydrostatic drive systems. Swashplate angle is controlled externally from a lever or cable or, in some instances, an external remote control device. Testing generally consists of output displacement versus control position, which is a linear relationship with efficiency affecting total linearity.

Response time and hysteresis testing are no longer required, with the exception of fluid compressibility response time. This applies to direct-actuated control pistons. Where the control lever operates a pilot source, response time and hysteresis testing is required.

REMOTE HYDRAULIC CONTROL

Remote hydraulic control is used when a remote pressure signal is available to control either flow or pressure. The remote device response has, of course, a definite effect on pump response. When testing a component of this type, external response times are generally neglected to avoid confusion. This holds true for component testing only. When testing a component within a system, all factors must be considered.

A remote pressure control is what strokes the control piston. In controls of this type, a spring with a greater force is used to provide a wide pressure control range versus pump stroke. For pump on–off controls, such as pressure compensators, a lighter spring is used on the control piston. As pressure is applied, spring rate affects the crack pressure versus full flow pressure. In remote hydraulic control pumps, the higher the spring rate, the better the control unless differential-pressure control piston designs are used. In this case, the spring force is replaced by a controlled hydraulic return force.

REMOTE ELECTRIC CONTROL

Remote electric control offers many control possibilities to the pump designer. Simple manual input from a vehicle operator to microprocessor control of variable-volume pumps are now used. Remote electric control sometimes sounds more complicated than it really is. The important thing to remember is that the pump is no different than other designs already discussed. Again, the control piston is either stroked with hydraulic pressure against a spring or against a differential pressure signal. The only difference is that an electric signal is sent to either a proportional valve or servovalve, which in turn operates the control piston.

This pump design is tested to determine electrical input versus displacement, control valve leakage, and response time. Repeatability, linearity, and hysteresis are generally functions of the electronic control design. Open-loop electronic signals generated from potentiometers, programmable controllers, or microprocessors can provide very stable control, but where greater accuracy is required, a closed-loop feedback signal can be used. This signal controls the pump based on required displacement versus actual displacement.

1.1.4 Other Pump Types

The most common pumps used in hydraulic systems have been discussed, but there are other types available that commonly are used in specific applications. These designs are different from gear-, vane-, and piston-pump designs and because of this, deserve a brief description.

AIR-OPERATED PUMPS

An air-operated pump is depicted in Fig. 1.8; this pump provides a hydraulic or pneumatic output pressure proportional to an input air pressure. The ratio of these pressures is based on the surface area of the air input piston versus the surface area of the output piston. Actual versus theoretical output pressure is determined by this calculated ratio minus any mechanical inefficiency of the unit caused by internal component friction or leakage.

The output flow is determined not only by piston displacement, but by piston leakage and compressibility factors for hydraulic or pneumatic output.

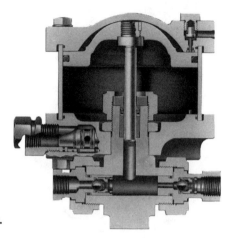

Fig. 1.8. Air-operated pump.
(Courtesy Haskel, Inc.)

These devices are usually configured with one or two drive pistons, depending on the rated capacity of the unit. Pressure output can be as high as 100,000 psi for certain models, while normal maximum pressures are in the 40,000–50,000 psi range.

Testing an air-operated pump can consist of input air pressure versus output pressure (in static- and high-flow conditions), minimum and maximum operating pressure and flow, temperature range, viscosity range, suction characteristics, and internal and external leakage of component parts.

HAND-OPERATED PUMPS

Hand-operated pumps can be single-piston low-volume high-pressure pumps for pressures in excess of 100,000 psi. Manual pumps can also be of a wide variety of designs used for fluid transfer, lubrication, or any number of applications. Generally, the most important test is output flow and pressure versus input power. The hand pump shown in Fig. 1.9 delivers a fixed volume of fluid for every full stroke on the handle. As pressure output increases, so does the input force required.

Tests applied to this type of pump include displacement (in.^{3}) per full stroke, pressure output (maximum), suction characteristics, handle force versus pressure, internal and external leakage, viscosity range, and temperature range.

CENTRIFUGAL PUMPS

Centrifugal pump design is based on the principle that as an impeller turns, it imparts velocity on a stream of fluid, thus providing

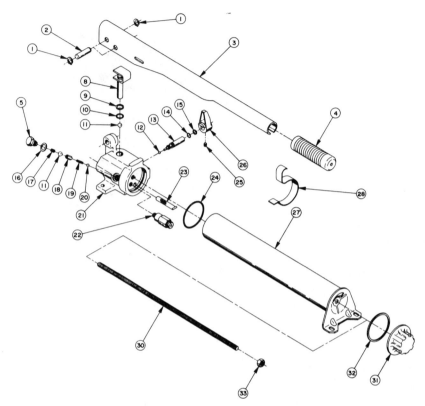

Fig. 1.9. Hand-operated pump. *(Courtesy OTC Division, Sealed Power Corp.)*

flow. As shown in Fig. 1.10, you can see the large clearance between the impeller and flow passage in the pump housing. This design can provide very high flows, while outlet pressure is low and suction capability is poor. This pump can be applied successfully in low-pressure applications, such as supercharging a high-pressure pump or for any high-volume-flow requirement.

Because the design is not a positive displacement type, efficiency is quite different than as previously discussed. As pressure increases, flow drops drastically to the point of no flow at all. The impeller will just rotate freely, supplying pressure, but no flow. A very important consideration when applying this type of pump is fluid viscosity or specific gravity; pump efficiency can be affected tremendously by these variables.

Tests performed on this pump are flow versus outlet pressure, maximum–minimum suction, horsepower across all flow and pressure points, drive speed, and viscosity or specific gravity effects on performance.

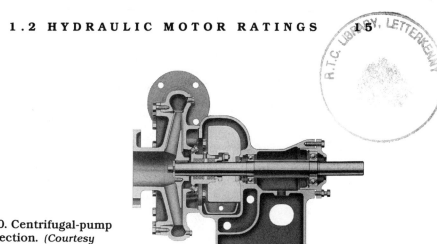

Fig. 1.10. Centrifugal-pump cross section. *(Courtesy Goulds Pumps, Inc.)*

1.2 Hydraulic Motor Ratings

Hydraulic motors are nearly identical in design to pumps. Rather than drawing fluid in and pressurizing it to do work, the motor takes the pressurized fluid in to create rotary motion and then pushes the used fluid out at low pressure. The same basic designs are used, mainly gear, vane, and several piston designs.

TORQUE VERSUS PRESSURE

One critical rating is the ratio of output torque (ft-lbf or in.-lbf) produced versus the input pressure. In a motor's normal speed range, this is quite constant, with the exception of reduced torque at higher pressures due to increased internal leakage (efficiency).

Torque is a measure of output power based on the relationship of displacement, speed, and input pressure. The following equations are commonly used:

$$\text{Torque (in.-lbf)} = \frac{\text{Pressure (psi)} \times \text{Displacement (in.}^3\text{/rev)}}{2\pi}$$

$$\text{Torque (in.-lbf)} = \frac{\text{Horsepower} \times 63025}{\text{RPM}}$$

$$\text{Torque (in.-lbf)} = \frac{\text{Flow (gpm)} \times \text{Pressure (psi)} \times 36.77}{\text{RPM}}$$

$$\frac{\text{Torque (in.-lbf)}}{100\,\text{psi}} = \frac{\text{Displacement (in.}^3\text{/rev)}}{0.0628}$$

$$\text{Horsepower} = \frac{\text{Torque output (in.-lbf)} \times \text{RPM}}{63025}$$

OUTLET PRESSURE

Motor outlet pressure directly affects output torque. To determine the actual torque a motor generates, the outlet pressure must be subtracted from the inlet pressure before calculating input pressure versus torque. In open-loop systems, return-line pressure may vary, depending on system cycles, so a worst case situation must always be accounted for. If 150 psi is the maximum return pressure seen, and motor inlet pressure is 3000 psi, the actual pressure effect at the motor inlet is 2850 psi. In closed-loop systems, return pressure is constant, usually about 150–300 psi, based on the hydrostatic-system control design. Again, this constant motor outlet pressure must be accounted for.

Some motors are rated for either single-direction rotation or bi-directional operation. Single-directional units may not be designed to withstand full system pressure, so maximum outlet port pressure ratings are applied. Otherwise, full system pressure may be applied to either motor port.

FLOW VERSUS SPEED

Motor speed is directly proportional to input flow (gpm) at the motor inlet, with the exception of the efficiency effect. In fixed-displacement motor designs, speed is controlled by an outside source, such as a flow control or stroke adjustment of a variable-volume pump. Therefore, the theoretical formula for predicting output speed is

$$\text{Speed (rev/min)} = \frac{231 \times \text{Flow (gpm)}}{\text{Displacement (in.}^3\text{/rev)}}$$

Actual output speed is again affected by efficiency, mechanical design, bearing rating, and viscosity.

At low minimum and maximum speed ranges, most motors have increased leakage that causes erratic rotation of the output shaft. This is similar to pump output ripple, as discussed earlier. The internal pressurized surfaces create an erratic flow within the pump, which causes this effect. Standard motors generally are rated with a minimum speed to overcome the effect. On the other hand, a group of motors designated "high-torque, low-speed motors" provide smooth rotation at low RPM but have limited high-speed rating. Motors are tested to determine these upper and lower limits to overcome erratic rotation at the low end or excessive upper limit speed that exceeds the mechanical design of the motor. Again, several factors come into effect: viscosity, temperature, bearing rating, and mechanical design of the motor.

EFFICIENCY

Internal leakage is the most critical component in predicting actual motor output—both speed and torque. The critical factors in determining efficiency are the fluid viscosity characteristics and the mechanical design limits of the motor. Volumetric efficiency is the ratio of input flow versus used flow to produce rotation and is expressed as

$$\text{Volumetric efficiency} = \frac{\text{Used flow (gpm)}}{\text{Input flow (gpm)}} \times 100$$

"Used flow" relates to theoretical speed versus actual speed. This is calculated by the ratio of theoretical displacement at speed versus actual speed output.

BEARING RATING

Hydraulic motors are used for a wide variety of tasks, some of which can impose excessive side loading to the outer bearings. Not only do motor designers need to be concerned about internal forces created by imbalance or rotational forces, but external forces must also be considered. Some motor designs that are commonly used in rugged applications are provided with outer bearings which far exceed normal load ratings. Other designs do not, but are easily adapted to overhung load adapters to compensate for side loads.

TEMPERATURE

Owing to seal limitations or viscosity characteristics, a motor (as all other hydraulic components) is rated at minimum and maximum operating temperature. Close tolerances of rotating components can also be affected by expansion or contraction due to fluctuating temperatures. Usually dissimilar metals are avoided to help negate this potential problem.

VISCOSITY

Again, viscosity plays an important role when motor performance is analyzed. Excessive viscosity creates higher return-line pressures, which decrease torque output. Insufficient viscosity causes excessive leakage and potential lubricity problems, which can cause premature motor failure.

CASE DRAIN PRESSURE

Open-loop piston motors occasionally have case drains, unless case pressure is diverted to the low-pressure outlet. Some motors do not drain leakage, but allow this leakage to blow by the rotating components, seeking the lowest-pressure path. Pumps with case drains are rated at maximum and minimum pressure to prevent housing failure and provide proper operating parameters.

INTERNAL COMPONENT WEAR

Motor wear is analyzed by disassembling the test motor after certain tests are performed. These tests can include shock cycling, contaminate ingression, excessive temperature, or just life testing at normal operating conditions.

The internal wear surfaces such as pistons and piston bore, piston shoes, and bearing plate and bearings are all inspected to find increase in tolerance clearances or any uneven wearing due to side loads or contamination.

This test helps design engineers determine proper material, material finish, and clearance dimensions for all dynamic components within the motor.

1.2.1 Fixed-Displacement Motors

This motor design, as discussed, rotates at whatever speed inlet flow, minus efficiency, allows. Also, it provides whatever torque inlet pressure, minus outlet pressure, allows.

GEAR MOTORS

Figure 1.11 depicts a fixed-displacement gear motor. As you can see, the figure is nearly identical in design to a gear pump; the only common exception is the design of internal porting for timing pur-

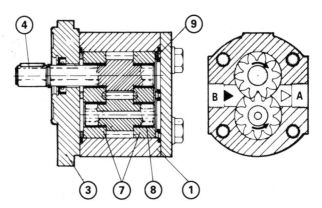

Fig. 1.11. Gear-motor cross section. *(Courtesy Rexroth Corp.)*

poses. Timing the pressurized fluid to enter the rotating group is one of the most difficult tasks when designing a gear motor. Improper timing can cause noise, inefficiency, cavitation, or erratic torque output. Gear-motor designs consist of several variations: gear-on-gear, crescent-gear, gerotor, differential gear, and roller gerotors. The main differences produced by various designs include minimum and maximum speeds, efficiency, and pressure rating. (Refer to the gear-pump discussion for a detailed description of operational ratings other than those described for fixed-displacement motors.)

VANE MOTORS

Again, the design of a vane motor is identical to a vane pump except for timing and vane pressurization. Unlike a pump that requires rotation prior to pressurization of the vanes, pressure is immediately available at the inlet port of the motor. Vane-motor designs include standard vane, axial vane, and rotary-abutment rotor. (Refer to the vane-pump discussion for a detailed description of ratings other than those described for fixed-displacement motors.)

PISTON MOTORS

The most efficient hydraulic motor is the piston design, owing to the extremely tight clearances possible between the pistons and cylinder block bore. Higher torques and speeds are possible than with gear and vane designs. An example of higher-speed units is shown in Fig. 1.12; this is a bent-axis design similar to the bent-axis pumps described earlier. Low-speed motors are supplied in several

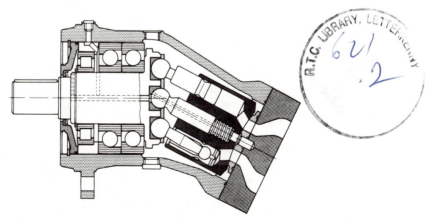

Fig. 1.12. Bent-axis piston-motor cross section. *(Courtesy Rexroth Corp.)*

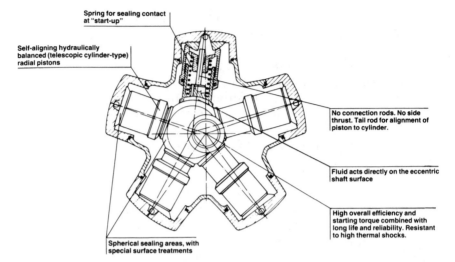

Spring for sealing contact
at "start-up"

Self-aligning hydraulically
balanced (telescopic cylinder-type)
radial pistons

No connection rods. No side
thrust. Tail rod for alignment of
piston to cylinder.

Fluid acts directly on the eccentric
shaft surface

High overall efficiency and
starting torque combined with
long life and reliability. Resistant
to high thermal shocks.

Spherical sealing areas, with
special surface treatments

Fig. 1.13. Radial-piston-motor cross section. *(Courtesy Rexroth Corp.)*

design versions. One of the more common designs used for higher torques is the radial piston depicted in Fig. 1.13. Fluid directed at the top of the piston assembly transmits a force against an eccentric shaft surface, turning the output shaft. Large piston surface areas and exceptionally low internal leakage allow this design to produce very high torques with speeds less than 1 RPM.

1.2.2 Variable-Displacement Motors

Variable-displacement motors are similar in design to variable-displacement pumps. This feature allows system advantages not possible with fixed motors. Higher operating speeds are possible with reduced torque. For example, a fixed-volume pump supplying a variable motor at full displacement produces the largest amount of torque. By decreasing motor displacement, speed increases, allowing higher operating speeds than could be realized with a fixed motor. This advantage is a nice feature when higher speeds do not require full torque, since torque decreases with decreased displacement.

Design of these motors is similar to the piston pumps described earlier. Swashplate angle or rotary group angle is varied mechanically, hydraulically, or electrically to accomplish displacement change. Testing for these controls is necessary to determine maximum and minimum input force, pressure, or voltage in relationship to actual displacement. In more critical applications it is necessary to determine response time on and off stroke. This response time again is usually affected by fluid viscosity, operating speed, and system pressure.

1.3 Valve Ratings

Proper control of a system is based on the consistent operation of all valves within the system. Without consistent valve operation, the system becomes unpredictable. While this may not be applicable to all systems, it is necessary that valve performance be known to a system designer. Valves are used to control fluid direction, pressure, and flow.

MINIMUM AND MAXIMUM FLOW

This rating is performed to determine two important factors. The first being the minimum flow required for operation, which is usually required to find the point where internal leakage is overcome as well as any mechanical flow requirements. For example, many components are pilot-operated and can require as much as several gallons per minute of flow to operate, such as large pilot-operated four-way valves. This is a case where valve response time is affected by insufficient flow. Minimum flow can also be several cubic centimeters per minute of flow, required to supply makeup for inefficiency due to spool leakage or bleed orifices. A good example is the leakage within a spool valve. The clearance between the spool and sleeve is very small, but much flow is lost from pressurized fluid escaping across the lands and returning to the reservoir. Until these minimum flow requirements are met, the valve may function erratically or may not function at all.

Maximum flow ratings generally determine at what point a valve will fail due to flow forces or at what point pressure drop becomes too excessive for most applications. In pressure control valves, excessive flow will cause erratic operation or pressure drop can exceed the minimum pressure set point of the valve. Spool valves are designed to shear the flow path during a power shift from a solenoid operator or to return the spool to a normal position with spring force. At some flow point, the flow forces may exceed the force of an operator or return spring. This maximum flow is usually published by valve manufacturers. However, the flow rate is based on tests using a particular fluid. A review of the actual fluid used as well as the system flow cycle is required before applying a valve. A common problem overlooked is return flow from a cylinder. While a valve may be sized to operate at pump output, cylinder return flow can double or triple actual return-line flow due to the internal volume ratio of the cylinder.

MINIMUM AND MAXIMUM PRESSURE

This rating denotes at what minimum pressure a valve will operate smoothly or control pressure and at what maximum pressure the valve is mechanically designed to withstand or control with some degree of safety margin. Minimum pressure in pressure or flow control valves generally applies to the minimum pressure that is required to balance the mechanical design of the valve. This balance is generally the relationship of the pressurized area versus the spring force. In order to provide stable valve operation, a spring force of some value is required to eliminate hysteresis or premature opening at the low end of the scale. Spool valves, however, usually require a minimum pressure in order to shift. In the smaller sizes that are directly actuated by a solenoid or are manually operated, this usually does not apply. In the medium to large valves, however, the spool is shifted by pilot pressure. This pressure is the minimum pressure required to overcome the flow forces seen during the maximum rating of the valve.

Maximum pressure rating of a valve generally applies to the maximum safe operating limit. Safety margins of 2:1, 3:1, and 4:1 are used and so the terms proof pressure and burst pressure are also used. Proof pressure tests determine the maximum pressure a component will withstand over the recommended operating pressure with no damage or degradation of performance. Burst pressure tests determine the point at which an internal or external component breaks. This test can be extremely dangerous and should only be conducted in a test chamber capable of stopping the equivalent of an explosion. Operator safety is the primary concern during this test.

FLOW VERSUS PRESSURE DROP

This rating provides data necessary to select a component that will fulfill the required function but not induce excessive pressure drop, which reduces system output force. The curve generated is taken at various points along the flow rating of the valve. Pressure drop is recorded at each point to provide this relationship. It is important to note the fluid used in the evaluation by the valve manufacturer when reviewing published data. Fluid viscosity has a direct relation to this rating. The higher the viscosity, the higher the pressure drop. This is especially important in mobile equipment owing to cold-weather conditions. Components have burst during cold-weather startup due to pressure drop reaching burst pressure because of viscosity increase.

In certain systems, pressure drop cannot be tolerated owing to the maximum output of the pump or the efficiency required. Actuators may require the full flow of the pump at full pressure. A reduction of pressure to the actuator owing to loss through a component can be avoided by oversizing the component. This oversizing, of course, also increases system cost. Pressure drop also generates heat and reduces system efficiency. A stackup of components improperly sized can result in a system that does not have the power output desired, or a system that runs hot due to wasted energy.

RESPONSE TIME

This rating predicts the speed at which a valve will respond to a given situation. In some systems, this is not critical; in other systems, it is mandatory. A relief valve with slow response may not open fast enough to vent a shot of oil caused by an actuator that has stopped suddenly. During this response period, if pump flow is high enough, system fluid can be compressed by the pump to the burst pressure of a component, thus causing a potentially dangerous condition. It is important to remember that a pump is dumb and will put out as much pressure as the system will accept. Unless some valve limits pressure, rupture will occur, or the system will stall out due to insufficient input power.

Response time is usually rated in milliseconds at a certain viscosity, pressure, and flow. Because of this speed, test equipment can become expensive and sometimes mechanically impossible. Response time testing can only be as accurate as the test equipment accuracy.

CYCLE TESTS

Many systems, such as those used in high-production assembly lines, operate continuously. Valve failure can cause costly downtime. In order to predict the cycle life of a component, it can be tested under specific parameters to duplicate those parameters encountered in the field. In some cases, a manufacturer will perform general endurance tests just to gather data on internal component wear. Solenoid valves can be cycled to measure spool/sleeve wear or solenoid life. Pressure or flow control valves can be cycled to determine deterioration of pressure or flow setting over a given period of time.

Unfortunately this test is one of the most difficult to duplicate when specific conditions exist. A valve can be tested at a given flow or pressure, but many outside influences can greatly vary the resulting data. These outside factors can be fluid lubricity, viscosity, viscosity breakdown, oil temperature, ambient temperature, contamination levels, operating voltage and current, or excessive mechanical force among dozens of other potential unknowns. A very common question manufacturers face is, "How many cycles will your component last?" The answer, if an answer is possible, is at best an educated guess unless tests exactly duplicate field conditions.

INTERNAL LEAKAGE

Most valves used within a system have internal leakage. Leakage varies depending on internal valve design. Some valves are rated at zero leakage or bubble tight, while others, such as spool valves, can leak several cubic inches per minute. Leakage rating can be critical in low-flow systems as well as circuits that require load-holding capability. Whatever the reason, a system should be evaluated as to the effects combined component leakage can cause. One critical circuit, which will be discussed in more detail in Chapter 3, is the leak test circuit. If a test engineer has to contend with component leakage during this type of test, the test results would be useless. Another critical circuit would be a personnel lift. In a neutral-circuit condition, certain components, such as pilot-operated check valves, are used to hold the platform in position while the power supply is off or in a bypass condition. Leakage through the check can cause the platform to descend or, at worst, cause catastrophic failure.

In any event, component leakage is defined as cubic centimeters or cubic inches per minute at a certain fluid viscosity. Combined valve leakage in a low-flow circuit could result in a slow or inefficient system. This rating, while not always critical, should be reviewed when selecting a component in a critical situation.

VISCOSITY

Again, it is important to point out that most parameters associated with a valve's performance are related to fluid viscosity. Flow ratings, pressure drop, leakage, and response time are all affected by the fluid used. Careful evaluation of a manufacturer's ratings versus your specific application are required to determine if a component is suited for the application.

TEMPERATURE

The affect of temperature on a component, whether it be internal or external, is an important consideration. A valve is usually rated at minimum and maximum temperature limits. This can be due to internal conditions or external ambient conditions; in general, this rating usually relates to fluid viscosity or seal design capability.

1.3.1 Directional-Control Valves

There are several valve designs which fall into this directional-control category. These valves are used to either shut off flow or divert flow to accomplish a specific function. Several methods of operating these valves is available, but the result is always the same. Valve operators include AC and DC solenoids, rotary actuators (electric, pneumatic, and hydraulic), pilot-operated (pneumatic and hydraulic), cam-roller, manual lever, among many others.

In the solenoid valve type, tests are conducted to determine minimum and maximum voltage of the coil as well as its output force, temperature limits, cycle life, and resistance.

Rotary actuators, which are typically used for remote control of ball and butterfly valves, are tested for output torque. The electric version has a torque output based on the geared-reduction ratio of the electric torque motor minus any mechanical inefficiency. The pneumatic actuator is tested for input air pressure versus resultant torque. The hydraulic actuator is tested for input hydraulic pressure versus resultant torque. (A complete description of rotary actuator tests follows in this chapter.) Another consideration is actuator response time, which translates to valve response. This response time is important when selecting an appropriate valve actuator for a system.

Pilot-operated valve actuators are tested for minimum and maximum pressure as well as shift response and vent response. A pilot piston may be slower venting than shifting, owing to orifices that limit flow into the pilot piston chamber. These orifices prevent the valve from shifting too fast and causing internal component damage. The orifice is sized for a certain maximum flow at the highest potential pressure. If two pilot operators are used for a three-position configuration, the vented-side piston chamber must vent oil out the same orifice in which the oil entered. Because it is now at a lower pressure, flow through the orifice is lower, thus causing slower response. In a single pilot-operated valve with spring-return, the response can be even slower because the spring must force oil back through the orifice on the opposite side. Spring force is generally much less than force available from a pilot piston.

Cam-roller and manually operated valves are tested for actuator mechanical force required to shift the valve. This force can vary at different points along the spool travel owing to flow forces and mechanical design of the valve. These operators can also be used for simple on–off conditions, as well as for throttling a valve to control output flow and pressure. In the latter case, the spool generally has

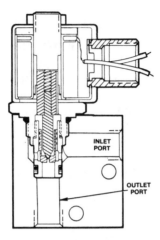

INLET
PORT

OUTLET
PORT

Fig. 1.14. Two-way normally closed poppet-solenoid-valve cross section. *(Courtesy Fluid Controls, Inc.)*

notches or grooves to improve the linear relationship of flow or pressure versus spool position.

TWO-WAY VALVES

Figure 1.14 depicts a two-way, normally closed, poppet-style solenoid valve. It is designed to control on–off conditions in a circuit with flow in one direction only. When the solenoid is energized, the plunger lifts to open a small vent orifice, venting pressure in the spring chamber of the poppet. This creates a hydraulic force that lifts the poppet to allow flow to exit through the poppet seat.

This type of valve design is tested for minimum and maximum flow and pressure, pressure drop, and seat leakage. Poppet valves generally are rated at zero leakage to several drops per minute leakage, so seat leakage tests can be difficult to perform if not set up properly.

THREE-WAY VALVES

Figure 1.15 depicts a single-solenoid, three-way spool valve. This design is used to divert flow between two ports, or it can be used as an on–off valve with the load vented in one position. Depending on the valve design, all three ports may be rated at full system pressure, or one port may have to be connected to return.

Spool timing can sometimes be important when a three-way valve uses two seats for low-internal-leakage characteristics. This is an important fact required when testing leakage of this design. Often, it may appear that the valve test is acceptable when test procedures are set up neglecting this problem.

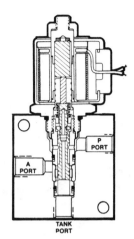

Fig. 1.15. Three-way spool-
solenoid-valve cross section.
(Courtesy Fluid Controls, Inc.)

FOUR-WAY VALVES

Figure 1.16 depicts a basic detented, manually operated, three-position, four-way valve with all ports blocked in neutral. You can see the round notches on the spool lands which can be used for decompression or for limited throttling. Also, because the spool O.D. is consistent, pressure forces on the lands are negligible. Flow forces and the return springs are the main factors in determining the spool force required for valve shifting.

This figure also shows the most common internal porting configuration used. The pressure inlet enters at the center of the spool, with the two work ports (A and B) on either side. At each end of the spool is the tank connection and the operator drain cavity. In this case, return-line pressure can be higher than some designs that combine the return passages with the operator drain cavity.

In Fig. 1.17 you can see the operation of a small direct-actuated solenoid valve. As shown in the manual valve, it too is a four-way, three-position valve with all ports blocked in neutral. To show the difference between AC and DC operators, one of each type is shown in this cross section. Also, note that the return-line passages and the operator cavities are connected. This means that return-line pressure must be kept low. Solenoid-operated tube assemblies rarely can withstand more than minimal return-line pressure.

In Fig. 1.18 you see a pilot-operated, four-way, three-position solenoid valve with all the ports blocked in neutral. This configuration is used in medium to large spool valves, because the solenoid force of a direct-acting valve is insufficient to overcome the forces encountered at higher flows.

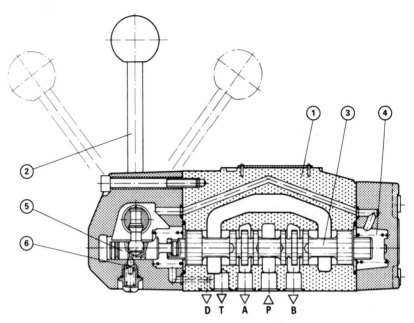

Fig. 1.16. Detented, manually operated, three-position, four-way directional valve. *(Courtesy Rexroth Corp.)*

Basically, the pilot valve is a direct-acting solenoid valve with *A* and *B* ports vented to tank in neutral. The main high-flow valve is a four-way, three-position, hydraulically piloted valve with all ports blocked in neutral. The main spring chambers are vented in neutral because the pilot valve is ported this way. Return-line pressure is balanced on both sides of the spool and the opposing spring forces hold the spool steady in all conditions encountered in neutral.

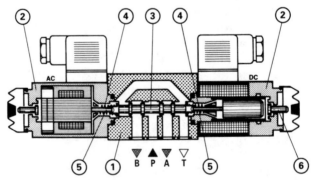

Fig. 1.17. Direct-operated four-way solenoid valve with AC or DC solenoids. *(Courtesy Rexroth Corp.)*

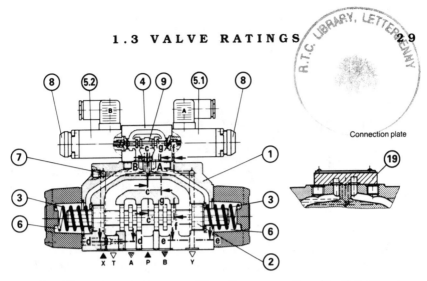

Connection plate

Fig. 1.18. Pilot-operated, three-position, four-way solenoid valve. *(Courtesy Rexroth Corp.)*

All three valve designs shown are tested for minimum and maximum flow and pressure at each port, response time, spool leakage, and pressure drop. Spool land length and position at shift can cause various pressure drop or leakage values, so testing should be conducted at each valve position.

CHECK VALVES

The check valve is the most simple of the directional control valves. It consists of a ball or poppet held on a seat with a spring, as shown in Fig. 1.19. The check valve shown is a cartridge design that can fit within a manifold, a valve body, or any other system component.

Tests performed on this valve are maximum and minimum flow and pressure, crack pressure, leakage, and pressure drop. Leakage is usually minor, from zero leakage to several drops per minute.

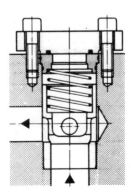

Fig. 1.19. Check valve cross section (cartridge design). *(Courtesy Rexroth Corp.)*

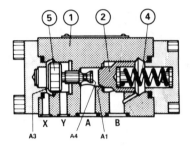

Fig. 1.20. Pilot-operated check-valve cross section. *(Courtesy Rexroth Corp.)*

PILOT-OPERATED CHECK VALVES

When a system requires a check valve to allow reverse flow, a pilot-operated check valve, as depicted in Fig. 1.20, can be used. This is a poppet-type check valve with a single pilot piston. Free flow is allowed from port A to B; but to flow from B to A, the piston must be pressurized at the X port to force the poppet off the seat. The pilot piston area is always greater than the seat I.D. area. This allows the piston force to be greater than the force holding the poppet closed when both are at equal pressure. In some cases, this ratio is not sufficient, so pilot-operated check valves can be supplied with a two-stage pilot piston, which usually includes a small ball or poppet within the main poppet to vent the spring chamber, thus reducing force on the main poppet.

Testing includes the same tests as described under check valves, and, also, pilot pressure versus system pressure testing to determine pressure opening ratio. Some designs also include a pilot piston seal to reduce system leakage or to isolate the pilot source from the main line. In this case, pilot piston leakage is also tested.

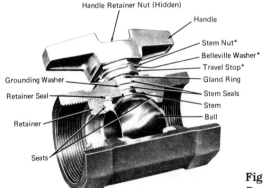

Fig. 1.21. Ball valve. *(Courtesy Rockwell International.)*

Fig. 1.22. Butterfly valve. *(Courtesy Rockwell International.)*

MANUAL SHUTOFF VALVES

Several valve designs, such as ball valves, butterfly valves, globe valves, and gate valves, fall into this group. Figure 1.21 depicts a ball-valve design, and Fig. 1.22 depicts a butterfly-valve design. In most cases these valves are used for on–off service rather than for throttling service, in which they sometimes are used.

Common tests applied to these valves include minimum and maximum flow and pressure, pressure drop, pressure drop versus handle position, internal seat leakage, and external leakage at the stem packings.

1.3.2 Pressure-Control Valves

Several categories of control valves fall under the heading of pressure control. These valves regulate inlet pressure and outlet pressure, or they balance pressure to limit the amount of power a system or subcircuit can produce. Some circuits use a pressure-control valve only as a safety valve, for limiting maximum system pressure. Other circuits are designed around a pressure-control valve and in these instances precision regulation is often important.

Pressure regulation is the key rating in this category. The controlability or regulation tolerance band in a given situation is an important datum. Repeatability of set point defines the valves capability to repeat the same setting at the same flow and viscosity.

Pressure-control valves typically have hysteresis between opening pressure (crack pressure) and closing pressure (repeat pressure). This is caused by the different forces acting on internal components in both cases. For example, a simple direct-acting relief valve has a spring holding a ball or poppet on a seat. At crack pressure, the force of the spring on the ball is equal to the force against the surface area of the ball on the I.D. of the seat. As pressure increases, the ball rises off the seat to allow flow to pass through the seat I.D. to tank. While this is happening, the spring is compressed further as the ball is pushed back by increasing pressure due to flow. This is called crack pressure versus full flow pressure or pressure rise. Crack pressure is defined as the point at which the valve begins to open allowing oil to pass. This is generally a very low flow of fluid or merely visible drops. By the time full flow is passing through the valve, increased spring force as well as pressure drop through the valve combine, resulting in pressure rise.

Reseat pressure of this valve now occurs when pressure drops below the spring force and the ball can shear the flow forces through the seat I.D. As you can see, we have a different set of parameters for crack pressure and reseat pressure. This requires consideration when applying a pressure control valve within a system. Reseat pressure can be as low as 70% of crack pressure. In a system that requires maximum pressure to function, pressure-control cycles should be studied to determine if this pressure loss before reseat will create erratic system performance.

The basic function of a pressure-control valve is to sense line pressure with a spring-loaded surface area, and open a flow path sufficiently to bleed oil to prevent higher pressures than set point. Several common designs accomplish this task nicely. The main difference between these designs is that they are either direct-acting or pilot-operated, and that the surface area used to sense pressure (or react to it) is designed in different configurations.

The most basic design is, of course, the direct-acting relief valve, which is used for low-flow applications or to vent the spring chamber of a larger flow capacity valve. There are several variations of this design including ball or poppet style, sliding spool style, and differential-area piston style.

The pilot-operated valve is basically a poppet on a seat held by a light spring. The spring chamber and pressure port are connected by a small orifice. Also connected is the vent relief, which actually senses the pressure and vents the spring chamber at set pressure, causing the poppet to open. This design can be provided in several versions including poppet style and sliding spool style.

The following valves depicted are all pilot-operated poppet style. This will highlight the fact that the same basic valve design can provide different functions by changing orifice or sensing-line locations.

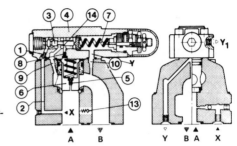

Fig. 1.23. Internally piloted, internally drained, relief-valve cross section. *(Courtesy Rexroth Corp.)*

RELIEF VALVES

Figure 1.23 depicts an internally piloted, internally drained, relief-valve design. Note the orifice (5) in the main poppet (6), which connects the A port to the spring chamber to balance pressure on both sides of the poppet. The spring acts only to hold the poppet down on the seat. Pressure is sensed by the vent relief valve from the A port through another orifice (1). Once set pressure is reached, the vent-relief spool pushes against the vent spring (7), and opens the spring chamber to the B port through a third orifice (8). It is this orifice that controls response time of the main poppet assembly.

Figure 1.24 depicts an externally piloted, internally drained, relief-valve design. The basic function is the same as described previously, except that pressure can be remotely controlled. An external pressure signal at the X port now actuates the vent-relief spool to cause main poppet reaction. Some valves of this type are designed to actuate by the venting of pressure at the X port, rather than the presence of a set pressure signal. In this design, the vent valve on the main relief can be used as a maximum-pressure-limiting valve, while the remote signal can be used for infinite control up to that maximum limit.

It is important to remember that relief-valve set pressure is affected by pressure at the B port, or return line. Many valve designs are not rated for full system pressure on the return port, so caution

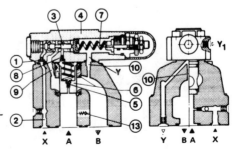

Fig. 1.24. Externally piloted, internally drained, relief-valve cross section. *(Courtesy Rexroth Corp.)*

is advised during relief-valve testing. It is also important when setting a relief valve to subtract return-line pressure from inlet pressure to establish the actual set point of the valve. If a relief valve is set at 3000 psi, and return-line pressure is at 200 psi, the valve will relieve at 3200 psi.

SEQUENCE VALVES

A sequence valve is similar to a relief valve except that the spring chamber of the vent relief is separately connected to tank, which allows the valve to maintain regulated pressure regardless of return-line pressure. Figure 1.25 depicts this valve with this separate drain at either port Y_1 or at Y with a plug between the spring cavity and the B port.

When testing this valve, the drain line should be unrestricted. Any line loss will add to the pressure setting, just as return-line loss adds to the setting of a relief valve.

Sequence valves are commonly used in circuits that require a minimum pressure (set point) to become active. Because of this, sequence valve B ports are normally designed to withstand full system pressure. This should be verified prior to the testing or selection of this type of component.

UNLOADING VALVES

The unloading-valve design depicted in Fig. 1.26 is used in systems that operate in a short pressure cycle, with longer periods of off time, a good example being an accumulator circuit, where the pump is only used to charge the accumulator. Once set pressure is reached, the poppet allows full pump flow to bypass at low pressure, reducing heat. The accumulator pressure is now sensed, and when this pressure drops below the low-end limit, the poppet closes and the pump recharges the accumulator.

For the valve shown, pilot pressure at port X is connected at the accumulator and is sensed external to port A where pump outlet

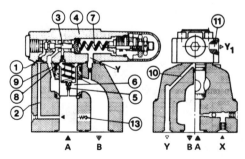

Fig. 1.25. Sequence-valve cross section. *(Courtesy Rexroth Corp.)*

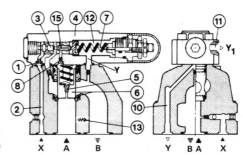

Fig. 1.26. Unloading-valve cross section. *(Courtesy Rexroth Corp.)*

pressure is connected. By connecting this sensing line remote, any pressure fluctuations seen at port A do not effect operation of the valve.

PRESSURE-REDUCING VALVES

Pressure-reducing valves are used to limit outlet pressure to a predetermined set point, and act somewhat in the reverse of the valves previously described. The valve is normally opened, allowing full flow to pass through. Once set pressure is reached, the valve throttles until downstream pressure is regulated at set point, regardless of inlet pressure.

These valves should be tested to determine their capability in maintaining reduced pressure in both dynamic and static conditions. As an example, a cylinder moves at full flow, and the pressure-reducing valve limits pressure as the cylinder acts against certain imposed loads. Once the cylinder bottoms, it is important that the pressure remains at the set reduced pressure. If the valve does not relieve this pressure, static conditions will allow pressure in this static line to build to main system pressure. Some valves are rated as reducing/relieving valves. This should be known prior to testing for valve performance. A third port, the drain port, is supplied to allow drain flow to bypass to tank. This bypass is required to keep seat leakage from building up downstream pressure above the set point when the valve is in a static condition.

COUNTERBALANCE VALVES

A counterbalance valve typically consists of a pressure-controlling piston and a bypass check valve for free-reverse flow, and usually is applied in a circuit where an overhung load could "runaway" without some backpressure as restriction. As shown in Fig. 1.27, the valve consists of the main poppet (2), control spool (3), pilot piston (4), follower piston (5), and dampening orifice.

Counterbalance valves are typically rated for operating character-

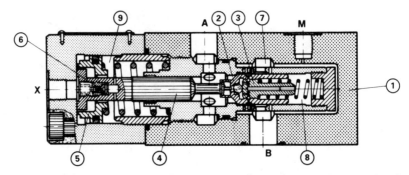

Fig. 1.27. Counterbalance-valve cross section. *(Courtesy Rexroth Corp.)*

istics at flow and load-holding capability (leakage) in static conditions. It is important to test these valves at actual field conditions because of response times and variable ambient conditions not simulated in laboratory tests. Failure of these valves in an actual situation could lead to a potentially dangerous situation.

1.3.3 Flow-Control Valves

System speed can be regulated by flow-control valves. The two basic types used are pressure-compensated and nonpressure-compensated designs. Nonpressure-compensated valves are actually either fixed or variable orifices. The orifice size restricts flow by creating a pressure drop. In hydraulic systems, the remaining pump flow not passed through the flow-control valve is either dumped across the system relief, or, in pressure-compensated pump systems, the pump de-strokes to the flow rate required by the system. The orifice can only be sized to control a flow rate at a specific inlet pressure. If pressure increases, so does flow.

Pressure-compensated valves regulate a preset flow rate regardless of inlet or outlet pressure, as long as a minimum operating pressure is met. Again, an orifice is used but the orifice controls a throttling piston that actually does the regulating.

Flow regulation is, of course, the main test conducted on this valve. Nonpressure-compensated valves are tested to determine flow rate versus orifice size versus inlet pressure at a particular viscosity. Pressure-compensated valves are also tested in this manner, but inlet pressure and outlet pressure are varied to determine regulation hysteresis with variable pressure conditions. Outlet pressure changes can affect a valve's performance, so tests should include this check.

Response time of a flow control can be critical, depending on application. For most machine controls, response of the valve does not play an important role; but consider an application such as a forklift truck. Most forklifts use a pressure-compensated flow control to regulate mast descent rate. Because the valve is pressure-compensated, a wide variety of loading on the mast does not affect flow rate. However, a pressure-compensated flow control is usually a normally open device. Flow rate is not controlled until flow reaches set point. Until this point is reached, the valve allows lesser flow rates to pass through unrestricted with the exception of normal pressure drop. In the forklift, an operator removes a loaded pallet from a high shelve, backs out the truck, and shifts the directional valve to allow the mast cylinder to lower the pallet to the ground. A flow control with slow response will allow the load to freefall until stabilization occurs. A fast-response valve would probably cause the load to bounce, creating an equally dangerous condition. Thus, the importance of response time for certain applications.

NONPRESSURE-COMPENSATED FLOW CONTROLS

As previously noted, these valves can be either as simple as an orifice or adjustable, as depicted in Fig. 1.28. The valve shown is a throttling valve that provides a variable orifice based on rotation of the adjustment knob. Several design variations are available to control flow: needle valves, tapered needle valves, sliding plate, etc., but all provide the same function. The opening between the stem and seat is designed to provide as linear a flow versus knob rotation as possible. This is one test conducted when designing such a valve.

Another important rating is the relationship of pressure drop ver-

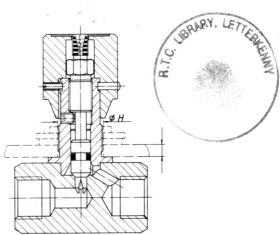

Fig. 1.28. Nonpressure-compensated flow-control valve. *(Courtesy Rexroth Corp.)*

sus flow. Since this valve is used to control flow via the pressure-drop method, this information is required for proper selection of a valve for a particilar application.

PRESSURE-COMPENSATED FLOW CONTROLS

Figure 1.29 depicts an adjustable pressure-compensated flow control. It, too, acts as a fixed orifice, but any change in pressure is compensated by the compensator spool (4). The adjustment knob depresses the control spool (5) so that a fixed orifice size is maintained. This orifice creates the signal that modulates the compensator spool to a stable flow condition. Stable flow occurs when the conpensator spool has created a restriction equivalent to the balance of inlet pressure versus outlet pressure forces assisted by the spring (6) and control orifice (5).

Tests conducted on this valve are minimum operating pressure, response time, regulation hysteresis with variable inlet and outlet pressure, and regulation repeatability.

A sharp-edged control orifice is used in many designs to reduce the influence of viscosity on flow regulation. This provides a more stable flow control during the temperature or pressure swings that are seen in many systems. In these instances, tests are also conducted to determine the flow-regulation tolerance over a temperature range. This can be a general test, or a specific set of operating conditions can be simulated.

It is also important to note that tests conducted with a test stand

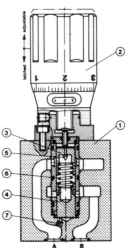

Fig. 1.29. Adjustable pressure-compensated flow-control valve. (*Courtesy Rexroth Corp.*)

do not always simulate actual field conditions. In many instances an actual system mock-up is required to pinpoint certain operating parameters not detected during normal testing.

PRIORITY FLOW CONTROLS

Priority-flow-control valves are similar to the standard pressure-compensated valves with one exception: Full pump flow is passed through the valve, one outlet port providing regulated flow, the second outlet port passing the remaining flow available from the pump.

Two variations are present in this valve type. One type will allow use of the secondary flow in another system subcircuit, provided this circuit operates at a pressure below the regulated outlet. The second type cannot operate properly unless secondary flow is diverted back to tank at low pressure. The only advantage of the second type over a normal flow control design is that less heat is generated from unused flow. A standard flow-control inlet port builds pressure up to the relief-valve setting until excess flow dumps across the relief at maximum system pressure. The bypass flow control dumps excess flow to tank at a pressure equal to or slightly higher than the pressure required at the regulated port.

In both types, regulated flow is tested with variable secondary port pressure. Caution should be used when testing the low-pressure secondary-flow design so that back pressure does not exceed recommended maximum on that port.

FLOW DIVIDERS

Flow dividers are used to divide input flow rate equally or proportionally out two outlet ports. Two common design methods include opposing-piston style or gear style. The opposing-piston method is very similar to the pressure-compensated flow-control operation. Orifices are used to provide a proportional split of flow through two connected, modulating pistons. The gear method involves the use of a set of gears that rotate at a speed proportional to inlet flow, just like a gear motor. The displacement of the gears (generally controlled by gear width) provides the proper flow proportion.

Tests on these valves consist of flow-division tolerance with varying inlet and outlet pressures. Loading of both outlet ports, simultaneously and individually, is recommended.

Note: Gear-style flow dividers can increase pressure above inlet pressure. When testing at high pressures, make certain that overpressure protection on both outlet ports is provided. In certain conditions, one side can act as motor, driving the other side like a pump.

1.3.4 Proportional Valves and Servovalves

This category of valve design encompasses a unique type of valve used for infinite control of system speed, force, and fluid direction. These valves are used within many system types. These systems can range from a simple proportional control manlift to the control of high-performance military aircraft and space-vehicle control. Both valve types require electronic control systems, which are often baffling to the hydraulic or mechanical engineer. Of course, most engineers do not like to admit this. Nevertheless, once the unfamiliarity is eliminated through personal experience with a system, the engineer can begin to realize the capability of this control device.

Proportional and servo systems are used in either an open-loop or closed-loop electronic configuration. In an open-loop mode, a control signal is sent to the valve and the valve stabilizes at the point the system matches the signal. For example, if the servo is controlling a hydraulic motor to obtain a conveyor belt drive speed, speed will be proportional to the input current or voltage. However, the open-loop system will not compensate for any fluctuations due to viscosity change or uneven belt-load conditions. The variable conditions will affect motor efficiency and will cause more or less fluid to slip by the rotating group. With the servo in a steady-flow condition, motor speed will increase or decrease as efficiency changes.

In a closed-loop configuration a tachometer pickup could be used to monitor actual motor or conveyor belt speed. The tachometer output would then be hooked into the feedback terminals of the servo electronics. Once speed is selected by an input to the servovalve, the electronics reference the tachometer output at that point. If measured speed varies, the servo will reposition itself until the desired (referenced) speed is once again matched.

This valve group has a set of definitions unique to itself. These terms should be understood and reviewed when defining the dynamics of a servosystem. The incorrect selection of a servovalve for your system can often become a very expensive mistake. The following list of terms is reprinted courtesy of Moog, Inc.

GENERAL TERMINOLOGY

Closed-Loop Control System An automatic control system in which the system input is compared with a measurement of the system output, and the resultant error signal is used to drive the system toward the desired output.

Servomechanism A continuously acting, bidirectional closed-loop control system.

Servovalve A device used to produce hydraulic control in a servo-mechanism.

Electrohydraulic Servovalve A servovalve that produces hydraulic control in response to electrical signal inputs; sometimes called a *transfer valve.**

Electrohydraulic Flow Control Servovalve A servovalve designed to produce hydraulic flow output proportional to electrical current input.

VALVE NOMENCLATURE

Hydraulic Amplifier A fluid valving device that acts as a power amplifier, such as a sliding spool, or a nozzle flapper, or a jet pipe with receivers.

Stage The portion of a servovalve that includes a hydraulic amplifier. Servovalves may be single stage, two stage, three stage, etc.

Output Stage The final stage of hydraulic amplification used in a servovalve, usually a sliding spool.

Port A fluid connection to the servovalve; for example, a supply port, a return port, or control port (sometimes called *load port** or *output port** or *strut port**).

Three-Way Valve A multiorifice fluid control element with supply, return, and one control port arranged so that valve action in one direction opens the control port to supply and reversed valve action opens the control port to return.

Four-Way Valve A multiorifice fluid control element with supply, return and two control ports arranged so that valve action in one direction simultaneously opens control port #1 to supply and control port #2 to return. Reversed valve action change is control port #1 to return and control port #2 to supply.

Flow Force Compensation A design technique for shaping the fluid passages about a variable orifice so as to reduce steady-state flow forces.

Load Compensation A valve design condition that yields relatively constant flow gain over a wide range of load pressure drop; sometimes called *gain compensation.**

ELECTRICAL INPUT CHARACTERISTICS

Torque Motor The electromechanical transducer commonly used with the input stage of a servovalve. Displacement of the armature of the torque motor is generally limited to a few thousandths of an inch.

*These terms are not recommended.

Input Current The current that is required for control of the valve, expressed in milliamps (ma). For three and four lead coils, input current is generally the differential coil current, expressed in ma.

Rated Current The specified input current of either polarity to produce rated flow, excluding any null bias current, expressed in ma. For three- and four-lead coils, rated current must be associated with a specific coil connection (series, differential, or parallel).

Quiescent Current The DC current present in each coil of a differential coil connection, the two coil currents having opposing polarities such that no electrical control power exists.

Electrical Control Power The electrical power dissipated in the valve coils that is required for control of the valve, expressed in milliwatts. For differential coil connection, the *total electrical input power* supplied to the servovalve is the sum of the electrical control power and the power due to quiescent current.

Electrical Null* The zero input current condition.

Coil Impedance The complex ratio of coil voltage to coil current. It is important to note that coil impedance may vary with signal frequency, amplitude, and other operating conditions owing to back emf generated by the moving armature. Therefore, coil impedance should be measured with explicit operating conditions.

Coil Resistance The DC resistance of each torque motor coil, expressed in ohms.

Coil Inductance The apparent inductive component of the coil impedance, expressed in henrys. For a valve torque motor having more than one coil, the total coil inductance will include mutual coupling effects.

Dither A low amplitude, relatively high-frequency periodic electrical signal sometimes superimposed on the servovalve input to improve system resolution. Dither is expressed by the dither frequency (Hz) and the peak-to-peak dither current amplitude (milliamps).

STATIC PERFORMANCE CHARACTERISTICS

Control Flow, also called *Load Flow** or *Flow Output** The fluid flow passing through the valve control ports, expressed in cubic inches per second (cis) or gallons per minute (gpm). In testing a four-way servovalve, flow passing out one control port is assumed equal to the flow passing in the other. This assumption is valid for no-load valve testing with a symmetrical load (e.g., with an

*These terms are not recommended.

equal area piston having insignificant friction and mass) and for static testing with loaded flow.

Rated Flow The specified control flow corresponding to rated current and specified load pressure drop, expressed in cis or gpm. Rated flow is normally specified as the no-load flow.

No-Load Flow The servovalve control flow with zero load pressure drop, expressed in cis or gpm.

Loaded Flow The servovalve control flow when there is load pressure drop, expressed in cis or gpm.

Internal Leakage The total internal valve flow from pressure to return with zero control flow (usually measured with control ports blocked), expressed in cis or gpm. Internal leakage will vary with input current, generally being a maximum at the valve null (*null leakage*). In a two-stage servovalve, internal leakage will include both hydraulic amplifier flow (sometimes called *tare flow**) and bypass flow through the output stage.

Total Valve Flow The sum of the control flow and the internal leakage flow, expressed in cis or gpm.

Load Pressure Drop The differential pressure between the control ports, expressed in psi. With conventional three-way valves, load pressure drop is the differential pressure between the valve control port and one-half the net supply pressure.

Valve Pressure Drop The sum of the differential pressures present across the control orifices of the output stage, expressed in psi. Valve pressure drop will equal the supply pressure minus the return pressure minus the load pressure drop.

Power Output The fluid power which is delivered to the load, expressed in hp.

$$\text{hp} = \frac{\text{Control flow, cis} \times \text{Load pressure drop, psi}}{6600}$$

Polarity The relationship between the direction of control flow and the direction of the input current.

Threshold The increment of input current required to produce a change in valve output, expressed as percentage of rated current; sometimes called *resolution.** Threshold is normally specified as the current increment encountered when changing the direction of application of input current.

Hysteresis The difference in the valve input currents required to produce the same valve output during a single cycle of valve input current when cycled at a rate below that at which dynamic effects are important. Hysteresis is normally specified as the maximum

*These terms are not recommended.

difference occurring in a complete cycle between plus and minus rated current, and is expressed as percentage of rated current.

Flow Curve The graphical representation of control flow versus input current. This is usually a continuous plot of a complete cycle between plus and minus rated current values.

Normal Flow Curve The locus of the midpoints of the complete cycle flow curve. This locus is the zero hysteresis flow curve; however, valve hysteresis is usually quite low such that one side of the flow curve can be used for the normal flow curve.

Flow Gain The slope of the control flow versus input current curve at any specific point or in any specific operating region, expressed in cis/ma or gpm/ma. The incremental flow gain may vary from point-to-point owing to valve nonlinearities. The nominal flow gain will generally show three operating regions: (1) the null region, (2) the region of normal flow control, and (3) the region where flow saturation effects occur. When this term is used without qualification, it is assumed to mean normal flow gain.

Normal Flow Gain The slope of a straight line drawn from the zero flow point of the normal flow curve, throughout the range of rated current of one polarity, and having a slope chosen to minimize deviations of the normal flow curve from the straight line. Flow gain may vary with the polarity of the input current, with the magnitude of load differential pressure, and with changes in operating conditions.

No-Load Flow Gain The normal flow gain with zero load differential pressure. No-load flow gain will vary with supply pressure and other operating conditions.

Rated Flow Gain* The ratio of rated flow to rated current, expressed in cis/ma or gpm/ma. When rated flow is specified for a loaded flow condition, the rated flow gain should be qualified similarly.

Flow Saturation The condition where flow gain decreases with increasing input current. Flow saturation may be deliberately introduced by mechanical limiting of the valve range, or may be the result of increasing pressure drops along internal fluid passages.

Flow Limit The condition wherein control flow no longer increases with increasing input current.

Symmetry The degree of equality between the normal flow gain of one polarity and that of the reversed polarity. Symmetry is measured as the difference in normal flow gain of each polarity, expressed as a percentage of the greater.

---

*These terms are not recommended.

Linearity The degree to which the normal flow curve conforms to a straight line with other operational variables held constant. Linearity is measured as the maximum deviation of the normal flow curve from the normal flow gain line, expressed as a percentage of rated current.

Pressure Gain The change in load pressure drop per unit input current with zero control flow (control ports blocked), expressed in psi/ma. Pressure gain is usually specified as the average slope of the curve of load pressure drop versus input current in the region between ±40% of maximum load pressure drop.

Pressure Threshold* The change in input current required to produce a specific change in the load pressure drop with zero control flow. Sometimes used as a combined threshold and pressure gain measurement.

NULL CHARACTERISTICS

Null The condition where the valve supplies zero control flow at zero load pressure drop.

Null Region The range of input current about null wherein effects of lap and bypass leakage in the output stage predominate. Normally the valve null region extends through a range of about ±5% rated current from null.

Lap In a sliding-spool valve, the relative axial position relationship between the fixed and movable flow metering edges with the spool at null. For a servovalve, lap is measured as the total separation at zero flow of straight line extensions of the nearly straight portions of the normal flow curve drawn separately for each polarity, expressed in percent of rated current.

Zero Lap, also called *Closed-Center* The lap condition where there is no separation of the straight line extensions of the normal flow curve, generally corresponding to precise alignment of the flow metering edges.

Overlap The lap condition that results in a decreased slope of the normal flow curve in the null region.

Underlap, also called *Open-Center* The lap condition that results in an increased slope of the normal flow curve in the null region.

Deadband* The null region associated with a spool overlap condition.

Null Bias The input current required to bring the valve to null under any specified set of operating conditions, excluding the effects of valve hysteresis, expressed as a percentage of rated cur-

---

*These terms are not recommended.

rent. Hysteresis effects may be discounted by taking the arithmetic average of the null bias currents measured on a symmetrical hysteresis loop.

Null Shift The change in null bias required as a result of a change in operating conditions of environment, expressed as a percentage of rated current. Null shift, sometimes called *centershift,** may occur with changes in supply pressure, temperature, and other operating conditions.

Null Pressure The pressure existing at both control ports at null, expressed in psi; sometimes called *centering pressure.**

Null Pressure Gain* The slope of the pressure gain characteristics at null, expressed in psi/ma.

Null Flow Gain* The slope of the control flow versus input current relationship at null, expressed in cis/ma. Null flow gain may be between 0 and 200% of the nominal flow gain due to the lap condition.

Null Leakage The total valve internal leakage flow at null, expressed in cis.

DYNAMIC CHARACTERISTICS

Frequency Response The complex ratio of control flow to input current as the current is varied sinusoidally over a range of frequencies. Frequency response is normally measured with constant input current amplitude and zero load pressure drop, and is expressed by the amplitude ratio and phase angle. Valve frequency response may vary with the input current amplitude, temperature, supply pressure, and other operating conditions.

Amplitude Ratio The ratio of the control flow amplitude to a sinusoidal input current amplitude at a particular frequency divided by the same ratio at a specified low frequency (usually 5 or 10 Hz). Amplitude ratio (AR) may be expressed in decibels where $db = 20 \log_{10} AR$.

Phase Angle The time separation between a sinusoidal input current and the corresponding variation of control flow, measured at a specified frequency and expressed in degrees (deg = time separation, sec × frequency, Hz × 360).

Proportional valve and servovalve testing can be either simple or extremely complex depending on the test required and the system architecture. In the simple end of the test spectrum these valves are tested for the operating characteristics listed at the beginning of this valve test section. One important test is contamination tolerance. In some cases, these valves can operate in systems with 25-

---

*These terms are not recommended.

micron filtration or worse. In other cases, 3-micron absolute filtration is not sufficient to allow adequate service life for the required application.

Servovalve testing generally is listed in four different categories: electrical, static tests (point-by-point data), static tests (continuous data), and dynamic tests.

Electrical testing includes coil DC resistance, coil inductance, and insulation resistance.

Point-by-point data static tests include null point adjustment, valve polarity, pressure gain characteristics, internal leakage, no-load flow characteristics, hysteresis, threshold, loaded flow characteristics, null shift with supply pressure, null shift with return pressure, null shift with quiescent current, null shift with temperature, and null shift with acceleration.

Continuous data static tests include blocked load characteristics, internal leakage, no-load flow characteristics, and threshold and loaded flow characteristics.

Dynamic tests include amplitude ratio, phase angle, frequency response, change in frequency response with supply pressure, and transient response.

PROPORTIONAL VALVES

The proportional valve is less expensive and less accurate than the servovalve. Proportional systems, if calibrated accurately, can provide quite exceptional results. However, in systems requiring fast response, excellent repeatability, and precise control under varying conditions, servovalves are the obvious choice.

The heart of the proportional valve is the proportional solenoid. Unlike a simple on–off coil as used in directional valves, the proportional coil provides plunger position proportional to input current. A proportional valve (Fig. 1.30) resembles a standard four-way solenoid valve. The coil is either available with or without a feedback device. Generally, when feedback is required (closed-loop system), a linear variable differential transformer (LVDT) is used. The LVDT sends a signal back to the valve electronics to verify spool or coil plunger position. (See Chapter 2 for a detailed description of LVDT operation.) Feedback is used to provide improved hysteresis characteristics of the valve. Hysteresis in this case applies to the repeatability of spool position at the same input current signal.

The proportional coil is used to position a spool at any point along the full spool stroke. The spool has tapered, notched, or grooved edges to provide metering control at various points throughout the stroke. In general applications, these edges are designed to achieve

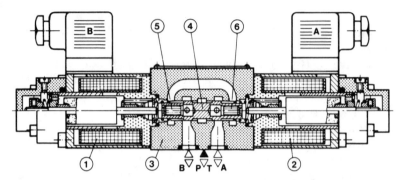

Fig. 1.30. Proportional-valve cross section. *(Courtesy Rexroth Corp.)*

as close a linear characteristic of pressure drop versus spool position as possible. The valve basically is a multiported, electronically controlled, nonpressure-compensated throttling valve. The spool is positioned at a point that reaches and maintains the desired flow or pressure. The degree of accuracy attained is dependant on system dynamics as well as the coil's ability to maintain spool position at that point. Typical proportional coils are available with 0.125 to 0.625 in. travel. This degree of travel does not allow for fine control if the spool is not held steady. Proportional coils with feedback are normally rated at about ± 3–5% hysteresis. Considering the fact that the spool must totally close off flow in one position and allow unrestricted flow in the opposite position, shows the level of control required over such a short spool stroke in the control range.

Proportional valves generally are available in a direct-acting design. Higher-flow designs such as depicted in Fig. 1.31 use a direct-

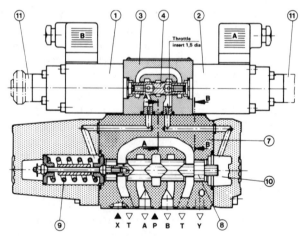

Fig. 1.31. Two-stage proportional-valve cross section. *(Courtesy Rexroth Corp.)*

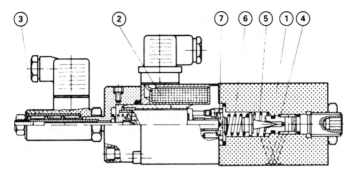

Fig. 1.32. Proportional relief valve. *(Courtesy Rexroth Corp.)*

acting proportional valve as a pilot control device of a larger pilot-operated control valve. This is needed because coil current required by a proportional valve is much greater than that of a servovalve. The flow forces generated within a high-flow spool valve are greater than the practical limitation of current required for sufficient coil force output.

Aside from variable directional control, proportional valves can be used in special configurations to modulate pressure, or act as a single function throttling or flow control. Figure 1.32 depicts a pilot-operated relief valve controlled by a proportional-coil-controlled pilot stage. Pilot pressure is controlled by orifice variation of the pilot stage. Orifice variation is proportional to the coil plunger position. Figure 1.33 depicts a proportional-flow-control valve. Again, the coil

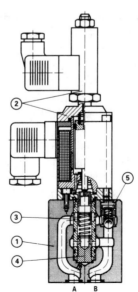

Fig. 1.33. Proportional-flow-control valve. *(Courtesy Rexroth Corp.)*

plunger is used to adjust an orifice opening. The valve then acts as a normal pressure-compensated flow control. The control piston hydraulically balances itself to provide a constant flow rate based on the sensed differential pressure across the orifice.

SERVOVALVES

There are various designs of servovalves. To obtain the maximum system performance in an exacting situation, careful selection of the proper valve is required. The servovalve industry is unique, in that few manufacturers are capable of producing a product with the repeatability and control required. This product is extremely specialized.

A servovalve can be supplied in a single-stage, two-stage, or three-stage design. The valve ports are controlled either by a spool (similar to a four-way valve) or flappers, which throttle the main ports. The servovalve receives an electronic signal from a manual or automatic output, and converts the signal to a proportional port opening.

The main difference between proportional valves and servovalves is controllability and response time. All servovalves use some type of feedback signal to sense the variation between the input signal and valve position. The high response characteristics can also provide a system that oscillates continually owing to the natural fluctuations within the system. A valve cannot respond immediately to a sensed change, so a condition may exist where the actual output and desired output are shifting back and forth trying to match each other.

The main valve spool (in two-stage designs) is hydraulically shifted to a position that creates desired system output. Output is either measured as flow or pressure. The hydraulic pilot signal is generated by a device that responds to an input electric signal. The signal is proportional to the desired valve position within the operating range of the valve. These devices are flapper, spool, or jet-pipe designs, which vary a differential pressure signal to both sides of the main spool. Spool position is then monitored with a position feedback device.

Internal valve feedback is generally obtained by a mechanical, hydraulic, or electric signal. In mechanical designs, the pilot spool and main spool are either directly connected or indirectly connected by force springs. The second method allows adjustment of the ratio between the two spools. Hydraulic feedback is generally accomplished as a pressure signal acting to balance the main spool. The pressure signal is generated by the variation of outlet pressure versus sensed internal valve pressure. In some designs, spool force is also sensed to obtain an additional source of feedback signal. Electric feedback is usually accomplished through the use of an LVDT sensing spool position. In this instance, the valve electronics is used to process the input signal data and actual position data to control desired hydraulic output.

In a single-stage design the spool is directly controlled by a torque motor or similar device. The motor adjusts spool position, which creates a fixed orifice across the main ports. This orifice is used to control pressure or flow rate. Other devices include force motors or proportional solenoids.

In the two-stage design depicted in Fig. 1.34, the feedback source is a wire attached to the torque motor armature and moved by the main stage spool. The spool position is actually measured as wire force. The pilot stage consists of a double nozzle controlled by a flapper attached to a torque motor. A command signal actuates the torque motor and the flapper shifts to a position that produces a differential hydraulic pilot source between the two nozzles. This imbalance shifts the main spool until the spool shifts to the point at which the feedback wire is in line with the torque motor armature. This alignment maintains the spool position keeping the system in balance until either the input signal is changed or the load changes causing the servovalve to compensate for this change.

In a three-stage design (Fig. 1.35) the main spool is hydraulically balanced by the pilot signal from a two-stage servovalve configuration. In this case, the pilot section consists of a torque motor (1) and a jet flapper (2) used as a hydraulic amplifier. The flapper is used to create differential pressure between both sides of the second stage spool (chambers 10 and 14). This positions the spool to allow another source of differential pressure on both sides of the main

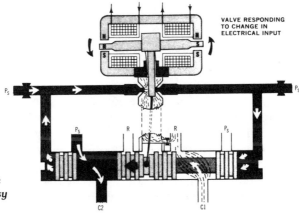

Fig. 1.34. Two-stage servovalve. (Courtesy Moog, Inc.)

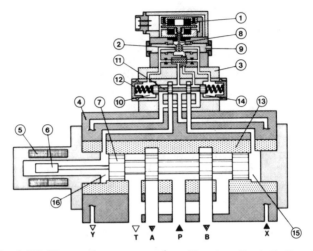

Fig. 1.35. Three-stage servovalve. *(Courtesy Rexroth Corp.)*

spool (chambers 16 and 25). The third stage (main) spool position is monitored by an LVDT for electrical position feedback [items 5 (coil) and 6 (core)].

1.3.5 Cartridge Valves

This category fits two distinct valve designs. The first design concept consists of any type of valve described previously that is housed within a machined cavity. These valves are commonly used to minimize external plumbing and to reduce overall system package size. Several manufacturers offer valves that fit into standard cavities. These cavities are machined into a metal block with all interconnecting passages also within the block. Figure 1.36 depicts a block assembly with several valves which make up a complete circuit.

Testing of each component follows the described tests for each valve type previously discussed. However, two other tests are now required. The first is the additional potential leakage paths around the body O.D. of each cartridge valve. Often valves are tested individually and then inserted into the block. Often the external seals are sheared when the valve is installed into the block owing to poor entrance angles, sharp edges, chips, nonconcentric bores, and wrong sized O-rings. Leakage testing after the block is assembled will detect this potential system problem.

The complete circuit should also be tested to ensure that component interaction performs as required. This relates to the proper selection of components to construct a system that fulfills all the func-

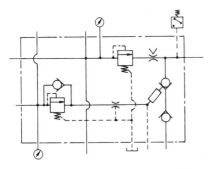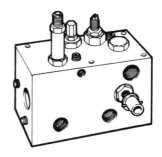

Fig. 1.36. Cartridge-valve circuit assembly. *(Courtesy Fluid Controls, Inc.)*

tions of the system. Sometimes a cycle of the system is overlooked and a component may have to pass more flow or pressure or less flow or pressure than it is capable of operating at.

The second cartridge-valve design concept is the pilot-operated logic valve used in high-flow, high-pressure systems. These valves have been used in Europe for quite some time and are now gaining acceptance in the United States.

A typical valve of this type is depicted in Fig. 1.37. In this case the valve is used as a high-flow relief valve with a vent relief used to control set-point pressure. The logic valve consists basically of a poppet or spool (3) in a housing (1) with a spring. The logic assembly is designed for high pressures (6000 psi and above) and high flows (hundreds of gpm), while the pilot section works at low flow. This valve can be used for on–off control, pressure control, flow control, or multiple functions, depending on the pilot actuator. These valves replace large solenoid-operated spool valves, allowing individual port timing as well as reducing internal leakage considerably.

Testing again depends on the valve function, but specific tests include seat leakage, O-ring leakage, and external leakage.

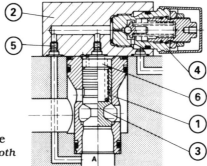

Fig. 1.37. Cartridge-logic-valve cross section. *(Courtesy Rexroth Corp.)*

1.3.6 Pneumatic Valves

Pneumatic valves basically operate the same as hydraulic valves, except orifices and flow passages are designed somewhat differently. All the tests previously described for each valve type apply for the most part. Cycle testing, internal and external leakage, minimum and maximum ratings for flow, pressure, operating voltage, etc., all apply.

The main difference in pneumatic components is the method used in rating flow and pressure capacity. Flow is usually rated at a C_v factor. C_v is defined as the number of U.S. gallons of water that pass through a given orifice area in 1 min at a pressure drop of 1 psi. This rating is applied to a valve to predict the pressure drop at compressed flow. The compressibility of air is a key factor, because actual flow (ACFM) versus free air flow (SCFM) varies, depending on system pressure. The following formulas are commonly applied to determine flow and pressure relationships for pneumatic components:

$$C_v = \frac{Q}{22.48} \frac{GT}{(P_1 - P_2)P_2}$$

where

Q = Flow in SCFM (14.7 psia @ 60°F)

P_1 = Inlet absolute pressure (psi + 14.7)

P_2 = Outlet absolute pressure (psi + 14.7) (Note: P_2 must be greater than $0.53 \times P_1$)

G = Specific gravity of flowing medium (air, $G = 1$)

T = Absolute air temperature (460 + °F)

$$Q = C_d A_o \sqrt{\frac{2\Delta P}{p_w}}$$

$$Q = \underbrace{\left(C_d A_o \sqrt{\frac{2}{p_w}}\, 0.26 \right)}_{C_v} \sqrt{\frac{\Delta P}{SG}}$$

or

$$Q = C_v \sqrt{\frac{\Delta P}{SG}}$$

where

Q = flow (in.^{3}/sec)

C_d = Orifice discharge coefficient (0.611 for a sharp-edged orifice)

A_o = Area of orifice (in.^{2})

ΔP = Pressure drop across the orifice (lbf/in.^{2})

p_w = Density of water

SG = Specific gravity of fluid

$$\text{ACFM} = \text{SCFM} \times \frac{P_1 T_2}{P_2 T_1}$$

Boyle's Law:

$$P_1 V_1 = P_2 V_2$$

or

$$\frac{V_1}{V_2} = \frac{P_2}{P_1}$$

Charles' Law:

$$\frac{V_1}{V_2} = \frac{T_1}{T_2}$$

where

P_1 = Original absolute pressure

P_2 = New absolute pressure

V_1 = Original volume

V_2 = New volume

T_1 = Original absolute pressure

T_2 = New absolute pressure

ACFM = Actual cubic feet per minute

SCFM = Standard cubic feet per minute

1.4 Cylinders and Rotary Actuators

Transmission of the force generated by a hydraulic system is accomplished by cylinders for linear travel, rotary actuators for limited rotary motion (usually under 360° total travel), and hydraulic motors for continuous rotary motion. Cylinders and rotary actuators fall under a different category of tests than the previously discussed hydraulic motors. These cylinders and rotary actuators have a fixed internal volume and surface area that are acted on by flow and pressure. Once the actuator has "bottomed out" or reached the end of its limited travel, flow must be vented or reversed to allow the actuator to return to the previous position.

1.4.1 Cylinders

Figure 1.38 depicts a standard hydraulic double-acting, single-rod cylinder. Total fluid volume is determined by

Internal volume (in.³)
= Piston surface area (in.²)
− Rod surface area (in.²)] × Stroke (in.)

Cylinder force is determined by

Cylinder force (lbf) = Piston surface area (in.²) × Pressure (psi)

As this component is used to transmit power, most tests have to do with mechanical integrity and efficiency. Efficiency is the rating of how much force or pressure is lost in accomplishing a theoretical output. In the case of cylinders, the following ratings are applied to define cylinder performance:

Breakaway Pressure The minimum pressure required for the cylinder to begin moving. This is caused by internal friction forces, by seals, and by the weight of the rod and piston.

Internal Leakage The rate of fluid lost across piston seals from the pressurized side of the cylinder to the vented side.

Spring-Return Force In single-acting, spring-return cylinders, the force of the spring is designed to overcome frictional forces

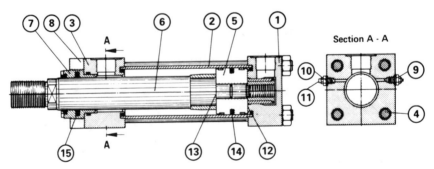

Fig. 1.38. Double-acting-cylinder cross section. *(Courtesy Rexroth Corp.)*

and any residual return line pressures on the opposite side of the piston.

Output Force Versus Pressure This defines the difference between theoretical force versus actual force at a certain pressure.

Mechanical integrity can encompass both hydraulic and mechanical capabilities. This rating is used to determine a cylinder's strength in holding or withstanding forces generated from the load or from the hydraulic supply. These forces can be seen at normal operating conditions or from shock and high-stress conditions.

Maximum Flow This rating applies to the maximum speed a cylinder can withstand. Seals must prevent leakage and the end caps and tube must be designed with sufficient strength to withstand the force of the piston against the end stops. In order to reduce this shock, end cushions provide hydraulic deceleration.

End Cushioning This option is tested to determine the reduction of speed versus time at the end of stroke and also the reduction of hydraulic-pressure spikes. The kinetic energy generated by this force must be reduced to a condition both mechanically and hydraulically within the rated capacity of the cylinder.

Maximum Operating Pressure The limit at which a cylinder can operate without internal or external damage or deformation with a degree of safety margin.

External Leakage The measurement of leakage at any external seal. This test is conducted at both static and dynamic operating conditions, with special attention to the rod seal area.

Buckling This failure can occur when the cylinder mechanical components are overcome by the force of the load. Most cylinder manufacturers rate the bore and rod diameter combination at a maximum load rating. Calculation of buckling is generally according to the Euler formula, since the piston rod is generally seen as the buckling member:

Buckling load $K = \dfrac{\pi^2 \times E \times J}{s_K^2}$ (at this load, the rod buckles)

$$\text{Maximum operating load } L = \frac{K}{S}$$

where

s_K = Free buckling length (cm)

E = Modulus of elasticity (kp/cm^2) (2.1×10^6 for steel)

J = Moment of inertia (cm^4) ($d^4 \times \pi/64 = 0.0491 \times d^4$ for circular section)

S = Safety factor (3.5)

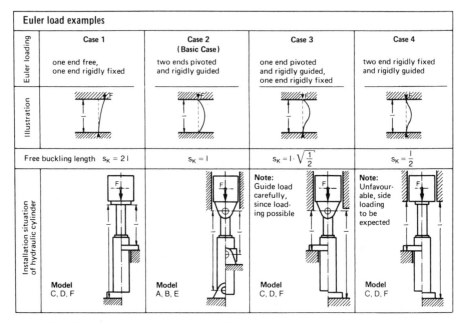

Fig. 1.39. Euler table (free buckling length). *(Courtesy Rexroth Corp.)*

The length to be stated as free buckling length must be taken from the Euler table in Fig. 1.39. Reinforcement by the cylinder tube is not taken into account in the calculation.

Bearing Load and Wear   The bearing surfaces that are subjected to the loads applied from external forces are tested to determine wear and maximum permissible load capacities. In many applications a cylinder is subjected to extreme side loads from forces not parallel to the rod force vector. This pushes the rod against one side of the rod bearing, and the piston against the opposite side of the tube. Eventually these surfaces wear, aided by contamination, and internal and external leaks appear. In more severe cases, mechanical deformation of bearing surfaces occurs resulting in component failure.

Contaminate Ingression   Rod seals are tested to determine their ability to prevent ambient contamination from entering the system when the cylinder retracts. Cylinders can be used in harsh, dirty environments such as those found in the mobile equipment industry. Effective rod seal design prevents these contaminates from entering and causing component failure within the system.

Cycle Life   Cylinder manufacturers conduct cycle tests on cylinders at varying ambient conditions to predict operating life in a given situation. These conditions can include typical factory environments as well as extreme temperature, atmospheric contamination, speed, and shock.

Mounting Style Strength The mechanical integrity of the cylinder mount is important data when applied to a system. Improper installation can cause metal fatigue and excessive side loading to the rod or tube assembly. These data can be obtained through calculation or by actual simulation testing.

1.4.2 Rotary Actuators

Similar in design to cylinders are piston rotary actuators that convert the linear motion of a piston (or pistons) to rotary motion. This is accomplished by several methods including rack-and-pinion (Fig. 1.40), piston-and-chain, and scotch-yoke designs. A vane actuator is also available that has similar features to a vane pump, with the exception that travel is limited by internal stops. Also, a piston and helix design operates like a cylinder. As the piston moves, intermeshing rotary helical gears between the piston and output shaft, turn the shaft.

All designs provide limited rotary motion and are basically tested in the same manner. Output torque is relatively constant in most designs except for the scotch-yoke design, where torque drops off slightly in midstroke, owing to the design of the piston–shaft interconnection. A pin mounted to the piston pushes a slotted yoke attached to the output shaft. Torque is greatest at both ends of travel because the torque arm is farthest from the output shaft centerline at these points.

Testing of rotary actuators includes most of the tests described for cylinders. One unique rating applied to these devices is backlash. This rating identifies the reverse movement of the rotary output shaft after pressure has been vented. Mechanical linkage connecting linear and rotary-movement components can cause backlash because of the buildup of manufacturing tolerances and

Fig. 1.40. Rack-and-pinion design rotary actuator.
(Courtesy Flo-Tork, Inc.)

stresses on certain internal parts while pressurized. This can be a critical factor in applying a rotary actuator to certain applications where precise positioning is necessary.

1.5 Filters

The ratings applied to filters fall into two separate categories: housing ratings and element ratings. Housing tests are conducted to determine mechanical design limits such as maximum pressure or maximum flow, fatigue tests, pressure drop, and leakage. When a bypass valve is contained within the housing, seat leakage, crack pressure, and full flow pressure drop are tested for. These data define the valve's performance and effectiveness in preventing unfiltered oil from bypassing the element when differential pressure across the element exceeds the set point of the valve.

Element testing is the critical test when working with filters. System contamination is the main cause of component failure. Defining internal system contamination as well as external contaminate ingression is extremely difficult, especially in applications such as mobile equipment and production test systems. Once the potential contamination level is determined, a filter is selected to reduce this level of contamination to a point that is compatible with component function and life expectancy of the system.

The following terms are commonly used to define and determine a filter's compatibility within a specific application and capability in eliminating system contamination.

Absolute Filtration Rating The largest diameter that will pass through an element under specific test parameters. This rating also determines the largest pore in the element structure.

Nominal Filtration Rating This is a measure of element efficiency, specifically a number used to define an average micron retention value. Exact industry test procedures have not been established for this rating, so actual filter performance varies from manufacturer to manufacturer. This rating method is being replaced with the more predictable beta rating.

Mean Filtration Rating This is a measurement of the average size of the pores of a filter element.

Beta Filtration Rating The ratio of particles of a given micron size between the upstream fluid (influent) and the downstream fluid (effluent). The formula used to express this ratio is

$$B_x = \frac{N\,(\text{upstream})}{N\,(\text{downstream})}$$

where

X = particle size in microns

N = number of particles larger than X micron value
p. 61

For example, let us assume we are testing to determine a filter's effectiveness at 10 micron particle size. Our tests have determined that 5000 particles larger than 10 micron were counted upstream, and 100 particles larger than 10 micron were counted downstream. Therefore, the beta rating is

$$B_{10} = \frac{5000}{100} = 50$$

Filter efficiency can then be determined from the following equation:

$$\text{Efficiency} = \left(1 - \frac{1}{\text{beta}}\right) \times 100\%$$

So, continuing with our example

$$\text{Efficiency} = \left(1 - \frac{1}{50}\right) \times 100\% = 98\%$$

Beta Ratio versus Efficiency Rating

Beta Ratio	Efficiency
1	0%
2	50%
20	95%
50	98%
100	99%
1,000	99.9%
10,000	99.99%

Dirt-Holding Capacity This is the weight of a specific contaminate that the element can retain at the maximum pressure drop recommended for that element. Generally expressed in grams, this capacity test determines how long an element can be used in an application before replacement is required.

Collapse Rating This is the differential pressure at which the element will mechanically fail. In most cases this is catastrophic failure where contaminates retained in the element can be released downstream.

Clean Pressure Drop This is the differential pressure across an element at initial installation. This rating is defined at a specific flow, temperature, and fluid viscosity.

Maximum Pressure Drop This is the maximum differential pressure across an element that is recommended by the manufacturer. This is usually a point below collapse pressure with a degree of safety margin. Bypass valves in the housing are usually sized to open at this point. Differential indicators (mechanical and electric) within the housing are used to signal an operator when this point is reached so that the element can be cleaned or replaced.

Flow Rate Flow rate is the flow at which the element produces a certain pressure drop. This rating can be used to define maximum flow capable of passing through the element, but more often is used to determine flow versus pressure drop.

Cleanliness Classification by Number of Particles Greater than 10 Microns in Size per 100 Millilitres

Contamination Standard	Class and Particle Count						
ISO Std. Class	18/15	17/14	16/13	15/12	14/11	13/10	12/9
Number of particles	77950	38975	19487	9744	4872	2436	1218
SAE Standard Class	6	5	4	3	2	1	0
Number of Particles	49592	25001	12456	6261	3121	1581	780

NAS 1638 Aerospace Industry Cleanliness Requirements of Parts Used in Hydraulic Systems—Maximum Contamination Limits (Number of Particles per 100 ml)

Size Range (microns)	Classes							
	00	0	1	2	3	4	5	6
5–15	125	250	500	1000	2000	4000	8000	16,000
15–25	22	44	89	178	356	712	1,425	2,850
25–50	4	8	16	32	63	126	253	506
50–100	1	2	3	6	11	22	45	90
over 100	0	0	1	1	2	4	8	16

	7	8	9	10	11	12
5–15	32,000	64,000	125,000	256,000	512,000	1,024,000
15–25	5,700	11,400	22,800	45,600	91,200	182,400
25–50	1,012	2,025	4,050	8,100	16,200	32,400
50–100	180	350	720	1,440	2,880	5,760
over 100	32	64	128	256	512	1,024

Note: In addition to the preceding ratings, filters are often subjected to mechanical and ambient extremes prior to establishing a rating for a particular application. For example, a filter may be exposed to high-level mechanical shock or vibration and then tested for degredation of performance. Likewise, temperature extremes are applied to determine the affect on element bonding media, elastomers, or element performance. Element deteriorization may surface as embrittlement, fluid absorption or disintegration.

Instrumentation

Now that we have reviewed the various tests applied to fluid power components, a review of the test equipment used to monitor these tests is necessary. In some cases, several types of equipment can be used to monitor the same test. This chapter is devoted to the introduction of this test equipment along with their theory of operation, operating characteristics, and application within a system.

A brief description of instrumentation accuracy levels will be discussed in this chapter. However, it is suggested that before applying any particular instrument, Chapter 4: Component and System Accuracy, should be reviewed for a more in-depth look at how these instruments can be misapplied during a test. This is the key element in producing test results that provide accurate data.

Unfortunately, cost is sometimes the restriction when reviewing instrument accuracy. Increased accuracy most always means increased cost. This must always be considered when starting a test system design.

2.1 Flow

Flow rate of a component or system is one of the most valuable pieces of test data used. Flow versus pressure drop, minimum and maximum flow capacities, leakage flow, efficiency, and response time are just a few of the tests that are dependant on accurate flow measurement. There now exist many flowmeter designs that may be used to monitor accurately (and sometimes not so accurately) fluid or air flow. Some meters use new and unfamiliar techniques to arrive at a flow rate, so they are not used owing to low confidence levels of test engineers. This is understandable because the meters that are commonly used have characteristics which sometimes produce erroneous data when least suspected. These characteristics are not usually the flowmeter manufacturer's fault, because each instrument has its limitations. The most common reasons for the erroneous data are lack of understanding of the instrument, misapplication within a system, or lack of technical communication between the manufacturer and user of the instrument.

Much information concerning flowmeter applications is available from the flowmeter manufacturers. Extensive testing by these manufacturers allows them to have accurate data as a reference when applying a meter to a particular application. A system can usually be supplied to fit most applications. In systems with extremely wide flow rates, two, three, or four meters can be paralleled with overlapping ranges to provide accurate readings. In other cases, where viscosity can vary greatly, or if multiple fluids are to be used, a manufacturer can supply a Universal Viscosity Calibration. With these data, the flow rate reading can be manually compensated based on the known viscosity of the fluid. The meter is calibrated at several viscosity points throughout its full flow range to provide the data for this calibration.

The following list consists of common terms and definitions associated with the measurement of flow.

Absolute Temperature Thermal value with relationship to absolute zero ($-460°F$). Absolute temperature is equal to measured temperature in $°F + 460°F$.

Annular Area The area (in square inches) of the opening between the outside diameter (OD) of the float and inside diameter (ID) of the tube in a rotameter.

Best-Fit Straight Line Calibration of a flowmeter based on the straight line of output pulses versus flow rate. The line is placed over the actual output curve in order to maintain the highest degree of accuracy without linearization.

Density The expression of mass per unit volume usually defined in pounds per cubic inch.

Dynamic Response The time (in milliseconds) in which a flowmeter responds to a step input of flow rate change.

Extended Range The turndown ratio of a flowmeter possible over the standard range accomplished by special calibration or additional electronics.

Flow Rate The expression of fluid or gas flow defined by the following common terms:

ACFM – actual cubic feet per minute: This is the compressed volume of a gas at pressure and is calculated by

$$\text{ACFM} = \text{SCFM} \times \frac{14.7\,\text{psig}}{P} \times \frac{T}{60°\,\text{F}}$$

or

$$\text{SCFM} = \text{ACFM} \times \frac{P}{14.7\,\text{psig}} \times \frac{60°\text{F}}{T}$$

where

   P = measured gas pressure in absolute terms (measured pressure psig + 14.7 = psia)

   T = measured gas temperature in absolute terms (measured temperature °F + 460°F)

cc/min = cubic centimeters per minute
gph = gallons per hour
gpm = gallons per minute
in.³/min = cubic inches per minute
pph = pounds per hour
ppm = pounds per minute
SCCM = standard cubic centimeters per minute
SCFH = standard cubic feet per hour
SCFM = standard cubic feet per minute
SLPH = standard liters per hour
SLPM = standard liters per minute

Flow Straightener A length of straight tubing with baffles in order to provide laminar flow into a turbine flowmeter.

Frequency Response The time (in milleseconds) a flowmeter can react to a change in flow rate.

F.S. Full scale, the term used to define the operating range of a meter. This term is often used to define accuracy. A flowmeter with a full scale accuracy of ± 1% and a flow range of 1–10 gpm will have an accuracy of ± 0.1 gpm.

Hz Hertz, the term used to depict frequency; 1 Hz = 1 cycle per second.

Hysteresis The difference of flowmeter output, at the same flow value, when readings are taken during increasing flow versus during decreasing flow.

K Factor The number of pulses a flowmeter produces for a given flow rate.

Laminar Flow Smooth flow, where viscous forces are greater than inertial forces, generally below a Reynolds number of 2000.

Linearity The percent deviation output tolerance a flowmeter will produce over the entire flowrange.

Linearizer An electronic module that compensates for pulse output errors due to nonlinearities caused by fluid characteristics or meter design.

Magnetic Pickup A proximity device that provides an electric pulse each time a magnetic material passes within its sensing range.

NIST National Institute of Standards and Technology (formerly National Bureau of Standards, NBS).

Overrange The amount of flow a flowmeter can pass without damage or decreased service life.

Q Flow.

Repeatability The percent deviation a flowmeter will produce with identical flow characteristics.

Resolution The largest amount of output step changes possible as the measured amount is varied over its entire range. Generally defined as a percentage of full scale output.

Reynold's Number The ratio of inertial to viscous forces in a fluid, defined by the formula

$$Re = \frac{PVD}{u} \quad \text{or} \quad R = \frac{3160 \times Q \times Gt}{D \times u}$$

where

P = density of fluid

u = viscosity in centipoise

V = velocity

D = inside diameter of pipe

Q = flow rate

Gt = specific gravity of fluid

RF Pickup A modulated carrier pickup whose frequency is altered when a certain material passes within its sensing range.

Sonic Velocity The point at which flow velocity approaches the speed of sound. In flowmeters these shock waves cause erratic output.

Specific Gravity A number which defines the ratio of the weight of a fluid to the reference weight of water, both of the same volume. For gases, the weight of air is used as the reference point.

Flowmeter Type	Flow Range	Accuracy	Maximum Operating Pressure	Turndown Ratio	Pressure Drop	Cost
Turbine	0.25–50,000 gpm	0.25% of Reading to 1% full scale	6000 psi	10:1/35:1	High	Medium–high
Glass-tube rotameter	0.002 cm³/min–500 gpm	2% full scale	50–500 psi	10:1	High	Medium–high
Metal-tube rotameter	0.5–2500 gpm	2% full scale	1000–1500 psi	10:1	High	Medium–high
Positive displacement flowmeter	1 cm³/min–2000 gpm	0.25%–1% of Reading	3500 psi	100:1–2000:1	Medium–high	High
In-line spring and piston	0.05–300 gpm	5% full scale	3000 psi	10:1	Medium	Low
In-line vane	0.1–500 gpm	2%–5% full scale	300–2000 psi	10:1	Medium	Low
Orifice plate	0.35–11,500 gpm	1%–3% full scale	150–300 psi	5:1	Medium	Low
Mass	0–100 SLM	0.4%–1% of Reading	500–4500 psi	10:1	Low	High

Fig. 2.1. Flowmeter characteristic chart.

Standard Conditions Flowmeter manufacturers use 70°F and 14.7 psia for gases and 70°F for liquids as a standard reference point. Liquids are considered noncompressible.

Turbulent Flow Fluid that moves in random flow patterns, when inertial forces are greater than viscous forces, generally above a Reynold's number of 4000.

Turndown Ratio The range a flowmeter can accurately detect from maximum flow down to minimum flow.

UVC Universal Viscosity Curve.

Velocity The speed at which a fluid moves based on flow rate versus the ID of the pipe, generally expressed as fps = feet per second or fpm = feet per minute

Viscosity The measure of a liquid's molecules ability to flow or slide past each other. There are several methods used to determine a reference value of viscosity.

Figure 2.1 is a cross reference of flowmeter types commonly associated with fluid power testing. There are many overlapping classes of meters within the same category, so a detailed review of any one particular brand is necessary. This chart can be used as a general guideline in selecting a meter type for a particular test, but remember that flowmeter specifications do not always provide all information necessary for proper selection. A careful review of test requirements is always mandatory before purchasing a flowmeter system.

This chapter will provide a more in-depth review of each flowmeter type listed in the chart.

2.1.1 Turbine Flowmeters

Probably the most common meter used is the turbine flowmeter. This meter consists basically of a rotor supported by bearings and a pickup as shown in Fig. 2.2. As the rotor turns, the pickup senses a passing rotor blade and produces an electrical pulse output. The flowing media causes the rotor to turn at a rate proportional to flow. This pulse rate from the pickup is then translated by the electronic signal conditioner into usable engineering units such as gpm, in.^{3}/min, or cm^3/min. Generally the angle of the rotor blades is 20°–40° to the fluid flow. Smaller angles affect repeatability, while larger angles cause increased end thrust.

The K factor of a flowmeter is the relationship of flow versus pulse output. Each flowmeter is different owing to manufacturing tolerances, bearing type, and fluid viscosity, so this K factor is derived during calibration for each specific meter. Over the specified flow range of the meter, this K factor is relatively linear within a certain

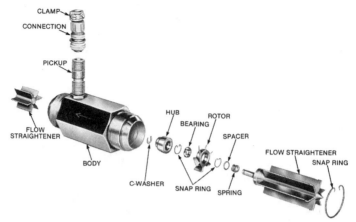

Fig. 2.2. Turbine flowmeter components *(Courtesy AMETEK, Inc., Cox Instruments.)*

tolerance band. This band is called linearity and is generally expressed as percentage accuracy of full scale or percentage accuracy of reading. (See Fig. 2.3.) When purchasing a meter of this type, it is very important to specify the fluid and viscosity so that a K factor is derived for the fluid used. Not only must the fluid viscosity change due to the temperature to be analyzed, but the viscosity change due to pressure, must be considered if the meter is installed in a system with a wide pressure swing. For a complete description of this phenomena, see Chapter 4, *Flowmeter Inaccuracy.*

Once the fluid viscosity is known, an accurate K factor can be produced. Depending on the accuracy required, calibration can be verified at multiple points along the flow range of the meter; these are generally 10-, 20-, or 30-point calibration checks. For example, a meter with a range of 1–20 gpm will be checked at every 1 gpm flow increment for actual pulse output. This will determine the linearity of that meter at that viscosity. If it is determined that the actual K-factor curve is outside the specified flowmeter accuracy, a linearization circuit can be applied to the meter electronics to compensate for this variation.

Turbine flowmeters typically have a flow range turndown of 10:1. This means that a meter with an upper limit of 20 gpm will go down as low as 2 gpm within the linear accuracy range. Again, with a linearizer or, rarely, where the fluid viscosity is acceptable at lower or higher flows, a meter can be provided with an extended turndown ratio of up to 100:1. To overcome magnetic drag on the low end, the

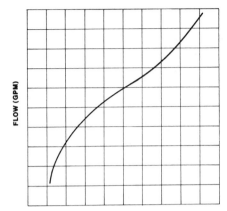

Fig. 2.3. Flowmeter linearity.

magnetic pickup can be replaced with a modulated carrier pickup that provides low drag, which equates to wider flow ranges. This requires different electronics in order to demodulate, filter, amplify, and shape the signal transmitted from the pickup. The magnetic pickup design produces a high-level output pulse that is easy to detect and, therefore, less electronic circuitry is required to produce usable information.

The meter requires laminar flow to provide accurate readings. Flow straighteners are installed before and after the rotor to accomplish this within the meter. It is also very important to install the meter in a section of straight tubing usually 10 diameters in length upstream of the meter and 5 diameters downstream. (This may vary slightly from one meter to the next.)

Another important consideration is flowmeter response time. Certain tests require instantaneous data to obtain high accuracy. Turbine flowmeters and their associated electronics can require up to several seconds before actual flow rate is detected. The turbine meters themselves have fast response, generally around 50 msec or so, but this depends on the system itself (the load) and how long it takes the column of oil to start moving and stabilizing. In addition to this time, add the response of the electronic conversion circuitry, which can run from about 100 msec to 3 sec depending on the method of pulse to readout conversion.

Typical turbine meter systems require about 20 cycles of pulse input before conversion signals stabilize. Then the actual response time would depend on the size or range of the turbine block and the actual flow being read.

TURBINE FLOWMETER ELECTRONICS

The electronics required for turbine flowmeters can be as simple as a pulse counter/converter, or may require several modules to compensate for error caused by temperature, nonlinearity, and so on. When selecting a turbine flowmeter, it is important to discuss all aspects of the application with a flowmeter manufacturer. Important considerations include operating pressure range, operating temperature range, minimum and maximum flow rate, fluid viscosity at several points, fluid density, pressure drop, accuracy required, response time, fluid lubricity, and filtration level. This information will allow the manufacturer to select the complete system he or she feels best suited for your application. All too often, an engineer assumes that published catalogue data will fit his or her system characteristics. All too often, this is a mistake.

The electronics system generally starts with a pulse rate counter which converts the signal (generally AC) into a workable engineering value. In cases where the meter is used within its standard range and the fluid viscosity is stable, the counter is generally sufficient. However, in high-accuracy systems other circuitry is used to maintain the ultimate accuracy possible.

A linearizer circuit is programmed to modify the sampled frequency and modify the input based on calibrated values for the particular fluid used. Some linearizer modules are available to change viscosity and specific gravity inputs so a wide range of fluids or wide viscosity-swing fluids can be used without flowmeter recalibration.

Temperature-compensation circuitry is also available to modify the sampled frequency input from the meter based on a programmed variation of linearity for the fluid used. This allows a meter to be calibrated at a mean temperature value, while being applied to a system with temperature variations outside the calibrated limit. Temperature is usually sensed by a thermocouple installed near the turbine meter. The thermocouple input to the compensation circuit modifies the flowmeter output proportional to the programmed deviation curve.

One problem that occurs on occasion is "zero bounce" while the meter is at zero or very low flowrates. This can be caused by the meter blades rocking back and forth or from unsuspected sources of outside pulses such as fluorescent light fixtures. The emitted frequency of the voltage cycle in AC fixtures can be detected by the pickup of a turbine meter. This "noise" can show up as an erratic flowrate at the low end of the meter scale, causing considerable confusion to the test engineer. Eliminating unwanted noise such as

this can be accomplished by bridging a capacitor across the two meter wires, turning the meter at a different angle to the emitted noise,
or using the extremely scientific method of wrapping certain areas
of the meter or display with aluminum foil.

2.1.2 Glass Tube Rotameters

Predecessor to the turbine meter, the rotameter was originally the
most widely used flowmeter. Unlike turbine meters, most rotameters cannot operate in high-pressure circuits, so they are usually
used in return lines. As depicted in Fig. 2.4 the rotameter consists
of a float within a uniformly tapered tube and a measurement scale.
The meter works on the variable-area principle. The float is pushed
upward into the tube until the opening between the float OD and
tube ID is sufficient to balance the upward force of the fluid with
the gravitational weight of the float. The higher the flow rate, the
larger the opening required and so the float rises proportionately.

The float can be provided in various designs depending on the
application. Some floats are viscosity sensitive in order to produce
wider flow ranges, while other designs include sharp edges at the
OD to negate most viscosity variations. The tube is generally constructed of glass or plastic and sometimes metal tubes are used for
higher-pressure service. In this case the float is magnetically cou-

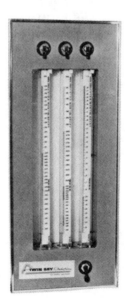

**Fig. 2.4. Rotameter-type flow-
meter.** *(Courtesy Twin Bay Industries.)*

pled to an outside indicator. Various float materials are used to provide higher and lower ranges for the same tube. Because the meter works on the fluid force versus float weight principle, increased or decreased float weight changes the reading.

The rotameter is a viscosity-sensitive device, so actual fluid viscosity at operating temperature and pressure should be known when ordering a meter.

2.1.3 Positive-Displacement Flowmeters

One of the most accurate flowmeters used for fluid power testing is the positive-displacement meter. Several design variations are available, which include gear type, helical gear, reciprocating piston type or wobble disk type. Figure 2.5 depicts a helical-gear-type meter with two intermeshing gears and a pickup. As fluid passes between the gears, causing them to rotate, the pickup provides a pulse output that is translated into a flow rate by the electronics package. Pulse rate is directly proportional to flow with little deviation caused by mechanical drag or viscosity. Internal tolerances of positive-displacement meters are extremely tight, so leakage is held to a minimum. Another feature of this meter is the extremely low-flow-rate capability possible. This allows positive-displacement meters to be very useful in leakage testing with highly accurate test results. Because of this low-end capability, another benefit produced is very wide flow rate ranges. Unlike turbine meters that eventually stall at the low end due to drag, the positive-displacement meter must keep turning. What goes in, must come out. Therefore, depending on how stable the electronic circuitry is at the low end due to filtering or pulse averaging time, a meter can provide measurement as low

Fig. 2.5. Helical-gear positive-displacement flowmeter. *(Courtesy Max Machinery, Inc.)*

as 0.0003 gpm with a 5000:1 turndown ratio, maintaining 0.5% of reading accuracy. Flow straighteners or straight lengths of tubing are not required.

These meters provide high accuracy, high turndown ratio, low flow capability, and viscosity change insensitivity and they have high-pressure capability. The few drawbacks include high price, higher pressure drops on the high end of the scale, and increased sensitivity to contamination.

POSITIVE-DISPLACEMENT METER ELECTRONICS

Because the viscosity's effect on the pulse rate is negligible in positive-displacement meters, no compensation is required for linearity, temperature, or pressure variations. A direct conversion of pulses/gallon is usually all that is required. This electronics package usually consists of a signal amplifier/converter and a pulse counter/converter. The counter senses pulses over a given time base to provide an average flow rate. This time rate affects the response time of the system. If the time base is too small, the readout could change too fast, resulting in difficult readings or data collection. A long time base results in a slow response system. Most packages available have an adjustment to provide an adjustable sample time to allow a stable reading with low response characteristics.

Some positive-displacement flowmeter systems can be provided with extremely fast response readings. Rather than using the number of pulses over a given time period method, they actually measure the time of one complete pulse. Response times for the full system can be as high as 15 msec.

2.1.4 Inline Piston and Spring Flowmeters

The inline piston and spring flowmeter (Fig. 2.6) is designed for general service and test systems where ± 5% full scale accuracy is sufficient. Visual indication of flow is provided by a "floating" indicator ring enclosed within an acrylic guard with a flow scale. The indicator is magnetically coupled to a sliding piston assembly, which acts against a spring. As flow increases, the piston is pushed against the spring until balance is achieved between spring force and the orifice opening in the piston. Orifice area changes with piston movement due to a metering cone, which increases area with increasing flow.

Viscosity has some affect on reading accuracy, as does fluid den-

Fig. 2.6. Inline piston and spring flowmeter. *(Courtesy Hedland, Division of Racine Federated, Inc.)*

sity. These meters are typically calibrated at one density value, but they can be calibrated with other fluid densities or a correction factor can be supplied for each meter.

2.1.5 Inline Spring and Vane Flowmeters

Inline spring and vane flowmeters can also be used where ± 5% full scale accuracy is acceptable. A swinging vane acting against a spring is forced back as flow increases. The vane is mechanically attached to a pointer, which indicates the rate of flow on a scale. The opening between the vane and the housing is designed to increase as flow increases, thus maintaining a low pressure drop regardless of flow. These units can be calibrated to the viscosity of most fluids.

2.1.6 Differential-Pressure Flowmeters

One method commonly used to detect flow is the differential-pressure technique. This principle is based on the fact that flow rate is related to the square root of the pressure differential across a restriction in the flow path. By measuring the drop across the restriction, flow can be determined based on this theory and through calibration results for the flowmeter itself. This design is very cost effective in larger line sizes. Typical differential-pressure flowmeter designs are orifice plate, venturi, flow tube, and nozzle types.

Figure 2.7 depicts an orifice plate flowmeter with removable orifice plates. The typical turndown ratio of this design is 4:1 because

Fig. 2.7. Orifice plate flowmeter.
(Courtesy Aqua Matic.)

the relationship of flow and pressure involves a square root. Because of this relationship, differential pressure drops off quickly as flow decreases. Accuracy is usually within 1–3% full scale. Recalibration is required to maintain accuracy if contaminates are present. The sharp-edge design of the orifice plate used to limit viscosity error can errode, thus changing orifice size and making the meter more sensitive to viscosity changes.

2.1.7 Mass Flowmeters

Mass flowmeters measure volumetric flow rates independent of density, pressure, and viscosity and are available to measure gas or liquid flow. Several methods are used to achieve this capability. One of the more common methods used is the Coriolis meter, where fluid flows through a U-shaped vibrating tube. The tube is vibrated at its natural frequency by a magnetic device at the bend of the U. This frequency is approximately 80 Hz and the tube movement is minimal, about 0.1 in. total. As flow enters the tube, the force of the fluid into the tube resists movement by pushing up on the tube, while existing fluid forces push down on the tube (Fig. 2.8). The higher the flow rate, the larger the resisting forces, causing the tube to

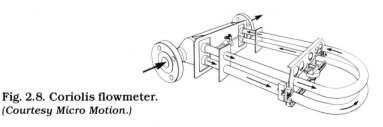

Fig. 2.8. Coriolis flowmeter.
(Courtesy Micro Motion.)

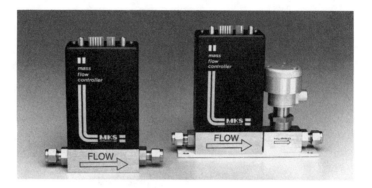

Fig. 2.9. Thermal-sensing flowmeter/controller. *(Courtesy MKS Instruments, Inc.)*

twist. The amount of twist is linearly proportional to flow rate. This change in movement is measured by magnetic sensors, and this signal is converted to a workable electronic signal for flow rate output.

Another type of mass flowmeter is a unit that uses thermal sensing as a detection technique. The capacity of a gas stream to convey heat is proportional to the product of mass flow and specific heat. Since the specific heat of gases varies only slightly with temperature and pressure, heat-transfer techniques of flow measurement will respond directly to mass flow.

A thermal-sensing flowmeter is depicted in Fig. 2.9. In this design, the flow sensor is a tube several inches long with a very small ID. The center of the tube is wound with two adjacent coils of electrical wire. These two coils uniformly heat the central section of tubing and act as resistance thermometers, which sense the temperature difference between the upstream and downstream coils. The coils are connected in series and power a part of a bridge circuit balanced at zero flow. When gas flows through the tube, the upstream coil is cooled more than the downstream coil and the bridge circuit becomes imbalanced. This imbalance is calibrated to provide a voltage output proportional to mass flow.

These meters are relatively accurate (± 1% full scale), have extremely low-flow sensing capability, and are excellent for detecting leakage flow rates. This design requires recalibration for a change of test media with different heat-transfer characteristics.

2.1.8 Paddle-Type Flowmeters

A paddle-type flowmeter incorporates a rotating paddle that provides a pulse output proportional to fluid velocity. The tip of the pad-

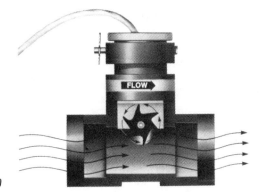

Fig. 2.10. Paddle-type
flowmeter. *(Courtesy
Beckman Industrial Corp.)*

dle protrudes into the flow stream (Fig. 2.10) and a magnetic pickup
is used to sense the pulse frequency. The frequency is then converted into actual engineering units based on the calibration of the
meter.

Accuracy of this type of unit is typically in the area of ± 1% full
scale.

2.1.9 Vortex Flowmeters

Vortex flowmeters can provide relatively accurate readings at high
flow rates. It is based on the principle that as fluid flows past an
unstreamlined object, called a "bluff body," the fluid boundary lay-
ers roll into vortices at frequencies proportional to the flow velocity
(Fig. 2.11). The vortex-shedding frequency is proportional to the
width of the bluff body. This ratio of proportionality is called a
Strouhal number.

The shed frequency is then sensed and converted to an output
linear to flow rate. Frequency can be sensed by a piezoelectric crystal
element that senses induced strain in the bluff body shedder bar.

Fig. 2.11. Vortex flowmeter
operating principle. *(Courtesy
Brooks Instrument Division, Em-
erson Electric Co.)*

Vibration causing error can generally be negated, because these units typically use two crystals out of phase with each other, which allows separation of mechanical noise and signal output with the electronic signal conditioner.

Other frequency-sensing methods are diaphragm pressure sensors beyond the shedder bar and ultrasonic beam sensors detecting vortices per time unit developed by a diametral strut.

2.1.10 Beaker Method

After all that is said about the many methods available to accurately detect flow rates and the sophisticated electronics required to perform these tasks, the beaker method is still one of the most widely used leakage flow-rate detection methods.

A graduated beaker and stopwatch are all the equipment that is required. Once a steady stream of fluid flows from an open line at the leakage test point, the beaker is placed under the stream and the stopwatch is started. Using 1 min as a test sample will provide exact engineering units in in.^{3}, cc, or gallons, based on the divisions on the beaker. This flow rate is unaffected by pressure, temperature, viscosity, or density. What you actually get is a slow, but cost-effective and accurate flow measurement.

2.2 Pressure and Vacuum

Pressure measurement is one of the most common tests applied to fluid power components. The ability of a pressure regulator to maintain set pressure, the proof and burst pressures of a component, the leakage of a component when using the pressure decay method, are just a few examples of the tests.

Pressure is generally defined in one of three terms: psig (pressure per square inch gauge), psia (pressure per square inch absolute), and psid (pressure per square inch differential). psig is defined as the pressure with atmospheric pressure as a zero reference point; psia is defined as a pressure with a perfect vacuum as a zero reference point. For example, at sea level 14.7 psi relates to the barometric or atmospheric pressure. Using this pressure, a pump output measured at 1000 psig would also be measured at 1014.7 psia, because the addition of barometric pressure is included. When calculating the pressure characteristics of a compressed gas, pressure is always figured in absolute in order to negate any outside effects. psid is basically the difference of two pressure readings used in pressure drop tests. A valve with 300 psig at the inlet, and 278 psig at the outlet at a certain flow, has a pressure drop of 22 psid.

Another common term is pressure in feet of head, generally used in defining the output of centrifugal pumps. This term describes the vertical lift capacity of a pump. For example, a centrifugal pump that is rated at 100 ft of head with water, will be able to pump water straight up a pipe 100 ft high before stalling out. Comparing feet of head to psig yields 1 ft of head = 0.4332 psig.

In lower-pressure measurement, a similar term is inches of water column or just inches of water. This term defines the pressure created by a vertical column of water. A typical use for this term is the measurement of liquid level in a storage tank; 1 in. of water = 0.03613 psig.

Vacuum measurement is used to define negative pressure below atmospheric pressure. psiv (pressure per square inch vacuum) is a common term as is in. Hg (inches of mercury). In hydraulic systems, pump suction capacity is defined as in. Hg. This term relates to the amount of negative pressure available to displace the weight of mercury in a tube; 1 in. Hg = 0.4912 psig.

There are several methods commonly used to measure pressure. All can provide accurate data, but the current trend is toward electronic measurement, which has advantages over all methods. These advantages include self-calibration, response time, ease of reading, temperature compensation, signal output for peripheral equipment, increased accuracy, and higher resolution. It is not uncommon, however, to find engineers requesting a pressure gauge next to a pressure transducer display as a backup. Unfamiliarity with the electronic principle of the transducer creates this lack of trust. In fact, some engineers believe that pressure gauges are more accurate than transducers, because they see the transducer display with a zero shift. They do not realize that the gauge does the same thing, except it does not have the sensitivity or resolution to show it.

2.2.1 Pressure Gages

Pressure gages have been used for many years as an accurate, reliable source of test data. There are many companies that produce gauges and much test data are available as a reference for selecting the proper gauge for critical applications.

A typical pressure gage is depicted in Fig. 2.12. A needle pointer moves across a calibrated dial face to depict sensed pressure. The most common method of converting pressure to pointer movement is the bourdon tube design as shown in Fig. 2.13. The tube itself is shaped like a "C," with fluid entering from one end and the other end sealed. As pressure increases, the "C" shape tends to straighten out, causing movement. Although this movement is small, the

Fig. 2.12. Pressure gage. *(Courtesy Marsh Instrument Co.)*

movement is amplified by some mechanical means (rack and pinion gears, etc.) in order to produce sufficient pointer travel. The accuracy of the translation of bourdon tube movement to pointer movement including mechanical friction defines the total gage accuracy. Typical gage accuracy can be anywhere from ± 5% to ± 0.1% full scale.

Other than range and accuracy, an important factor to consider when selecting a gage is the number of divisions on the dial face. The more divisions, the more accurate the reading. On very accurate test gages, a mirrored scale can be provided to prevent misreading due to parallax viewing. If an operator is standing at an angle to the gage, the distance between the pointer and the dial face can create a false reading. A mirror surrounding the divisions and behind the pointer, helps to alleviate this problem.

Fig. 2.13. Bourdon tube design. *(Courtesy Marsh Instrument Co.)*

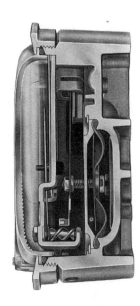

Fig. 2.14. Diaphragm gage design.
(Courtesy Dwyer Instruments, Inc.)

There are several mechanical methods used to convert pressure to dial movement, the bourdon tube method as just described, is the most common. For low-pressure measurement, the diaphragm method should be noted. As shown in Fig. 2.14, a diaphragm gage converts pressure into movement with a diaphragm connected to a mechanical linkage.

2.2.2 Manometers

An extremely accurate and relatively inexpensive instrument used for low-pressure measurement is the manometer as shown in Fig. 2.15. The device uses a U-shaped tube that is partially filled with a measuring fluid. One end of the tube is subjected to the test pressure, while the other end is either open to atmosphere for psig readings or connected to a secondary pressure source for psid readings. The measuring fluid is displaced as a balance of pressure versus the specific gravity of the fluid. The scale is divided to depict the weight of the fluid in terms of psi. Fluid on the pressurized side will be pushed down by the measured air pressure and forced up the vented side of the tube. The difference between these two columns of fluid as referenced to zero when both sides are balanced is the resultant pressure reading.

Other manometer designs, such as the well-type manometer, help ease readings by maintaining the reference side at a larger volume or by compensating the scale on the pressure side.

Fig. 2.15. Manometer. *(Courtesy Dwyer Instruments, Inc.)*

2.2.3 Pressure Transducers/Transmitters

A pressure transducer is a device that converts pressure to a corresponding current or voltage output. There are many variations in the technique used to accomplish this, and this sometimes causes confusion as to the type most suited for an application. The main differences, other than price, tend to be accuracy, output, pressure rangeability, temperature, resistance to vibration, and frequency response.

The following list contains common terms associated with pressure transducers.

Accuracy (Combined Error) Defines the overall accumulated accuracy rated as a percentage of full scale output. A transducer with ± 1% full scale accuracy with a range of 0–200 psig has an accuracy of ±2 psi. The combined effects of linearity, hysteresis, and repeatability, as well as other factors dependant on the transducer design, contribute to overall accuracy. Some manufacturers use the root sum of the squares (RSS) to predict total accuracy:

$$\text{Accuracy} = \sqrt{(\text{Linearity})^2 + (\text{Hysteresis})^2 + (\text{Repeatability})^2}$$

Dynamic Error Band This is the same as Static Error Band except with friction eliminated by light vibration.

Excitation This is the external voltage applied to a transducer in order to allow a proportional voltage based on pressure.

Frequency Response This is the speed at which a transducer follows pressure. Each transducer has a natural frequency of free oscillations of the sensing element. This frequency combined with the interface between the transducer and sensed media affect the frequency response. This rating is very helpful in determining a transducer's capability in tracking high response pressure fluctuations. For example, a transducer rated at 1000 Hz frequency response is capable of tracking up to 1 msec pressure fluctuations.

Full Scale Output This is the algebraic difference between the zero pressure output and full scale output.

Hook's Law This is the method on which strain gauge measurement is based. The gradient of Hook's line is defined by the ratio that is equivalent to the modulus of elasticity (Young's modulus).

Hysteresis This is the difference between transducer output at a specific pressure when pressure is raised to that pressure versus when pressure is lowered to that pressure.

Input Impedance This is the resistance measurement across the two excitation terminals of a transducer.

Insulation Resistance This is the resistance measurement between two insulation points when the transducer is subjected to a specific dc voltage at room temperature.

Linearity This defines a transducer's ability to follow a straight line of pressure versus output voltage over the full pressure range.

Null Voltage Output This is the voltage a transducer produces at zero pressure. Not all designs provide zero voltage at zero pressure.

Media Compatability This defines the transducer wetted components with various pressure media. This is important when working with exotic fluids and gases, as well as temperature extremes.

Overpressure Protection This is the pressure a transducer can withstand above its rated pressure without degradation of performance or failure.

Poisson Ratio This is the ratio between the strain of expansion in the direction of force and the strain of contraction perpendicular to that force.

Repeatability This defines a transducer's output deviation when sensing the same pressure at the same ambient conditions.

Sensitivity This defines the ratio of the change in output to the change of measured pressure.

Stability This is the ability of a transducer to maintain its operating characteristics within rated tolerance during its life cycle.

Static Error Band This represents the maximum deviation from best straight line drawn through the coordinates of 0% pressure range, 5% ± 1% output voltage ratio and 100% pressure range, 95% ± 1% output voltage ratio. This band includes the effects of linearity, friction, hysteresis, resolution, and repeatability and is expressed as a percentage of excitation voltage.

Type	Pressure Range (psi)	Linearity (Accuracy) (%)	Shock and Vibration Sensitivity	Frequency Response	Temperature Range (°F)	Temperature Effects	Cost
Unbonded wire	0–10K+	0.25%	Poor–good	0–2K+	−320–600	0.005%/°F	Medium
Foil	0–10K+	0.25–5%	Good	0–1K+	−65–250	0.01%/°F	Medium
Bonded semiconductor	0–10K+	0.5%	Excellent	0–1K+	−20–200	0.02%/°F	Low
Sputtered thin film	0–5K+	0.25%	Excellent	0–1K+	−320–525	0.005%/°F	Medium
Capacitance	0–200+	0.05%	Poor–good	0–100+	0–165	Needs control	High
Piezoelectric	0–10K+	0.1–1%	Poor	0–100K+	−450–400	0.01%/°F	Medium
Variable reluctance/inductive	0–10K+	0.5%	Good	0–1K+	−320–600	0.02%/°F	Medium
Differential transformer	0–10K+	0.5%	Poor	0–100+	0–165	—	Medium
Potentiometer	0–10K+	1%	Poor	0–50+	−65–300	Nonlinear 0.01%/°F	Medium
Diffused semiconductor	0–5K+	0.25%	Very good	0–1K+	−65–250	0.005%/°F	Medium

Fig. 2.16. Pressure transducer characteristic chart.

Temperature Compensation This defines the range that a transducer will perform to its specified tolerances with compensating circuitry, or just by the nature of its design.

Temperature Error (Thermal Sensitivity) This defines the variation of transducer output based on temperature deviation from room temperature.

Thermal Zero Shift This is the rating applied to define a transducer's zero output change based on temperature fluctuations. This shift affects the entire set range and shifts the total scale up or down accordingly.

Young's Modulus This is used to define the ratio of normal stress to strain.

Zero Balance This is the measured output at no pressure under room conditions. Absolute transducers use a value of 0 psia, while gage and sealed transducers use a value equal to atmospheric pressure.

Figure 2.16 is a cross reference of various pressure transducer types. This should be used for general reference, since new products are always coming into the market.

The following pages contain a more in-depth review of various transducer types. These specifications apply for both transducers (Fig. 2.17) and transmitters (Fig. 2.18). A transducer requires exter-

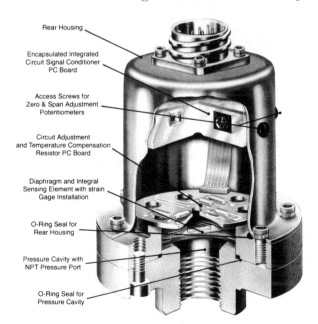

Fig. 2.17. Pressure transducer. *(Courtesy Viatran Corp.)*

Fig. 2.18. Pressure transmitter. *(Courtesy Viatran Corp.)*

nal signal conditioning, while a transmitter has the signal-conditioning circuitry built into the transducer housing.

STRAIN GAGE TRANSDUCERS

Strain gage transducers are based on the principle that solid material deforms as a result of imposed forces. A typical strain gage uses a conductive grid pattern of etched metallic foil mounted on a thin cross-section material. As the material stretches due to pressure, its resistance changes due to increased length and decreased cross-sectional area. This change in resistance is calibrated, and the output is proportional to pressure.

Strain gage transducers can use bonded metal foil, bonded semiconductor, and diffusion-bonded thin-film gage designs. A depos-

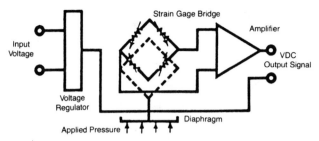

Fig. 2.19. Wheatstone bridge circuit. *(Courtesy Viatran Corp.)*

ited strain gage pattern can be formed on a diaphragm by means of thin-film deposition. By depositing the strain gage directly on the diaphragm, repeatability and hysteresis characteristics are greatly improved. A bonded semiconductor (silicon bar) is bonded to a diaphragm to form a relatively high output transducer. This type has the advantage of being compatible with many materials and good repeatability and hysteresis characteristics. A Wheatstone bridge circuit, as shown in Fig. 2.19, is used to supply electrical output. Excitation voltage is supplied to four strain gages from a regulated dc power source. As the element deflects, the resistance change unbalances the circuit, which changes the voltage across the bridge. This results in a very low mV/V output signal that is used directly as an output or that can be amplified for long transmission distances or to reduce noise interference.

Temperature compensation is required to null any effects of the modulus of elasticity of the strain element. Thermal expansion would otherwise cause the element to move, thus unbalancing the bridge circuit, providing a false output.

VARIABLE RELUCTANCE/INDUCTIVE TRANSDUCERS

Variable reluctance/inductive transducers use a diaphragm positioned between an E core, two coils, or a bourdon tube which drives an amature that rotates above an E core wrapped with two coils. As shown in Fig. 2.20, a diaphragm, when displaced by pressure, increases the gap between itself and one core, while decreasing the gap between itself and the other core. This upsets the balance of equal magnetic inductance between the coils when at zero pressure. Magnetic reluctance varies with change in gap and this determines the inductance value.

These coils are connected to a bridge circuit and are excited by an ac carrier voltage. The output voltage will vary based on the pressure applied to the diaphragm. This ac output can be fed into a carrier/

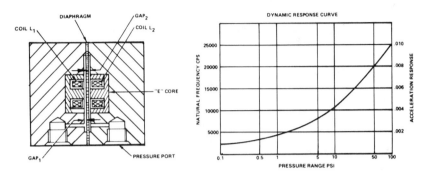

Fig. 2.20. Variable reluctance transducer design. *(Courtesy Validyne Engineering Corp.)*

demodulator that amplifies and rectifies (demodulates) the output to a dc output.

CAPACITIVE TRANSDUCERS

Capacitive transducers incorporate either two metallized quartz plates with a glass seal between them or a glass seal with silicon fused to both sides. The basic concept uses one side as the plate of a capacitor. As pressure deflects the diaphragm, it moves closer to the other plate increasing the capacitance of the device. This change of capacitance is not always linear and compensation may be required in the signal conditioner to linearize the output.

POTENTIOMETER TRANSDUCERS

Potentiometer transducers use a flexible pressure-sensing element to position a potentiometer, which provides an output based on pressure. The sensing element can be a diaphragm, bourdon tube, or any type of device that mechanically flexes under pressure. Output is not always linear with this device, so electronic linearization or calibration is provided to compensate for this error.

Excitation voltage is supplied to the potentiometer and the sensing element is connected to the wiper of the potentiometer. The wiper moves with pressure changes, thus producing a variable output voltage proportional to excitation voltage. Friction error is caused by the force required to mechanically position the wiper mechanism.

PIEZORESISTIVE TRANSDUCERS

Piezoresistive transducers work in a similar fashion to the strain gage type transducers described previously, but they use a crystalline silicon medium to sense pressure. This medium contains piezoresistors that are diffused into the material, overcoming problems such as temperature sensitivity and hysteresis due to thermoelastic strain and bonding degradation associated with wire strain gages. Silicon, being a perfect crystal, does not become perfectly stretched, returns to its original shape, and has excellent elasticity characteristics.

Typically, four piezoresistors are connected in a Wheatstone bridge circuit and as pressure causes the thin diaphragm to bend, the stress or strain is detected as the increase or decrease of resistance from each piezoresistor. Again, excitation voltage to the bridge is provided and output voltage is proportional to the excitation voltage and the amount or resistance change.

PRESSURE TRANSDUCER ELECTRONICS

Now that several types of pressure transducers have been reviewed, a brief description of the electronics required may be helpful. Also, note that there are many other transducer design techniques available that are not mentioned here. The types described are commonly used for fluid power testing, are ruggedly designed to withstand shock and vibration, and have strong signal output.

All transducers require excitation voltage, which either is provided from a regulated power supply or is supplied as part of the signal conditioner. A transducer cannot provide an output without an input. When choosing a transducer signal conditioner, check to make sure the excitation output and voltage input from the transducer are compatible.

The conditioner will generally have a zero and span adjustment to set the conditioner to the calibrated value of the transducer. Each transducer will be supplied with a calibrated mV/V output per some reference of pressure. This output value is appropriate for that transducer only, owing to error-causing mechanical and electronic effects. In some cases, shunt calibration is supplied on the conditioner, which can be used to simulate a pressure and is very helpful in setting the zero and span initially, as well as allowing easy periodic calibration checks. This is accomplished with a fixed resistor that is shunted across one arm of the strain gage bridge; it produces an unbalance equivalent equal to an "equivalent value." This value is derived from the shunt calibration output, the full scale output and the full scale display value and is expressed as

$$\text{Equivalent value} = \frac{\text{Shunt calibration output in mV/V}}{\text{Full scale output in mV/V}}$$
$$\times \text{Full scale transducer rating (psi)}$$

Type	Temperature range (°F)	Accuracy	Linearity	Response	Sensitivity (mV/°F)	Cost
Thermocouple						
J	32–1400	0.75–4%	Good	Slow	30–32	Low
K	32–1400	0.75–4%	Good	Slow	22–23	Low
T	−700– −300	0.75%–2%	Good	Slow	14–30	Low
E	32–1600	0.5–3°F	Good	Slow	39–44	Low
R	32–2700	0.5–5°F	Good	Slow	5–7	Low
S	32–2700	0.5–5°F	Good	Slow	5–7	Low
RTD	−400–1600	Good	Very good	Fast	—	Medium
Thermister	−150–650	Very good	Good	Fast	—	Low
Infared	32–6000	Good	Good	Medium	—	Medium
Bimetal thermometer	−100–1000	1–2% full scale	Fair	Slow	—	Low
Vapor tension thermometer	−40–450	1% full scale	Good	Slow	—	Medium

Fig. 2.21. Temperature monitor characteristic chart.

The signal conditioner can be supplied with an analog output for signalling peripheral devices, it can contain a digital readout, or it can contain both. In addition, some displays contain high/low switch points to signal when a set value is sensed and "peak hold" for locking in the highest pressure sensed. This is very useful and cost effective when trying to determine system shock data.

Pressure transmitters contain all electronics required except for operating voltage. An analog output can be used to provide a calibrated signal to a display or data-sensing device.

2.3 Temperature

Fluid viscosity is directly affected by temperature. The relationship of viscosity to fluid power component performance is an important factor when testing. Viscosity and temperature affect internal leakage, pressure drop, mechanical lubricity, response time, and more. Therefore, it is mandatory that test data be documented with fluid type and temperature to determine component performance at a specific point. Each fluid has its own unique viscosity versus temperature curve. This even applies to fluids of the same type and brand name. This curve should be reviewed and applied to test procedures.

There are several available methods to monitor fluid temperature as shown in Fig. 2.21. The following is a more detailed description of each type of monitoring device.

2.3.1 Bimetal Thermometers

Bimetal thermometers (Fig. 2.22) work on the principle of thermal expansion of dissimilar metals. These two metals are wound to-

Fig. 2.22. Bimetal thermometer. *(Courtesy Marsh Instrument Co.)*

gether in a helix or circular coil and are sized to produce a certain movement when subjected to a specific temperature change. The coil is mechanically linked to the gage pointer to depict actual temperature. This type of thermometer is useful when direct mounting at the temperature source is possible. Make certain that the proper stem length is chosen in order to ensure that the temperature-sensitive end of the stem is directly in the flow path of the fluid.

2.3.2 Vapor Tension Thermometers

Vapor tension thermometers (Fig. 2.23) are used for remote reading of temperature. The system consists of a closed, vapor-filled bourdon tube and sensing bulb connected by capillary tubing. The sealed vapor expands and contracts with temperature changes which causes increases and decreases of vapor pressure within the closed system. The gage itself consists of a pointer mechanically linked to the bourdon tube just as described under pressure gages. The system is, in fact, a pressure-sensing arrangement. Because of the design, this type of thermometer can be recalibrated via a face-mounted adjustment screw.

2.3.3 Thermocouples

The most widely used electronic temperature sensor for fluid power testing is the thermocouple (Fig. 2.24). Its operation is based on the Seebeck principle: When the junctions of two dissimilar met-

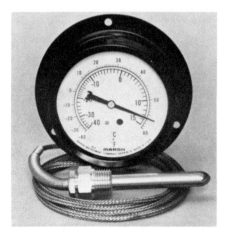

Fig. 2.23. Vapor tension thermometer. (Courtesy Marsh Instrument Co.)

Fig. 2.24. Thermocouple probe. *(Courtesy Minco Products Inc.)*

als forming a closed circuit are exposed to different temperatures, a net electromotive force is generated that induces a continuous electric current. If the circuit were open on one end, the circuit would produce a voltage based on the junction temperature and the composition of both metals used.

Unfortunately, the system becomes more complicated because of the interconnection of the thermocouple and the electronics required. The wire used between the two also produces a voltage effect, which must be compensated for. This interconnection should be reviewed when a thermocouple system is purchased. In order to negate this effect, a reference junction is created so that the thermocouple system measures the temperature difference between the reference junction and the measuring junction (the point of measurement). This is accomplished by an electronic circuit consisting of a regulated power supply, a combination of fixed resistors, plus a temperature-sensitive resistor designed for the particular thermocouple type.

The output voltage versus temperature is nonlinear, so electronic signal conditioners must compensate for this nonlinearity either by using polynomial equations or by referencing lookup tables based on NBS values for thermocouple voltage output. In addition, the voltage output is extremely small over the entire range, so system resolution must be extremely high. For example, a type K thermocouple voltage output changes 4 μV for a 0.1°C temperature change. This sensitivity then can be compounded by external noise, which can be picked up by the thermocouple wire. There are several methods of reducing or eliminating external noise; these methods include analog filtering, tree switching, integration (noise averaging), and guarding (shielding).

Thermocouples are manufactured with different materials that are rated for various temperature ranges and limits of error per ANSI Circular MC96.1-1975. These thermocouple types are as follows:

   T—Copper versus copper–nickel (copper–constantan)
      Temperature range: 0–350°C/32–700°F
      Standard error limits: 0–133°C = ± 1°C/133–350°C = ± 0.75%
      Special error limits: 0–133°C = ± 0.5°C/133–350°C = ± 0.4%
   J—Iron versus copper–nickel (iron–constantan)
      Temperature range: 0–750°C/32–1400°F
      Standard error limits: 0–203°C = ± 2.2°C/293–750°C = ± 0.75%
      Special error limits: 0–293°C = ± 1.1°C/293–750°C = ± 0.4%
   E—Nickel–chromium versus copper–nickel (chromel–constantan)
      Temperature range: 0–900°C/32–1600°F
      Standard error limits: 0–340°C = ± 1.7°C/340–900°C = ± 0.5%
      Special error limits: 0–340°C = ± 1°C/340–900°C = ± 0.4%
   K—Nickel–chromium versus nickel–aluminum (chromel–alumel)
      Temperature range: 0–1250°C/32–2300°F
      Standard error limits: 0–293°C = ± 2.2°C/293–1250°C = ± 0.75%
      Special error limits: 0–293°C = ± 1.1°C/293–1250°C = ± 0.4%
   R,S—Platinum versus platinum w/% rhodium (R = 13%, S = 10% rodium)
      Temperature range: 0–1450°C/32–2700°F
      Standard error limits: 0–600°C = ± 1.5°C/600–1450°C = ± 0.25%
      Special error limits: 0–600°C = ± 0.6°C/600–1450°C = ± 0.1%
   B—Platinum–6% rhodium versus platinum–30% rhodium
      Temperature range: 800–1700°C/1600–3100°F
      Standard error limits: 800–1700°C = ± 0.5%
      Special error limits: No rating available

The following ratings are for temperatures below 0°C:

   T—Copper versus copper nickel
      Temperature range: −200–0°C/−328–+32°F
      Standard error limits: −200–−66°C = ± 1°C/−66–0°C = ± 1.5%

E—Nickel chromium versus copper nickel
  Temperature range: $-200-0°C/-328-+32°F$
  Standard error limits: $-200--100°C = \pm 1.1°C/-100-0°C$
  $= \pm 1\%$
K—Nickel chromium versus nickel aluminum
  Temperature range: $-200-0°C/-328-+32°F$
  Standard error limits: $-200--110°C = \pm 2.2°C/-110-0°C$
  $= \pm 2\%$

In addition to ANSI standards applied for temperature limits of error, the thermocouple wire is manufactured per NBS standards, which detail thermoelectric voltage in absolute millivolts per temperature (per every 1°C).

2.3.4 Resistance Temperature Detector

The resistance temperature detector (RTD) is based on the principle that the resistivity of a metal changes with temperature. Common metals used are platinum, nickel, and copper owing to their resistivity characteristics. Figure 2.25 depicts a common RTD probe, which can be supplied in several configurations.

A low-voltage regulated power source is usually supplied to an RTD and three fixed resistors in a Wheatstone bridge circuit. The resistance variation is measured and, depending on the material, the output is somewhat linear. The curve is then linearized through the use of the Callendar–Van Dusen equation used to approximate the RTD curve.

The RTD is very sensitive, has fast response, and can be extremely accurate. However, an RTD is more fragile than a thermocouple. Another problem encountered is self-heating due to supply voltage. This induced heat changes the resistivity of the metal and thus creates an error. To correct this problem, either lower supply current or a larger RTD can be used to balance the system.

2.3.5 Thermister

Another device used, which is constructed of sintered semiconductor material, is the thermister. This device is also based on the

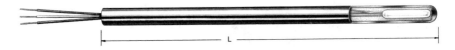

Fig. 2.25. RTD probe. *(Courtesy Claud S. Gordon Co.)*

resistance principle, where resistance decreases with increasing temperature. A thermister is relatively inexpensive, but owing to its nonlinear characteristics, more complex electronics are required to convert the output to actual temperature. This output can be some-what approximated by the Steinhart–Hart equation.

The definite advantage of a thermister is its excellent sensitivity. This is approximately 40 mV/°C, where the RTDs are at about 2 mV/°C and thermocouples are at about 0.05 mV/°C.

2.3.6 IC Sensors

Integrated circuit (IC) sensors are currently being used as a tem-perature monitor in certain applications. Their use is somewhat limited because until recently they have been too costly. These de-vices are available as current or voltage output devices and are rela-tively accurate and linear. They sense temperature by measuring the temperature variance in the base–emitter voltage (V_{BE}) of a silicon transistor. Typical output is -2.26 mV/°C, which requires output amplification.

2.3.7 Temperature Indicators/Controllers

Temperature measurement is generally used for one of two (or both) reasons. First, fluid or air temperature is required to check component performance based on fluid temperature or ambient conditions. Fluid temperature readings are needed for viscosity, so accurate temperature readings are needed for viscosity-sensitive tests. Second, fluid temperature must be controlled at specific tem-peratures so these readings may be taken.

Thermocouples, RTDs, thermisters, and IC sensors all require sig-nal conditioning and processing to convert the electric output of the device to a useful engineering unit. The mating of the proper signal conditioner is mandatory if accurate measurements are to be taken. Figure 2.26 depicts a typical temperature indicator/controller. Aside from containing the signal processing and temperature display, it also provides output for control of external heating or cooling sub-circuits.

Controllers typically used for hydraulic testing either turn a device on or off for heating and cooling or provide a variable output for modulating a device. For cooling, an on–off cycle would be a solenoid valve on the water feed line of a water/oil heat exchanger. This is usually sufficient to maintain fluid temperature to $+$ 5°F of set point. If higher accuracy is required, a modulating water valve can

Fig. 2.26. Temperature indicator/controller. *(Courtesy Fenwal Inc.)*

be controlled by a variable output signal to open and close over a proportional range. These valves generally require an interface device between the controller and valve actuator for proper operation. For example, a valve can be pneumatically operated over a 3–15 psi signal, 3 psi being closed and 15 psi being fully open. Within the pressure range the valve controls water flow rate. The controller, with an output of 4–20 mA or 0-5 Vdc must interface into an electric to pneumatic proportional regulator to provide the 3–15 psi output required by the valve.

System heating in most cases is generated by an on–off immersion or circulating heater. The heater is applied when assistance is required to bring temperature up to test temperature within a given period of time. The heater is usually started by a motor starter–relay. Some systems are capable of generating sufficient heat for all tests required, and cooling circuits are used to maintain temperature at an upper limit.

For these reasons, a temperature controller used for fluid power testing will either have one or two set point capability, either on–off or proportional output. A controller can be supplied with internal relays with sufficient current capacity to activate directly starter coils or solenoid valves for on–off applications. The options available for proportional control should be discussed to outline the flexibility of control.

Rate of Automatic Reset This function allows a controller to sense rapid changes in temperature and allow for compensation due to these variable temperature changes. Reset is used to monitor this rate and eliminate any offset between set point and sensed temperature.

Heat and Cool Gains When a fluid is to be heated and cooled to maintain a certain temperature, the gain is adjusted to provide an on–off cycle proportional to the cycle of the system thermal dynamics. If temperature drops below the proportional band, heating is on continuously and above the band, cooling is on continuously.

Proportional On–Off This allows a small band width at a set point to provide proportional output on either side of set point. For example, if temperature is increasing to set point, a cooler can start to prevent temperature from overshooting set point. This feature, when used at the other extreme, will start a heater to prevent system fluid from dropping too low in temperature.

PID Control This is a popular feature used to maintain accurate temperature control. PID stands for proportional, integral, and derivative. Proportional control is based on a set proportioning time to compensate for increase/decrease temperature rates. Integral control adjusts the proportional bandwidth with respect to the set point to compensate for system error due to overshoot of the controller. This overshoot is based on the time temperature stabilizes due to the time proportioning function of the controller. Derivative control senses the rate at which temperature is increasing or decreasing and adjusts the controller cycle time to avoid overshoot.

2.4 Speed (rpm)

When testing pumps or motors, it is necessary to obtain accurate speed data, since flow, input torque, and output torque are related to speed. Speed readings are extremely accurate when measured with magnetic pickups. These sensors, combined with a multiple tooth gear, are used quite extensively in all areas of speed measurement. Other methods such as mechanical linkages and photoelectric sensors with reflecting reference points are used, but, by far, the magnetic sensor is the leader in the fluid power test market.

There are two basic pickup designs available: passive and active (zero velocity). Passive pickups convert mechanical motion to an ac pulse output. As shown in Fig. 2.27, a passive pickup consists of a coil, pole piece, magnet, and housing. These components produce a magnetic field that, when altered by a ferrous gear tooth passing by, generate an ac voltage in the coil. The ac voltage is proportional to the rate of change of magnetic flux and is proportional to speed generally. The frequency of the ac output signal is exactly proportional to rpm.

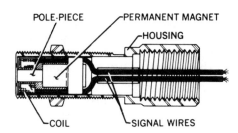

Fig. 2.27. Passive tachometer
pickup design. *(Courtesy Airpax.)*

This frequency is then read and translated into rpm by a frequency signal conditioner. A typical gear used for pickup can have
48–120 teeth for most common applications. For example, a 60-
tooth gear is used and it is desired to read rpm on a display. One
revolution will generate 60 pulses, so with a speed range of 500–
5000 rpm we will have a frequency output of 30,000–300,000 Hz.
The signal conditioner will be set to divide the frequency pulses by
60 to convert input frequency to actual revolutions. Also, the signal
conditioner must sample input frequency compared to time to convert frequency to rpm (a time-based unit).

Several important considerations should be reviewed when
matching a pickup to a gear. The distance between the pickup is
critical in order to provide a good signal as shown in Fig. 2.28. A
sinusoidal output pulse is the desired signal, therefore, gear width,
gear spacing, gear shape, and gear height all can provide unwanted
characteristics. Several pickup designs are available to help tailor
the output signal based on gear configuration.

Another important consideration is the surface speed of the gear,

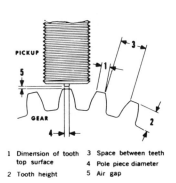

1 Dimension of tooth   3 Space between teeth
   top surface   4 Pole piece diameter
2 Tooth height   5 Air gap

For maximum output and approximately sinusoidal
waveform, the following conditions should obtain:

(1) equal to or more than (4)
(2) equal to or more than (3)
(3) approximately three times (4)
(5) as small as possible, .005″ typical.
Gear width equal to or more than (4)

Fig. 2.28. Pickup and gear distance. *(Courtesy Airpax.)*

since output voltage is dependent on a minimum surface speed. Pickup output voltage is also affected by the gap between the gear and pickup, the gear tooth size, and the load impedence connected to the pickup. This relationship can be expressed as

$$\text{Surface speed (ips)} = \frac{\text{rpm} \times \text{diameter (in.)} \times \pi}{60}$$

Your lowest speed requirement should be reviewed prior to purchasing a passive magnetic pickup system. Generally, in the speed ranges encountered for pump and motor tests, 100–200 rpm is about the lowest stable speed reading possible.

To overcome this low-end sensing problem, the active or zero-velocity pickup can be used. It is not dependent on velocity for an output. It senses the position of the gear tooth using Hall-effect-type principles and produces constant amplitude logic pulses from zero to very high speeds. Each time a gear passes, the output switches from zero to a positive 5 V. Unlike the passive pickup, where voltage output is dependent on a number of factors, the active pickup produces clean square-wave output pulses regardless of speed.

2.5 Torque

Measurement of rotational force (torque) defines the power required to drive a pump, or the power a hydraulic motor or rotary actuator can produce. Torque is generally defined in inch pounds or foot pounds when testing hydraulic motors. When testing pumps, torque is converted to horsepower and the following formulas apply:

$$HP = \frac{\text{torque (in.-lbf)} \times \text{rpm}}{63025}$$

$$HP = \frac{\text{Flow (gpm)} \times \text{pressure (psi)}}{1714 \times [\text{efficiency (overall)} \times 100]}$$

$$\text{torque (in.-lbf)} = \frac{HP \times 63025}{\text{rpm}}$$

$$\text{torque (in.-lbf)} = \frac{\text{flow (gpm)} \times \text{pressure (psi)} \times 36.77}{\text{rpm}}$$

$$\text{torque (in.-lbf)}/100 \text{ psi} = \frac{\text{displacement (in.}^3\text{/rev)}}{0.0628}$$

$$\text{torque (ft.-lbf)} = \frac{HP \times 5250}{\text{rpm}}$$

2.5.1 Moment Arm Torque Method

Torque can be measured in several ways, depending on the budget and accuracy requirements. One method used to test motors in a static condition is by attaching an arm to the output shaft and placing a force gage at a measured point on the arm, say at 1 ft from the motor shaft centerline. If the force gauge were calibrated in pounds of force, the actual reading would be read as foot pounds of torque.

A similar method involves mounting an electric-motor-driven dynamometer on bearings, for and aft, that are concentric with the drive shaft. Motor reaction torque, which equals pump driving torque, is measured at a torque arm on the motor housing.

2.5.2 Hydraulic Pressure Torque Measurement

It is common practice to drive pumps under test with a hydraulic motor, or to load motors under test by driving a hydraulic pump across a relief valve. In either case it is possible to determine torque by calculating the equivalent torque of the drive motor or drive pump. By reading an accurate pressure gage and knowing the torque efficiency of the drive motor or driven pump the following formula can be used to approximate torque:

$$\text{torque (in.-lbf)} = \frac{\text{pressure (psi)} \times \text{displacement (in.}^3/\text{rev)}}{2\pi}$$

This is a theoretical calculation. Efficiency should be taken into account based on pressure and speed of the drive motor or driven pump.

2.5.3 Electric Motor Current/Torque Conversion

When testing a pump, and driving it with an electric motor, it is possible to measure amperage draw of the motor to determine input horsepower. However, there are several problems associated with this method. The input voltage to the motor can fluctuate, thus producing an error when reading amperage. Also, motor efficiency can vary and usually torque output decays with service life. This relationship can be expressed as

$$\text{HP} = \frac{\text{volts} \times \text{amps} \times \text{efficiency}}{746}$$

In order to compensate for the amperage reading variance due to input voltage, measurements can be taken in watts. This is a value that is expressed as:

$$watts = volts \times amperes$$

$$amperes = \frac{watts}{volts}$$

Again, motor efficiency must still be taken into account to generate a close approximation of torque.

2.5.4 Slip Ring Torque Sensors

Slip ring torque sensor design allows constant torque readings during rotation. It operates on the strain gage principle and can be used with ac or dc excitation. Connection of the strain gage to the signal conditioner is accomplished with four silver graphite brushes rubbing against four silver slip rings mounted on the rotating shaft (Fig. 2.29).

The frequent problem encountered with this design is the wear and cleaning required between the slip rings and brushes. Brush wear requires changing the brushes. Wear can be decreased by lifting the brushes off the slip rings when torque readings are not necessary. The transducer can usually be equipped with either a manual or pneumatic brush-lifting mechanism for this purpose. Cleaning the contact surfaces between the slip rings and brushes helps maintain a satisfactory signal to noise ratio.

2.5.5 Rotary Transformers

Rotary transformer design eliminates the problems associated with the slip ring design. Ac excitation and signal output are trans-

Fig. 2.29. Slip ring torque sensor design. *(Courtesy Eaton Corp., Lebow Products.)*

Fig. 2.30. Rotary transformer
torque sensor design. *(Courtesy
Eaton Corp., Lebow Products.)*

fered between the rotating shaft and the stationary housing (Fig. 2.30) by two rotating transformers.

The primary and secondary windings of the two transformers are separated, one on the rotating member, the other on the stationary. Magnetic flux lines are produced by applying a time-varying voltage to one of the coils. This noncontact transmission of excitation and signal provides an accurate method of torque monitoring with little maintenance required. Again, the strain gage is used as the actual torque sensing element.

2.5.6 Torsional Variable Differential Transformer (TVDT)

The TVDT measures torque by monitoring the deflection of three tubular pieces of magnetic material mounted on a nonmagnetic shaft. An air gap between each piece, at a 45° angle to the shaft, provides an increasing or decreasing gap dimension proportional to torque. This gap is monitored by two transformers, which are affected by the magnetic reluctance of the gaps.

Ac voltage is applied to the primary sides of each transformer. The secondary windings are connected to a signal conditioner, which deciphers the variation in voltage output. Voltage output magnitude is proportional to torque.

2.5.7 Phase-Shift Torque Systems

Phase-shift torque systems use two magnetic pickups sensing pulses produced from two external tooth gears mounted on a common shaft. Shaft material is designed to allow torsional movement during torque situations. When the shaft is rotating at zero or very low torque, the output pulses from each pickup are in phase with each other. However, as torque is applied, the pulse outputs sepa-

Fig. 2.31. Torque table. *(Courtesy Eaton Corp., Lebow Products.)*

rate at a greater distance with increasing torque. This separation of pulse output is called phase shift.

The electronics require measuring this time difference between both pulses. This difference (or phase shift) is proportional to torque and is converted to a dc signal to power the torque readout.

2.5.8 Flanged Reaction Torque Sensors

Reaction torque sensors monitor turning force of a device. They are not designed to rotate except for the slight wind-up inherent to their design (typically 0.5–1°). They are generally mounted between two opposing torque devices such as a motor and pump with the output shafts rotating freely inside the sensor housing. The sensor actually monitors the twisting force of the motor and pump housing. When torque is applied, the sensor produces an output proportional to torque. Again, the strain gage method is used.

2.5.9 Torque Tables

This is another reaction torque sensing device that measures turning force (Fig. 2.31). The test device is mounted directly to the table and the output shaft is connected to a rotating fixture. The table is supported by four flexure straps at each corner of the table. Each strap contains a strain gage, each one part of a four-arm bridge.

As torque is applied, the outputs of the strain gages are monitored and the resultant variation is processed and calibrated to the corresponding torque value.

2.6 Force

Force measurement is used to define power exerted in a linear direction of travel and is usually expressed in pounds of force. For expressing low force values, ounces or grams of force are used.

Again, depending on the accuracy required and the budget, several methods can be used for this measurement.

2.6.1 Spring Rate Method

This is probably the least expensive method available. Springs are designed with a predictable rate of force. This rate is determined by wire diameter, material, coils per inch, free length, and compressed length. If these values are known, it is possible to calculate actual force required to move a spring to a specific compressed length. If force measuring equipment is available to provide spring length versus force data, accurate results are possible, provided the spring is designed so that material fatigue does not affect repeatability.

This measurement technique can be assembled for a specific purpose, or commercially available units can be used. These devices are referred to as spring force gages and operate on this principle. Force gage mechanisms are calibrated to convert spring movement to actual force displayed on a gage.

2.6.2 Hydraulic Pressure Method

If a hydraulic cylinder were used to generate force, the hydraulic pressure used to start movement could be converted to pounds of force by the formula: force (lb) = pressure (psi) × cylinder surface area (in.^{2}). A reliable method would be required to determine the exact time the measured component started moving to record accurate starting force.

Conversely, if you wanted to measure force exerted by a component, a cylinder could be used to generate a pressure proportional to force. In this method the cylinder volume would be filled with fluid and a pressure gage would be connected to the cylinder outlet port. After the cylinder was positioned to the proper stroke, and the test component was attached to the cylinder, the cylinder outlet would be closed through valving. As the test component generated force, pressure would be generated in the trapped cylinder chamber, and this pressure would be indicated by the gage. This technique is only viable if measurement of force at a given point is acceptable. Exact positioning is also difficult owing to the compressibility of the fluid and mechanical tolerances.

2.6.3 Load Cells

Load cells are the preferred linear force measurement device used where budget permits. Load cells come in several designs as de-

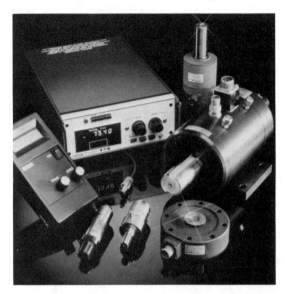

Fig. 2.32. Several load cell types. *(Courtesy Eaton Corp., Lebow Products.)*

picted in Fig. 2.32, which depicts a rod load cell designed specifically for cable tension force. Figure 2.33 shows a general purpose tension/compression load cell, and Fig. 2.34 shows a load beam cell design.

All load cells depicted operate on the strain gage method with mechanical design of the load cell producing predictable and repeatable motion proportional to applied force. The strain gages are connected in a four-arm Wheatstone bridge arrangement, providing a voltage output proportional to force. Variations in load cell design provide

Fig. 2.33. Tension/compression load cell. *(Courtesy Eaton Corp., Lebow Products.)*

Fig. 2.34. Load beam cell. *(Courtesy Eaton Corp., Lebow Products.)*

configurations to fit into different mounting configurations, allow for various load limits and uneven applications of load forces.

There are several considerations that should be reviewed when applying a load cell. The maximum combined force, in both linear and nonlinear motion, should be calculated. These forces should be determined in both a static or a dynamic operation to verify that the load cell is not subjected to combined forces over its rated capacity.

Response of the load cell and associated electronics should be reviewed when high-response component testing is required. The combined response time of the strain gage bridge, electronics, and transducer structure, as well as the input types, all contribute to this problem.

Another potential problem is the relationship of the natural frequency (ringing frequency) between the load cell and force being measured. The laws of motion state that when a force is applied to a material, the material in achieving equilibrium exerts an equal force which results in oscillation. It is important that the natural frequency of the load cell and force be analyzed to prevent resonance produced by matching the natural frequencies of both.

Mechanical problems associated with measuring force include fatigue loading, off-axis mounting between the force and load cell, impact loading (high force shock cycles), and mechanical stress. These factors can cause premature failure of the load cell or produce inaccurate data due to degradation of performance or improper mounting techniques.

2.7 Position

Position of a component under test can be important information in determining operating characteristics. Examples of position tests include spool position versus pressure drop, rotary actuator torque versus position, cylinder or actuator piston leakage testing, solenoid force versus position, and manual valve performance versus position.

There are several methods use to derive accurate position data. Some applications may be difficult to measure due to system resolution required. For example, a common test for hydraulic cylinders is to measure piston leakage by monitoring rod drift while the cylinder is under pressure. Stroke length can vary and rod drift acceptance levels can be as low as 0.001 in. If a cylinder has a stroke of 10 in., the degree of resolution or accuracy of a position transducer with a stroke of 10 in. would have to be 0.0001% of reading over its full stroke. This would be asking too much from any commercial transducer. Techniques can be used to allow this reading to be taken accurately. In this case, a transducer with a stroke length of 0.1 in. can be allowed to follow the cylinder and lock in position and zero at a reference point. A transducer with $\pm$ 1% of reading accuracy would then provide the $\pm$ 0.001 in. resolution required.

2.7.1 Linear Variable Differential Transformers (LVDT)

The LVDT (Fig. 2.35) is a precise measuring device that provides position feedback. A moving rod is coupled to a magnetic core, which moves axially within a primary and two secondary coil windings of a transformer. Two secondary windings are placed on either side of a primary, which is supplied with a dc excitation voltage. In a centered (or null) position, zero voltage between the two secondary windings is sensed. This is because the output of each winding is equal, but 180° out of phase with each other. The resultant output is thus zero.

As the rod is moved, the output of each secondary becomes unbalanced in both voltage and phase. This imbalance is translated to provide position output and direction of movement. Position output is proportional to secondary voltage and phase determines position from null position.

When applying an LVDT to a system, it is important to consider the accuracy of measurement tolerances desired. For example, an LVDT can be used in cylinder tests where cylinder dimensions, as well as rod drift when testing for piston seal leaks, can be monitored. If a cylinder retracted length was 12 in. and stroke length was

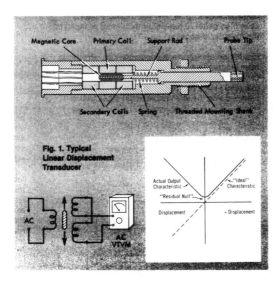

Fig. 2.35. Linear variable differential transformer (LVDT). *(Courtesy Daytronic Corp.)*

20 in., an LVDT with an 8 in. length would be sufficient to check cylinder external dimensions. However, an LVDT with 0.25% full scale accuracy with an 8 in. total travel translates to an accuracy of ± 0.020 in. Now, if the same LVDT were used to monitor piston leakage during hydrostatic drift testing, the best detection you could expect is ± 0.020 in. This is totally unacceptable in most applications, since 0.001 in. movement is sometimes cause for rejection. The obvious solution here is to use an LVDT with less overall travel for seal leak testing or a device with higher resolution for monitoring both tests.

2.7.2 Magnetostrictive Linear Displacement Transducers (LDT)

The LDT is probably the most accurate linear position transducer regardless of stroke length. These transducers can be provided with stroke lengths from 1 in. to 30 ft depending on the application. Hydraulic cylinder manufacturers also offer this device built into a cylinder to provide an actuator with precise position feedback. This eliminates external mounting of position transducers, thus protecting the sensing mechanism from ambient conditions.

The principle used to produce this exceptional accuracy is called magnetostriction. When the magnetic field is changed around a ferromagnetic material, a strain is induced in the material. The Tem-

Fig. 2.36. Magnetostrictive linear displacement transducer (LDT). *(Courtesy Temposonics/MTS Systems Corp.)*

posonics LDT (Fig. 2.36) uses a patented system, which can be provided with either an analog or a digital output. In their design, a torsional strain pulse is induced in a specially designed magnetostrictive tube by the momentary interaction of two magnetic fields. One of the fields emanates from a permanent magnet that passes along the outside of the tube. The other field is produced by a current pulse launched along a wire inside the tube. The interaction between the two fields produces a strain pulse that travels at ultrasonic speed down the tube that serves as a waveguide and is detected by a coil arrangement at the end of the device. The waveguide is spring loaded within a stainless-steel 3/8-in. OD rod. The position of the permanent magnet outside the rod can be pinpointed by measuring the lapsed time between launching the electric pulse within the tube and the arrival of the resulting strain pulse at the end of the tube.

The time interval for these pulses is related to modulus and density of the magnetostrictive waveguide. The pulses are produced by an oscillator, generating continuous pulses at a precise rate. The returning stress pulses are detected and a pulse width signal (based on the magnet's position) is developed. The pulse width is determined by a quartz crystal oscillator.

Rated accuracies of the Temposonics unit are ± 0.05% stroke or better linearity and ± 0.001% or 0.0001 in. (whichever is greater) repeatability. Output resolution on analog models is rated as stepless continuous output, while digital output systems are supplied up to 16 bits (65,536 parts) with an optional 18 bits (262,144 parts) resolution output. In addition to providing position measurement, velocity measurement can be easily attained with optional electronics.

2.7.3 Rotary Encoders

These devices can be used for either rotary or linear position measurement. There are several methods used to provide a digital or analog output proportional to position. Some encoders are limited to travel of less than 360°, while others can be used for continuous rotation.

One method used contains an optical sensor, monitoring light either reflected or interrupted by a rotating grid disk. The disk contains opaque and transparent segments divided along the disk face. An LED provides a light sensor and a photodetector senses when light is passed or blocked through the disk segments. The greater the amount of segments, the greater the resolution of the instrument. Some encoders provide only a single pulse (incremental encoder), while others sense several bits of information simultaneously (absolute encoder) to provide a digital word of position information. The grids are produced with high-resolution photolithography on a glass, plastic, or metal disk. Multiturn units can be provided with a pulse per one revolution, so that number of revolutions as well as position within that revolution can be monitored.

Another method used is the variable differential capacitance encoder used to measure angular position (Fig. 2.37). Rotation is limited to 120° total travel on the model shown. It is very useful and economical for accurately measuring angular movement, such as required in rotary actuator testing.

This device contains two opposing sets of stator plates and rotors. The rotors, each attached to the input shaft, move between two static plates. This principle provides a variable capacitor with voltage varied based on the intermeshing areas of the rotor and stator plates. The plates are separated by an air gap, which serves as the dielectric (insulating material). This design allows near frictionless rotation and long life. Differential output voltage from both variable capacitors serves as the transducer position versus voltage output.

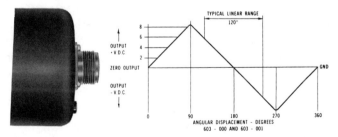

Fig. 2.37. Variable differential capacitance encoder. *(Courtesy Trans-Tek, Inc.)*

Rotary encoders are occasionally used to measure linear travel of long stroke actuators. In this case the encoder is attached to the actuator at a right angle with a cable, similar to a fishing rod and reel. As the actuator moves, the encoder turns proportionately to distance traveled. Each turn represents a linear measurement and a count of revolution pulses provides total travel information.

2.7.4 Vernier Gages

An inexpensive mechanical measuring device can be used to measure position or distance traveled. Vernier dial gages and calipers can be used, provided a solid reference surface can be attained. For example, measuring cylinder rod drift due to piston seal leak can be accomplished by first stalling the cylinder at a point, locking a vernier device on the cylinder or fixed surface, and placing the vernier pickup on the cylinder rod. Rod movement will be read directly in inches. Reliable readings are possible provided the operator reads the device properly and sets the zero reference during stable mechanical conditions.

2.8 Viscosity

The demand for specialized hydraulic fluids creates an overwhelming selection of choices for the fluidpower designer. While some systems work with "standard" type fluids, many employ fluids that are created for a specific purpose. Aircraft system fluids, for example, are designed for wide temperature swings and fire resistance. Their components must be compatible with a temperature range -65 to $+275°F$ and higher. This is quite a wide range for a fluid to be usable. Some industrial systems are designed for use with water-based fluids to reduce potential fire hazards. Other fluids are nearly impossible to work with but their advantages sometimes outweigh the effort required to employ them.

There are several methods used to describe the viscosity characteristics of a fluid. One unit of measure is centipoise. Centipoise is actually a smaller unit of measure than the poise, which is defined as the force per unit area required to move one parallel surface at a speed of 1 cm/sec past another parallel surface separated by a fluid film 1 cm thick. Metrically speaking, force is measured in dynes, so the formula is

$$1 \text{ poise} = 1 \frac{\text{dyne second}}{\text{square centimeter}}$$

$$1 \text{ centipoise} = 0.01 \text{ poise}$$

The centipoise unit defines absolute viscosity or dynamic viscosity. It is used to define a fluid's resistance to flow or the internal friction of the fluid. It is denoted by a quantity μ, called the coefficient of viscosity, and is expressed as

$$\mu = \frac{F/A}{V/L} \quad \text{or} \quad \text{poise} = \frac{\text{shear stress}}{\text{rate of shear}}$$

where

F = opposing force

A = area of parallel plates

V = velocity

L = length of travel

Kinematic viscosity takes into account both absolute viscosity and density. Basically, this method divides the coefficient of absolute viscosity by the density of the liquid. The unit of measure is called the stoke, most often defined in centistokes, and is expressed as

$$\text{centistoke} = \frac{\text{centipoise}}{\text{density}}$$

$$\text{centipoise} = \text{centistoke} \times \text{density}$$

Density of a liquid refers to its mass per unit volume. Typically, the unit of pounds mass per cubic foot ($1bm/ft^3$) are used:

$$\text{density} = \frac{\text{specific weight}}{32.2 \text{ ft/sec}^{2*}}$$

Specific weight is the weight of a fluid due to the gravitational pull of the earth and is expressed in pounds-force per cubic foot (lbf/ft^3). Specific weight equals density times gravitational acceleration. The only place where specific weight and density are equal is where the local acceleration of gravity is equal to the standard acceleration of gravity (32.1759 ft/sec^2).

Specific gravity is a reference unit that defines the ratio of the weight of a certain volume of fluid to an equal volume of water whose temperature is 60°F. This rate is expressed as

$$\text{specific gravity} = \frac{\text{weight of fluid}}{\text{weight of water}}$$

---

*Or the gravity constant for the specific location the measurement is taken.

Another common measure of viscosity is Saybolt Universal Seconds (SUS). It defines the measure of time 60 cm^3 of oil takes to flow through a 0.176 cm orifice 1.225 cm long. Obviously, when a fluid is cooler, it is thicker, and it is thinner at warmer temperatures. This affects the rate of time, so SUS is always expressed as SUS at a given temperature.

Viscosity index ratings of a fluid refer to the slopes of viscosity and temperature. A fluid that changes viscosity slightly with temperature variation has a high VI number, while a low VI number refers to a fluid that changes greatly with temperature variation. Originally the VI scale used 0 and 100 as the upper and lower values to define this characteristic. New fluids, improved additives and refining techniques have now increased the VI scale to well over 100.

Pour point of a fluid defines the temperature at which the fluid will begin to flow. This is important when designing a system exposed to extreme temperature. It is common practice to use a fluid with a pour point of at least 20°F lower than the lowest temperature to which the system will be exposed.

Viscosity is an extremely important factor to be considered when selecting a component for a system—probably one of the most important. It can drastically affect component performance, and some agree that next to contamination problems it rates second on the list. Decreased efficiency, increased internal leakage, and borderline pump suction capacity and system response are just a few factors greatly influenced by viscosity swing.

It is unfortunate that most engineers think that viscosity is sufficiently defined—that absolute viscosity, kinematic viscosity, and viscosity at temperature seem to provide all the information necessary. This is not the case. Fluids are also affected by pressure. The change in viscosity of some fluids is larger with pressure change than temperature change. Where do you find the pressure/viscosity rating of a fluid? In most cases you do not find it at all. In most cases engineers and even technical support people from oil manufacturers are unaware of this effect. For a somewhat detailed description of this effect, see Chapter 4, Flowmeter Inaccuracy. Described there is an accumulation of information I obtained from readers of Hydraulics & Pneumatics magazine after the editors were kind enough to publish my quest for the answers.

2.8.1 Falling Ball Viscometers

The falling ball viscometer is used to measure absolute viscosity. Its basic principle is based on the concept that an object will take a measurable time to be pulled by gravitational forces through the fluid under investigation. These devices range from a simple two-ball design to a device that will allow temperature and pressure control to determine viscosity characteristics at all operating conditions.

The two-ball design (Fig. 2.38) uses one ball in a reference fluid, whose viscosity is known. The test fluid is drawn into a second chamber containing the second ball. Once this chamber is filled, the balls are both set at a common zero reference point. At this point, the operator tilts the gage between 30° and 45°. Once the reference ball reaches the end of the scale, the device is moved to a horizontal position where a reading of the second ball position is taken. The scale allows reading of viscosity in centistokes. The reading is taken at ambient pressure and at a temperature of 80°F.

At the other end of the viscometer design range is a device that allows oil to be measured at controlled pressures and temperatures (Fig. 2.39); it consists of a stainless-steel measuring barrel enclosed within a pressure housing, a heating element, and an insulating jacket. After the unit has been filled with the test fluid, temperature stabilization usually takes about 4 hr. At this time, the ball is rolled back and forth to mix the fluid; then the ball is rolled against the barrel seal, where it is held by a solenoid. The cell is then rotated to its upright position and a button labeled "roll" is depressed. This simultaneously deenergizes the solenoid and starts a digital timer. At the bottom of the barrel is a contact assembly that stops the timer when the ball strikes it. The roll time of the ball is proportional to

Fig. 2.38. Two-ball viscometer design. *(Courtesy Louis C. Eitzen Co., Inc.)*

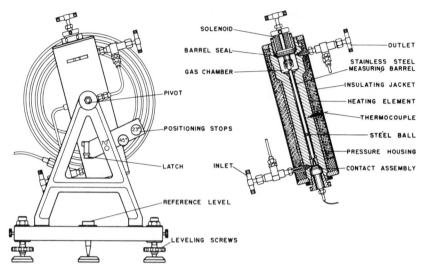

SOLENOID
BARREL SEAL
GAS CHAMBER
PIVOT
POSITIONING STOPS
LATCH
INLET
REFERENCE LEVEL
LEVELING SCREWS
OUTLET
STAINLESS STEEL MEASURING BARREL
INSULATING JACKET
HEATING ELEMENT
THERMOCOUPLE
STEEL BALL
PRESSURE HOUSING
CONTACT ASSEMBLY

Fig. 2.39. Falling ball Ruska viscometer design. *(Courtesy Ruska Instrument Corp.)*

the viscosity of the fluid. Calibration of the meter is accomplished by using fluids of known viscosities and densities. This device can operate at pressures to 10,000 psi, temperatures to 300°F, with 0.1% repeatability at viscosities of 0.1–3000 centipoise.

2.8.2 Electromagnetic Viscometers

Another device similar in concept to the rolling ball viscometer is the electromagnetic viscometer (Fig. 2.40). Rather than allowing gravity to pull a ball through the sample fluid, a bobbin is pulled through the fluid by force generated by a coil. The unit allows fluid to be drawn into the sampling tube by energizing coil A and C simultaneously. Coil A draws the bobbin to the start point and coil C opens a check valve, allowing fluid to be drawn by the motion of the bobbin. Coil C is deenergized allowing the check valve to close. Coil A is deenergized and coil B, when energized, draws the bobbin through the sampled fluid. The time it takes the bobbin to move through the fluid is sensed by the change in magnetic flux when the bobbin reaches its final detection position near coil B.

2.8.3 Vibrating Viscometers

The vibrating viscometer uses a vibrating probe that is immersed in the test fluid (Fig. 2.41). The probe is connected to a driver on

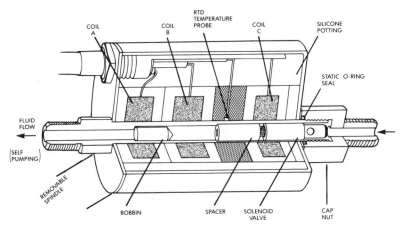

Fig. 2.40. Electromagnetic viscometer design. *(Courtesy Cambridge Applied Systems, Inc.)*

one end and a pickup on the other end. The driver consists of a coil excited at a frequency of 120 Hz and a drive armature attached to one end of the probe. The armature vibrates at 120 Hz from the pulsing magnetic field created by the coil. The opposite end of the pickup is attached to another armature within a coil, except that the stator of the pickup coil is a permanent magnet. This arrangement induces a 120 Hz voltage in the pickup coil.

Viscosity is determined by the amplitude of the probe vibration. Resistance to the shearing action of the probe is created by the viscosity of the sensed fluid. The higher the viscosity, the smaller the amplitude.

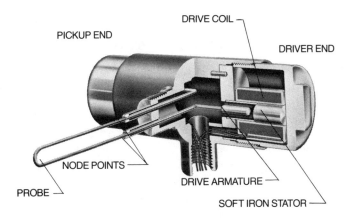

Fig. 2.41. Vibrating viscometer. *(Courtesy Automation Products, Inc.)*

2.8.4 Rotational Viscometers

A rotational viscometer rotates an element within a fluid container, or rotates the fluid container that has a stationary element. The element is designed to create a certain rotational torque in relation to shear strength of the fluid. This method relates to the basic absolute viscosity measurement principle, in which the shear rate is a function of the relative velocity of the surfaces and the distance between them.

Actual torque required for rotation can be detected in several ways. Some rotational viscometers use an electric motor to generate rotational movement. As the fluid becomes more viscous, the torque required increases. At this point motor armature current increases proportionately with the shear forces required. This current increase is measured and converted to viscosity.

Other viscometers use a linear traveling element to detect the same fluid characteristics. In this case, because full rotation is not generated, the plate (element) is timed from starting point to finish point. Time is proportional to viscosity and displayed accordingly.

2.8.5 Hole-in-Cup Viscometers

The hole-in-cup viscometer is based on the SUS principle of viscosity definition; that is, the unit consists of an oil reservoir with a precision hole at the bottom. As discussed earlier, the SUS rating is based on the time 60 cm^3 of fluid takes to flow through a 0.176 cm orifice 1.225 cm long. The simplicity of this device is also reflected in its cost. Fluids of a known viscosity are generally used to set up an equivalency chart for reference of various test fluids.

2.9 Contamination

Fluid contamination is the major cause of component or system failure. Much research has been conducted to define acceptable levels of contamination, as well as the proper methods required to maintain these minimum levels within a system. Contamination can be induced by many sources. These can include casting particles, faulty cylinder rod seals, pump wear, ambient contamination entering reservoir breathers, and so on.

The first problem in defining these acceptable levels was to have an accurate method of measuring the degree and size of contaminates present. In earlier days of hydraulics, a filter patch was inserted into a flowing fluid and contaminates were collected over a period of time. This patch was removed and studied under a microscope to measure the size and quantity of various contaminates extracted from the system. The structure of the particles was then studied to define the source of contamination. Metal particles meant that components were wearing, or fitting threads were being sheared during assembly. Casting sand particles proved to be a good source for system contamination when valve manufacturers did not apply proper flushing techniques prior to assembly and test. Elastomer particles meant that seals were extruding or wearing. Failure of hose inner linings was also a source for this type of particle.

Ambient sources were determined to be a major factor, especially due to contaminates drawn in through cylinder rod seals in mobile equipment and airborne particles in factory environments. Not until recently has the reservoir filler/breather been deemed a major cause of system contamination. It is now recommended that breathers be sized at the required filtration level of the system. Breathers were typically installed at filtration levels generally associated with suction strainers. A common application for spin-on filters is their use as reservoir breathers. They are inexpensive, can be screwed onto a pipe nipple, and allow controlled filtration of airborne particles.

Alvin Lieberman Hiac/Royco was kind enough to furnish some excellent information concerning various instruments available. The following is a reprint of an article entitled "Optical Instrument Monitor Liquid-Borne Solids," originally published in *Chemical Engineering*, Vol. 95, No. 10, 22–26 (1978), through the permission of the author and *Chemical Engineering*. (Please note that since the article was published Hiac and Royco have merged.)

2.9.1. Introduction

Various optical instruments have been developed for measuring the number and size of particles suspended in various fluids. Such instruments have been devised for analyzing both gas and liquid streams; however, this article will be limited to those units used to determine the characteristics of particles suspended in liquids. Since many design variations exist, this discussion will not attempt to cover all of them; instead, attention will be given to several optical devices that illustrate general principles.

Optical instruments offer several advantages. They can produce real-time data; they can handle a broad range of liquids; they can operate online (within specified concentration limits); they can measure over a fairly wide dynamic size range; they can gather a large amount of reproducible data quickly; and they can be operated by relatively unskilled personnel.

However, certain limitations on optical instruments should also be kept in mind. They take size data based on optical properties that must be referred to a calibration base; they yield a size distribution based on particle population, rather than particle mass; their concentration restrictions may necessitate sample dilution; and they are subject to bias if air bubbles or immiscible liquids are present.

Commercially available optical instruments fall into two general classes. In the first, an assemblage of particles is scanned to reveal information on overall concentration. With more-sophisticated instruments of this type, spatial filtering of scattered light permits determination of mean diameter and variance of particle size distributions.

In the second class of instruments, the optical system views a volume so small that individual particles in a flow are observed and each one provides an indication of size. Both concentration and size data can be derived, since the instrument reports the number of particles over several size ranges measured in a given volume of liquid.

2.9.2. Photometers

The operating principle of these devices is quite simple. A light beam is directed through the sample (Fig. 2.42). The total amount of light scattered at specific angles, or absorbed by the opaque particles, can be related directly to the quantity of particulate material in the liquid. Empirical calibration to a turbidity unit scale is normally used to define the quantity of suspended solids, although the degree of scattering is dependent on the size, shape, and concentration of the particles.

Online photometric analyzers are used primarily to measure turbidity and low concentrations of solids in process liquids or water lines. Back-scattering systems are used to determine concentrations in slurries containing 10–15% solids. Angular scattering systems can be used to monitor solids contents up to 50,000 ppm. Absorption analyzers, which handle lower concentrations (to approximately 1,000 ppm), are used for monitoring water treatment systems and process lines.

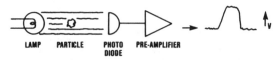

LAMP   PARTICLE   PHOTO   PRE-AMPLIFIER
                        DIODE

Fig. 2.42. Photometer operating principle. *(Courtesy Hiac/Royco Instruments Division.)*

DEPOLARIZATION MEASUREMENTS

Under development by the Precision Products Div. of Badger Meter, Inc. (Tulsa, Okla.) is a method for determining suspended particle concentration, based on the depolarization of scattered light. A depolarization ratio is determined by measuring the intensity of scattered light passing through a polarizing prism. Measurements are made along the optical axes both perpendicular and parallel to the plane of incident polarized light. The ratio of intensities represents the depolarization ratio.

This ratio can be related to concentration, since multiple scattering increases with the number of particles present. Primary scattered light becomes increasingly polarized with multiple scattering.

Reports indicate that this method gives good response to solids concentrations in waste liquids from a few to 5,000 ppm, with no undue bias from particle size or dissolved materials. Solids buildup on the optical windows is not believed to be a problem.

FRAUNHOFER DIFFRACTION ANALYZERS

The French company Industrielle des Lasers Cilas (Marcoussis) offers a particle-size analyzer (Model 227) that measures the intensity of Fraunhofer diffraction patterns produced by passing He-Ne laser light through a suspension. Although the pattern diffracted through polydisperse material is monotonic, smaller particles diffract the laser beam over a larger angle than do larger particles. The overall energy level at any point varies with concentration.

Diffracted light passes through a rotating mask, whose window diameters increase toward the periphery. A photodetector then determines the level of light energies at various radii from the laser beam's axis. From this information, the particle distribution can be deduced using a preprogrammed matrix calculation.

Size measurements can be made from 1 to 128 microns (μm). Cumulative weights are calculated at 1, 2, 4, 8, 16, 32, 64, and 128 microns. Data reported by the manufacturer indicate good results with cement and alumina samples. Sample concentrations should be in the range of 0.1 to 1 g in 500 mL of liquid. Measurement time of approximately 5 min is required.

The Leeds & Northrup Co. (North Wales, PA) has developed a line of low-angle light-scattering instruments (Microtrac) that operate on the principle of Fraunhofer diffraction, using laser light sources, optical filters, and microprocessors. These instruments measure particles in either liquid or air suspension, and may also be used online for process monitoring.

Although the intensity of the total light scattered (photo) by particles is proportional to the square of their diameters, the angle of diffraction is inversely proportional to their diameters. By placing a specially shaped filter in the diffraction plane of the collecting lens, one can relate light fluxes to scattering angles; these fluxes are related to particle volume.

The spatial filter is rotated to extract signals serially. (Collected light can be related to the second, third, or fourth power of the particle diameter.) Once these signals are detected and converted to digital form, a microprocessor computes the distribution of particle volume and surface area. Output data are presented in the form of the particle volume in each of 13 size bands. Information is also given on the percentage of particles passing each size interval, as well as their mean diameters, surface areas, and relative total volume. Measurements are made over two size ranges: 1.9–176 microns and 3.3–300 microns.

The product line has recently been extended to include a low-cost monitor that provides a continuous measure of solids concentration in wastewater and other industrial streams. This instrument measures true volumetric concentration, which can be converted to mass concentration with a one-time, single-point gravimetric calibration.

Similar instruments are being used in refining operations and chemical processes, including the manufacture of powdered resins, catalysts, and a variety of powdered and slurried chemicals.

2.9.3 Lab Methods to Detect Particle Groups

This brief summary of some operating laboratory methods is restricted to those used to characterize particle assemblages in liquids. Holographic analyzers and automated image analyzers are treated subsequently as single-particle devices.

1. Bagchi and Vold[1] describe a method for determining the average particle size of coarse suspensions from measurements of apparent specific turbidity. They show that the turbidity is inversely proportional to the average radius from about 12 to 50 μm for materials whose turbidity is independent of the wavelengths of light. The method has been used with materials in concentrations of 1–10 g/L.

2. Dobbins and Jizmagian[2] describe two methods for finding mean size by measurement of optical cross section. In the first

1. Bagchi, P., and Vold, R. D., *J. Colloid Science*, Vol. 53, No. 2, 1975, p. 194.
2. Dobbins, R. A., and Jizmagian, G. S., *J. Opt. Soc. of America*, Vol. 56, No. 10, 1966, p. 1,351.

method, when concentration is known, the relationship between mean scattering cross section and volume—surface mean diameter is fixed. In the second method, both the volume-surface mean diameter and concentration can be measured if transmittances are measured at two wavelengths. Data were obtained for mean diameters ranging from 1 to 5 μm.

3. Groves, Yalabik, and Tempel[3] discuss the operation of a centrifugal photosedimentometer using laser light. A transparent hollow disk is rotated, causing particles to settle under the influence of centrifugal force. The relationship between sedimentation time and optical density at any point along the radius is a characteristic of the particle size distribution.

A data logger is used to record light-transmission level, location along the radius, and settling time. This enables computer treatment of information to yield relatively rapid descriptions of particle size data. The method can size particles from 5–10 μm to less than 0.1 μm, and can handle sample concentrations up to one percent.

4. Jordan, Fryer, and Hemmen[4] use a hydrophotometer to find particle distribution in the size range from 2 to 50 μm via a sedimentation method. Optical density at a fixed point in a gravity settling column is measured as a function of time and is correlated to mass loading.

5. Robillard, Patitsas, and Kaye[5] describe the use of Mie scattering measurements at two wavelengths to determine the mean diameter and refractive index of particulate material. The method requires comparison of experimental data with computer-generated curves derived from Mie theory, followed by selection of best-fit parameters. Size distribution is inferred from the intensity ratio of the scattering extrema. This work follows and extends earlier work by T.P. Wallace[6] It is not applicable to multimodal distribution.

2.9.4 Nonimaging, Single-Particle Counters

Among these are three optical single-particle counters commercially available in the United States. All pass a liquid sample through an illuminated sensing volume; the amount of light detected depends on the degree of absorption or scattering as the particles tra-

3. Groves, M. J., Yalabik, H., and Tempel, J. A., *Powder Technology*, Vol. 11, No. 3, 1975, p. 245.

4. Jordan, C. F., Fryer, G. E., and Hemmen, E. H., *J. Sed. Petrology*, Vol. 41, No. 2, 1971, p. 489.

5. Robillard, P., Patitsas, A. J., and Kaye, B. H., *Powder Technology*, Vol. 10, No. 6, 1974, p. 307.

6. Wallace, T. P., and Kratohirl, *J. Polymer Sci.*, Vol. 8, Pt. A-2, 1970, p. 1,425.

Particle Area vs Pulse Size

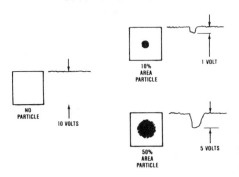

Fig. 2.43a. Single-particle counter operating principle. *(Courtesy Hiac/Royco Instruments Division.)*

verse the sensing volume. The sensing volume is small enough so that single particles are detected, even when they occur in relatively high concentrations. From the amplitude of the light pulse, particle size information can be derived (Fig. 2.43a). By sorting and counting the light pulses, the instrument can ascertain the distribution and concentration of particles.

Climet Instrument Co. (Redlands, CA) manufactures a liquid particle analyzer (Model CL 220), which detects the white light scattered by individual particles as they traverse the sensing zone. Scattered light is collected at scattering angles from 15° to 105°. Suspended particles—sized in terms of their equivalent optical diameter as referred to latex spheres—are detected and can be resolved between 2 and 200 μm. A sample flow rate of up to 500 mL/

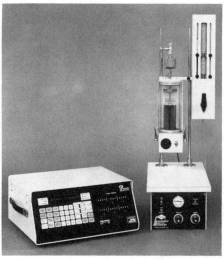

Fig. 2.43b. Single-particle counter system. *(Courtesy Hiac/Royco Instruments Division.)*

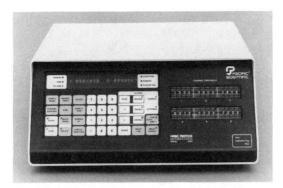

Fig. 2.43c. Microprocessor-based particle counter. *(Courtesy Hiac/Royco Instruments Division.)*

min can be used, and by varying the ratio of sample fluid to filtered recirculating fluid, one can measure particle populations between 1 and 10^8 per milliliter.

HIAC Instruments Div. of Pacific Scientific Co. (Montclair, CA) manufactures a design that observes a telecentric white-light beam and detects reduction in light level caused by particle passage. The firm makes a series of sensors that can detect particles from 1 μm to a maximum of 9 mm. Each sensor has a dynamic range reported to be 60:1, based on the smallest detectable particle (as referred to an equivalent optical diameter for a latex sphere) and the minimum dimension of the internal passageway. The smallest sensor handles particles 1 to 60 μm, in concentrations up to some 30,000/mL at a flow rate of 4–6 mL/min. The largest one handles particles to 9 mm in concentrations up to 0.2/mL at a flow rate of 100 L/min (Fig. 2.43b and 2.43c).

Royco Instruments, Inc. (Menlo Park, CA) manufactures two instrument packages. One observes a focused beam and detects reduction in white-light level caused by particle passage. Two flow cells are available with this package, both of which can detect particles of less than 2 μm with a dynamic range of at least 100 to 1. Sizing is based on calibration with either latex spheres or the AC Fine-Test Dust.[7] The smallest cell has a concentration capability of 12,000/mL at flow rates to 35 mL/min; the larger cell can handle concentrations of 3000/mL at flow rates to 150 mL/min.

The second package detects near-forward scattering produced by particles passing through a laser beam. The same flow cells can be

7. American National Standards Institute Calibration Method, ANSI B93.28, 1973.

used with this package, permitting identical concentration and flow rate capabilities. Minimum size sensitivity is 0.5 μm, and a dynamic range of 50 : 1 is achieved.

Spectrex Corp. (Redwood City, CA) markets a device that uses light scattered from a laser beam to detect particles larger than approximately 5 to 100 μm in concentrations to 1000/mL within a glass container. The laser beam can scan a volume of 10 cm^3 in 15 sec. Since out-of-focus particles and the bottle walls are not detected by the collection optics, only particles within the in-focus point for both the illumination and collection optics are reported.

A microprocessor attachment automates the complete size analysis, prints and plots size and mass distribution, and provides a settling scan.

IMAGING SINGLE-PARTICLE COUNTERS

Image analysis systems. Many manufacturers produce image analysis systems. These are basically video-camera devices that are focused on a microscope stage or on a photograph. Signals from the video tube are transmitted to a computer processor for data reduction. A wide range of information can be developed by the computer, including length and area size data, shape factors, and statistical analyses.

Differential scattering systems. Science Spectrum, Inc. (Santa Barbara, CA) has developed instruments that can locate a single particle and retain it electrostatically in position so that differential scattering intensity can be recorded as a function of scattering angle. In this way, particle composition and shape information can be retrieved by matching measured data with those in a computer file. Particles as small as 1–2 μm can be studied, but the measurement time is on the order of seconds per particle. Data produced include average size, size distribution, shape, and other particle suspension characteristics.

2.9.5 Lab Methods for Analyzing Single Particles

Numerous techniques for characterizing single particles have been reported in the literature. Included among these are:

1. Bartholdi et al.[8] describe a scanning system in which a particle is illuminated by an argon-ion laser beam as it passes through a flow chamber. Scattering between 7.5° and 21.5° is measured by a 128-element photodiode array. Measurement is initiated by a delay

---

8. Bartholdi, M., et al., *Optics Letters*, Vol. 1, No. 6, 1977, p. 223.

signal once the particle has passed through a Coulter orifice. Approximately 500 μsec per particle are required to record a complete scatter pattern for particles larger than approximately 10 μm.

2. Breitmeyer and Sambandam[9] detail an inline holography analysis, where holograms are made of blood cells in saline solutions to determine orientation and cell shape. An argon-ion laser was used and photographs were observed for data reduction.

3. Eisert et al.[10] discuss a microphotometer that uses a flow-through system for orienting nonspherical cells along the flow axis. A focused laser beam at the axis of the hydrodynamically focused flow path is arranged so that its diameter is smaller than that of the cell. If constant flow velocity is maintained, the pulse width will indicate the cell length in the flow system. Sizing from approximately 5 to 300 μm lengths at rates up to 50,000/sec is reported.

4. Kaye[11] describes the use of a low-angle laser-light-scattering instrument. Scattered light is measured at angles as low as 1.5° from volumes of approximately 10 μL, illuminated with a 5-mW He-Ne laser. Particles as small as 0.1 μm can be detected. The sample flow rate is extremely low, but concentration capabilities are high.

5. Mullancy et al.[12] have employed a forward-angle photometer. Hydrodynamic focusing maintains a sample stream of 50 μm diameter that is illuminated by a He-Ne laser beam of 100 μm diameter. Light scattered at between 0.5° and 2° is collected. Sensitivity from approximately 3 μm up to 20 μm has been obtained with a flow rate of 1 mL/min at concentrations up to 50,000/mL.

6. Ricci and Cooper[13] outline the use of a flying spot laser that scans a flowing slurry stream with a two-dimensional raster scan. Particles interrupt the focused beam that would otherwise illuminate a photodiode. Particles from 5 to 1000 μm are measured in concentrations up to 2%, in streams flowing at rates up to 5 gal/min. The laser beam, reduced to 10 μm diameter, scans at 3 cm distance at a frequency of 500 Hz and a velocity of 50 m/sec.

7. Salzman et al.[14] describe a flow system, multiangle light-scattering instrument. Particles are characterized by their light-scatter patterns. A He-Ne laser is focused at the center of the sample stream.

9. Breitmeyer, M., and Sambandam, M. K., *J. Assn. Adv. Med. Instrum.*, Vol. 6, No. 6, 1972, p. 365.

10. Eisert, W. G., et al., *Rev. Sci. Instr.*, Vol. 48, No. 8, 1975, p. 1,021.

11. Kaye, W., *J. Colloid Int. Sci.*, Vol. 44, No. 2, 1973, p. 384.

12. Mullaney, P. F., et al., *Rev. Sci. Instr.*, Vol. 40, No. 8, 1969, p. 1,029.

13. Ricci, R. J., and Cooper, H. R., *ISA Trans.*, Vol. 9, No. 1, 1970, p. 28.

14. Salzman, G. C., et al., *Clin. Chem.*, Vol. 21, No. 9, 1975, p. 1,297.

As each particle is illuminated by the beam, a 250-μsec pulse of scattered light is collected by a 32-photodetector array laid out in a concentric pattern that encompasses scattering angles from 0.3° to 20°. The scatter pattern is then stored in a computer for analysis by a mathematical clustering algorithm.

8. Uzgiris and Kaplan[15] describe application of a laser Doppler velocimeter to determination of electrophoretic mobility distributions. A standard LDV system was used to observe the frequency shift in scattered laser light due to applied voltage on the suspension of particles in a sample. Mobilities ranging from 1 to 5 μm/s/V/cm were recorded.

2.9.6 Material Handling Requirements

Batch systems. Since essentially all of the optical devices use flow systems, it is necessary to transfer material from a container through the instrument. One may assume that the necessary preliminary steps of adequate particle dispersion and uniform mixing have been completed. Since the instruments measure particles as local nonhomogeneities in the liquid, it is necessary to avoid artifact introduction. This requires elimination of bubbles, inclusions of immiscible liquids, and contamination from airborne dust or surface debris on the container. In addition, if concentration data are required, then the need for careful volumetric control is obvious.

For the single-particle counting devices, if the maximum recommended concentration is exceeded, coincidence errors can occur. In this case, it may be necessary to dilute the sample. If dilution is required, then aside from the normal care in maintaining good volumetric control, dilution ratio selection must be chosen high enough to avoid coincidence errors, but not so high as to reduce concentration of large particles in typical distributions to a level that does not permit accumulation of statistically significant data.

If a suitable dilution ratio cannot be found for all size ranges, then it may be necessary to measure small particles at a high dilution ratio, and large particles at a low ratio. For the most part, dilution is a manual procedure; however, there are many commercially available automatic dilutors manufactured for clinical chemistry procedures. These typically dilute by a factor ranging from 10 to 500, with high precision and accuracy. Diluted samples of 10–20 mL are typically produced; the concentrated material is aspirated in microliter

15. Uzgiris, E. E., and Kaplan, J. H., *J. Colloid Interface Sci.*, Vol. 55, No. 1, 1976, p. 148.

quantities and then dispensed along with clean diluent in milliliter quantities.

Inline systems. If an optical particle counter is connected inline, then the problems of solids dispersion and mixing, sedimentation during sample storage, careless handling, etc., associated with container preparations are eliminated. However, a number of other problems are present. First, it is necessary that sample stream flow rate be controlled; an isokinetic sample probe may be required; a means of ensuring gas bubble elimination may be needed; an inline dilution system may be required.

Sample stream flow rates can be controlled by using pressure-compensated flow control valves with suitable protection to avoid orifice clogging, or by using a metering pump; both are downstream of the particle counter. Bubble control can be attained to some extent either by maintaining system pressure so high as to ensure that all gases remain in solution in the liquid or by incorporating a low-pressure vacuum deaerator immediately upstream of the particle counter inlet.

At present, no commercially available inline dilution system is available for dispersion sampling. Some one-of-a-kind developments have been made. These are basically modifications of batch dilutors. Clean diluent is pumped at a fixed flow rate into a mixing chamber, where a smaller sample flow is mixed with it. The diluted sample is then fed to the particle chamber. These systems are limited by the available quantity of clean diluent, particle losses in the mixing chamber, and flow measurement stability.

Some preliminary work has been reported on fixed-ratio sample-diluent flow systems feeding into motionless mixers for minimum particle loss before presentation to the optical instrument. Clean diluent is supplied by recirculating the diluted stream through a suitable filter.

2.9.7 Data development

Data production. Optical instruments have a number of common features in terms of the data they produce. In each device, the measurements are dependent upon the optical properties of the particulate material being different from those of the substrate; the primary data output is related to a projected area of the particles; a truncated size measurement is always made, with the low end varying with instrument sensitivity. These comments apply to both multiple- and single-particle analyzers in which size information is produced.

Multiple-particle optical instruments will usually produce data describing a particle-size number distribution. With judicious data-processing circuitry, either differential or cumulative distributions can be defined. Single-particle optical instruments will produce data based on the number of particles in several size ranges.

Depending on the resolution of the instrument, the width of each range can be made small enough so that essentially all particles in the range can be represented as having the mean diameter of the range. In this way, mathematical manipulation can be performed to convert from a number base to a volume or an area base with minimum error.

Data processing. The conversion from number base to volume or area base includes certain assumptions that must be taken in mind if comparison with data obtained by other means is desired. First, the size description is based on calibration to an idealized basis; next, a limited range of shape factors is assumed; next, uniform particle composition in the sample is assumed. Once these assumptions are accepted, conversion from optical diameters to aerodynamic diameters or sieve data can be accomplished for a range of materials. In some cases, empirical calibration may be required.

All of the optical instruments produce data that can be easily transmitted to a computer for storage and/or processing. The data can be a series of dc signal levels, or parallel or serial output digital pulses. Part or all of the data processing system can be included with the instrument.

Data can be provided that indicate cumulative or differential particle-size distribution curves, mean diameters with standard deviations, total quantity of particles per unit volume of liquid, and number of particles in several particle size ranges. With this easily acquired and easily processed data, application of the instrument output through a computer or microprocessor to a control function for process lines can be accomplished with relative ease. It is necessary to select a particular size parameter or ratio of sizes as the dependent variable in a control function, define an optimum operating range, and choose the process parameter that controls the variable of concern.

2.10 Sound Level

One important consideration in developing a component or system is the output noise level it produces during operation. In 1970 the Occupational Safety and Health Administration (OSHA) was formed to research and recommend new safety and health stan-

dards for the work environment. The regulatory committee was responsible for all areas that either affect safety conditions or pose
health hazards to workers. They determined that certain noise levels
caused temporary and sometimes permanent hearing loss. At the
very least these levels of noise were annoying to a worker causing
stress, mood swings, poor performance of tasks, and difficulty in
communication.

OSHA originally settled on a 90 dBA noise level exposure, as an 8-
hr time-weighted average maximum. The Environmental Protection
Agency (EPA) recommended that this level be reduced to 85 dBA
continuous exposure for an 8-hr shift. At present most manufacturers request the 85 dBA level for equipment operating in the factory
environment. In some cases machinery with higher levels of output
are isolated in remote areas or separate rooms, and operators are
required to wear hearing-protection devices.

Sound travels through the air (the transmitting medium) at a specific intensity and frequency. Intensity describes the level or loud-
ness of a sound and is normally expressed in decibels (dB) as a unit
of measure. The decibel is a dimensionless unit that is the ratio of
two numeric values on a logarithmic scale. The ratio consists of a
measured quantity versus a reference quantity and is expressed as

$$dB = 10 \log_{10} \frac{P_1}{P_0}$$

Owing to the nature of the logarithmic scale, it is not easy to add
or subtract additive noise levels. For example, the combined noise
level of two machines each producing 90 dB noise levels would be
93 dB. As a scale factor, 0 dB is the threshold of hearing, while 130
dB is the threshold of pain. A rule of thumb used is that for every 3
dB increase in sound level, sound output actually doubles. The reason for this method of level measurement is due to the extremely
wide range of audible detection.

Sound frequency (pitch) defines the wave motion in cycles per second (Hz). The wave consists of air pressure variation in a sinusoidal
waveform. The higher the frequency, the higher the pitch of a
sound. The normal hearing range can be as low as 20 Hz and as
high as 20,000 Hz. Frequencies higher than 20,000 Hz are termed
ultrasonic. The range of 500–2000 Hz is referred to as the speech
zone frequencies.

When no sound is present, the atmospheric pressure is undis-
turbed; this pressure is generally used as a reference pressure to the
pressure increase caused by a sound. As sound waves move through

the air, the pressure level of the air is disturbed. Audible sound pressure is as low as 0.00002 N/m^2 (the threshold of hearing) and as high as 20 N/m^2 (the threshold of feeling).

Sound level meters are designed to measure sound pressure levels on three weighting networks designated A, B, and C. These weighting scales affect the meter reading, so that the scale to be used is dependent on the characteristic of the sound to be measured. The weighting actually is designed to increase or decrease the response of the meter, either sampling average sound level output longer periods of time or sampling at higher speed to detect peak levels of output.

A meter will allow readings at various frequency ranges, termed octave bands. This allows measurement of sections of a sound to determine the source of various outputs. This is especially useful when trying to pinpoint the source of excessive machine noise or when determining the type of noise-reducing material necessary to isolate a particular output. The frequency range desired is allowed through a band pass filter, which cuts out frequencies above and below the desired measuring range.

When noise measurements are taken, it is extremely necessary to consider ambient noise conditions. When manufacturers rate their equipment, it is generally at perfect acoustic surroundings. When the equipment is used, there are many contributing factors that can vary the output. Improper installation can cause transmission of mechanical vibration, creating new "sounding boards" for the resonant frequencies. Factory environments produce noise, and this output is sometimes neglected or forgotten when designing to strigent noise regulations.

Another factor is the distance the reading is taken from the noise source. As a rule, the area where a machine operator is operating is the measuring point. A component rating is generally taken at 5 or 10 ft from the source when ratings are developed.

2.10.1 Sound Level Meters

A sound level meter is the most frequently used meter to determine noise level of fluidpower equipment (Fig. 2.44). This meter consists of a calibrated microphone, amplifier, attenuator, weighting networks, and indicating meter. The microphone converts sound pressure levels to an electrical signal. Important factors in the selection of the proper microphone include sensitivity, frequency response, dynamic range, directivity pattern, low internal

Fig. 2.44. Sound level meter.
(Courtesy Quest Electronics.)

noise generation (noise floor), impedance, and temperature sensitivity.

There are several types of microphone designs including:

Piezoelectric A design that uses crystals to generate electric current by applied mechanical stress. In this category there are crystal microphones and ceramic microphones. Crystal units have a good frequency response but are susceptible to high temperatures and humidity and have poor aging characteristics. Ceramic units have a smooth frequency response and are resistant to ambient extremes. The only drawback is that its upper frequency limit is not as high as a condensor design.

Condensor A design widely used for sound measurement, it is also refered to as a electrostatic or capacitor microphone and uses the variation in capacitance in response to sound level changes to generate an output. Two types of condensor microphones are the air-dielectric and the electret. The air-dielectric uses a diaphragm to move with sound pressure. The diaphragm is one plate of the capacitor, while the second plate is stationary. The voltage charge across the capacitor is proportional to the distance between the two plates. The electret design is similar, except it uses a permanently charged plastic material between the plates rather than air.

Dynamic A design that uses an aluminum or plastic diaphragm with a small coil attached. The coil moves within an annular gap of a magnetic field from displacement caused by diaphragm movement. This movement induces a voltage in the coil that is stepped up through a transformer and then into an amplifier.

The sound level meter takes the output voltage from the microphone (which is ac) amplifies it, rectifies it to dc, and uses the dc to move the indicating pointer on the scale. The attenuator controls the overall amplification of the meter, while the weighting networks control the response versus frequency characteristics.

The American National Standard S1.4-1971, Specification for Sound Level Meters, states that a sound level meter must be able to measure the rms (root mean square) level of the sound pressure, have a response that meets tight specifications of smoothness of response and linearity, and contains A, B, and C weighting networks. The A and B scales have a response (response rated in dB) that decreases with decreasing frequency, with the A scale decreasing the most. The C scale has a flat response over the frequency range of 25–8000 Hz. The C scale is generally used to supply a signal to auxilliary equipment for a more detailed analysis.

2.10.2 Frequency Analyzers

In some cases, a sound level meter may be insufficient to truly define a noise source. It is not capable of distinguishing frequency distribution of the sound. A frequency analyzer is able to depict various frequency components at various sound intensities. Two types of analyzers are used, the constant percentage bandwidth or the constant bandwidth design. This just defines the width of the frequency band sampled over the full spectrum.

2.10.3 Octave Band Analyzers

Octave band analyzers are used for a simpler approach to measuring complex sound characteristics. The device separates certain frequency ranges with filtration and each range is measured. Each range is generally plotted on a graph so that the relationship of sound pressure level versus octave band frequency can be studied.

2.10.4 Noise Exposure Monitors

A noise exposure monitor is used to accumulate sound data information over a period of time. This provides data on the "noise dose" that a person may be subjected to. This information is useful when comparing the exposure of a worker to the regulations set by OSHA.

There are three variations of this instrument: one accumulates the total amount of sound energy a worker is subjected to for a workday, another monitors the time a specific decibel level is exceeded, and the third measures the rate at which sound energy impinges on the exposed person over the short-term periods of time.

2.10.5 Noise Dosimeters

Noise dosimeters are small and portable units that can be placed in the pocket of a worker, which allows an analysis of noise dose to a worker who moves in various locations throughout the day. Sound levels between 90 and 115 dBA are generally monitored, with accumulated doses stored in a semiconductor memory circuit.

At the end of the time period, the dosimeter is plugged into a read-out device and the noise exposure is displayed as a percentage of the allowable limit.

2.10.6 Impact Noise Analyzer

An impact noise analyzer is generally used to monitor exposure to high peak levels, such as experienced around punch presses. The device will generally allow storage of peak instantaneous levels, average levels, and continuous indication of peak level. Analysis of collected data will allow calculation of impact noise duration.

2.11 Electric Test Equipment

Equipment available to measure electrical characteristics seems endless. The main values normally associated with fluidpower testing include voltage, current, resistance, inductance, capacitance, and combinations of these values, such as wattage and HP consumption. Common tests include minimum and maximum solenoid voltage, solenoid coil force and resistance, electric motor power required to drive a pump, and voltage or current versus output of a proportional valve or servovalve

Before reviewing the basic equipment used to monitor these values, it might be helpful to discuss a few basic electrical principles.

Current is generally expressed in amperes and refers to the flow of electricity. This is equivalent to hydraulic flow rate. The higher the amperage rating of a system, the more flow of electricity is possible. The ampere is defined as a flow of 1 C/sec. One coulomb is equal to 6.3×10^{18} electrons. Since the charge of one electron is too difficult to measure for practical purposes, we created a unit of measure more suitable. The actual definition of an ampere is the rate of current that must flow through an electrotytic cell in order to deposit silver at the rate of 0.001118 g/sec.

Voltage refers to the potential difference across two terminals of an electric circuit. This is equivalent to pressure in a hydraulic system. This potential difference is also referred to as electromotive force (emf). The emf of an electric circuit is the potential voltage drop across the entire circuit. A 440 Vac power source will generate twice as much power as will a 220 Vac power source at the same current.

Resistance is the measurement of current resistance in a conducting material and is expressed in ohms (Ω). An ohm is defined as the resistance of a column of pure mercury 1 mm^2 in cross section and 106.3 cm long, at a temperature of 0°C. Resistance of a conductor is dependent on the type of material and its length, cross-sectional area, and temperature. In hydraulics, this is equivalent to pressure drop, or loss in pressure due to pressure required to move a specified flow rate through an orifice.

The relationship of these values is expressed as

$$\text{volts} = \text{amperes} \times \text{ohms}$$

$$\text{amperes} = \frac{\text{volts}}{\text{ohms}}$$

$$\text{watts} = \text{amperes} \times \text{volts}$$

$$\text{ohms} = \frac{\text{volts}}{\text{amperes}}$$

Direct current (dc) and alternating current (ac) describe the two basic power sources used in fluidpower systems. Direct current is used in mobile equipment owing to the battery power available from the vehicle. It is also used in low-current control systems such as proportional valve and servovalve applications. A majority of the instrumentation used for testing relies on dc excitation and output signals for remote data acquisition.

Dc power can be supplied from a battery or through a rectified ac power source. Flow of dc current is always in one direction, and negative electron flow remaining in this direction. Both 12 and 24 Vdc are common system voltages.

Ac power is generally supplied from utility electric sources or from a generator in remote locations. Common ac voltages in the US are 480–440 Vac three-phase, 240–220 Vac three-phase, and 120–110 Vac single-phase. Most ac power in the United States is 60 cycles, meaning that the voltage polarity and current direction go through 60 cycles of reversal per second. In other countries, 50 Hz is a common voltage frequency, and this can cause problems when designing systems for overseas applications. Unlike dc current with a constant positive and negative, ac polarity reverses constantly. Thus the terms direct current and alternating current. Of course, these are not the only voltages and currents used in the United States, but they are the most common in industrial applications.

Various power supplies are used aboard ships and on aircraft, and instrumentation circuitry can employ almost every variation possible to perform a function. You will find that most components and equipment are supplied for the common power sources listed. Some solenoid coils can operate at either 50 or 60 Hz, which eliminates some problems associated with overseas customers. However, electric motors can only be operated at one frequency to develop the rated output power. When a 60-Hz motor is operated at 50 Hz, output speed is reduced proportionately. This creates testing problems when a system is built, because true output cannot be evaluated unless the electric motor is physically replaced with a motor with the proper frequency rating.

Ac power can be utilized in either a single phase or a polyphase configuration. This is dependent on the total power required by the device. Single phase is generally associated with lower-power devices. Most control circuitry in industrial use employs 115 Vac single-phase power. This also applies to instrumentation, data acquisition equipment, and programmable controllers. Depending on the surrounding environment, dc power is sometimes used as a control voltage. Some equipment also use 220 Vac or 440 Vac single-phase power such as low-power immersion heaters.

Measuring voltage and current of ac single-phase systems is a fairly simple operation. The power factor in a circuit with good sinusoidal characteristics is a function of the phase angle between the current and voltage components. Power factor (pf) is expressed as:

$$\text{pf} = \frac{\text{volts} \times \text{amperes} \times \cos\phi}{\text{volts} \times \text{amperes}}$$

where $\cos\phi$ is the cosine of the phase angle between the current and voltage.

The power factor in an ac circuit can be accurately determined, regardless of harmonic content, by the use of three, true rms voltmeters providing each voltmeter is capable of true rms response up to the highest order of harmonic present in the waveform. The connections for the three-voltmeter method are shown in Fig. 2.45. Each meter with the exception of V_1 must have negligible burden charac-

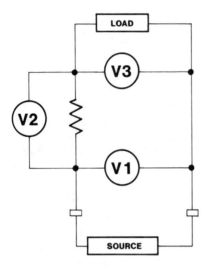

Fig. 2.45. Three-voltmeter measurement technique for determining power factor.

teristics. If high-impedance meters are not available, medium load instruments may be used, providing their loading effects are taken into consideration and calculated into the formula.

The formula for computing the cosine of the phase angle and thus the power factor is

$$pf = \frac{V_2^3 - V_1^2 - V_2^2}{2(V_1 \times V_2)}$$

The measurement of polyphase power (two or three phases) can become more complex if the load is unbalanced or if power-factor parameters are dissimilar across the various loading elements. If balanced, individual readings on each leg of the circuit may not be necessary. However, if power factor varies, an individual reading must be taken. In a balanced load, three-phase-system power factor is expressed as

$$pf = \frac{\text{true power}}{\text{volts} \times \text{amperes}}$$

The actual factor of unbalanced three-phase loads creates a difficult task. Usually, a weighted average of the power factors is calculated. This is one of the reasons it is difficult to extract accurate data when testing pumps and using motor current draw to calculate torque.

The most common component used in the marriage of electronics and fluidpower is the solenoid. This device is used to convert electric power into mechanical force through the use of electromagnetic induction. When current is passed through the coil, it generates a magnetic field. The amount of current and the design of the coil determines the resultant force potential. Coil design basically consists of wire diameter (current-carrying capacity) and ampere-turns; ampere-turns = $N \times I$, where, N = number of turns and I = current in amperes. The quantity of NI defines the amount of magnetomotive force (mmf).

Ampere-turns defines the magnetizing force, but the magnetic field intensity depends on the length of the coil. Solenoid intensity (H) can be calculated using the formula

$$H = \frac{NI \quad \text{(ampere-turns)}}{l \quad \text{(m)}}$$

where

H = intensity at the center of an air core (mks)

l = length between the poles at the end of the coil

Intensity is actually the mmf per unit of length.

The solenoid is designed to produce a certain force on a plunger within its inner core. The plunger is allowed free travel over a specified distance within the solenoid core. The plunger and its stop are separated by a short distance called the air gap. The air gap of a magnet is the space between its two poles. The shorter the air gap, the more intense the magnetic field in the gap for a given pole strength.

When the solenoid is energized, the plunger is forced in a linear direction by the potential force of the coil design. In direct-acting solenoids, the plunger pushes a spool or poppet directly. In pilot-operated designs, the plunger opens or closes an orifice or pushes a pilot spool to allow hydraulic or pneumatic forces to do the main work. There is a trade-off point where direct-acting and pilot-operated designs become more applicable to a particular situation. Because of the limited practical force possible by a solenoid, direct-acting solenoids are used in lower-flow applications where required spool forces do not become excessive.

Solenoids do create heat. This requires that duty cycle ratings need to be evaluated for coil application. Some coils are rated for continuous duty and others are not. This all depends on the ambient cooling potential as well as the design of the coil itself and how much force the coil is attempting to generate.

Solenoids are generally rated for their voltage, resistance (in ohms), frequency, insulation material, and dielectric strength.

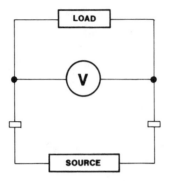

Fig. 2.46. Voltmeter connection.

2.11.1 Voltmeters

As shown in Fig. 2.46, a voltmeter measures voltage by detecting potential difference across two points in a circuit. The voltmeter has a very high resistance to limit the current flowing through the meter. Analog (moving coil) voltmeters use a coil mounted between the poles of a permanent magnet. When current flows through the coil, the magnetic field created by the coil reacts with the field of the magnet. This force actuates the pointer mechanism with position proportional to the amount of current in the coil.

If a voltmeter is not connected in parallel as shown, but in series, the high resistance of the meter itself would reduce voltage and cause false readings. When measuring dc circuits, polarity of the meter must be the same as the circuit. In ac circuits, the test leads can be reversed due to the nature of alternating current (provided the meter is compatible for both ac and dc voltage measurement).

Some meters include a multiple range switch to cover various voltage ranges. This switch connects the coil in series with various resistors, each of which is sized to deflect the pointer based on the full-scale range.

It is important to note that the resistance created by a voltmeter in a low-current circuit can reduce voltage. This effect is called loading down the circuit, and does give a false reading. This loading effect can be minimized by using a voltmeter with a much greater resistance than the resistance across which the voltage is measured. Figure 2.47 depicts one method used to measure voltage and amperage. In this case, the power expended in the voltmeter must be subtracted from the total measured power to determine true power. This compensation is applied with the formula

power (W) = (volts × amps) − voltmeter watt draw

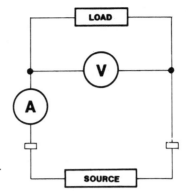

Fig. 2.47. Power measurement using voltmeter/ammeter.

Figure 2.48 depicts another method where the power expended in
the ammeter must be subtracted, using the formula

$$\text{power (W)} = (\text{volts} - \text{ammeter voltage drop}) \times \text{amps}$$

2.11.2 Ammeters

Current is measured with an ammeter, which, as shown in Fig.
2.49, is hooked in series with a circuit. The ammeter has a very low
resistance, so when inserted into a circuit, it adds little to the total
system resistance. The principle is the same as described for voltmeters; the pointer is moved proportionately to current.

Readings are not easily taken because the ammeter must be inserted into the circuit. Again, when measuring dc current, the meter must be hooked up in the same polarity as the circuit.

In certain low-current measurement applications a standard
shunt-type meter may produce an excessive voltage drop. These me-

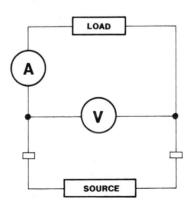

Fig. 2.48. Power measurement using ammeter/voltmeter.

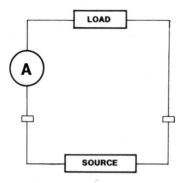

Fig. 2.49. Ammeter connection.

ters read voltage across the internal resistor, which has been installed in the line. This voltage is then converted to read amperage. To reduce this error, feedback ammeters are used, which typically have a voltage drop of 1 mV or less. The feedback design feeds the measured current into a high-gain amplifier, which amplifies the signal sufficiently to drive a voltmeter. Again, the voltage reading is converted to equivalent amperage.

In higher-voltage systems, an amp-clamp probe (Fig. 2.50) can be used. This ammeter does not require connection into the circuit, but instead is placed around the current carrying wire. The wire's magnetic field is used to indicate the amount of current.

2.11.3 Wattmeters

Wattmeters are used to represent the true power draw of a circuit or component. Wattage is actually a reading of voltage times amperage. This measurement compensates for errors caused by fluctuating voltage. For example, a 480 Vac power source can fluctuate owing to demands of several machines running from the same feed line. If you are trying to measure amperage draw of one particular machine, amperage draw is dependent on available voltage. As voltage decreases, amperage increases, and vice versa. So, by monitoring both amperage and voltage, and multiplying both, we have a unit of measure that is constant with constant load.

One important factor in measuring true power is the phase shift between voltage and current. If the sinusoidal waveforms of each are in phase, true power is measured. However, if the two components are out of phase, the phase angle must be determined and its cosine applied. This formula is

$$\text{power (W)} = \text{volts} \times \text{amps} \times \cos \phi$$

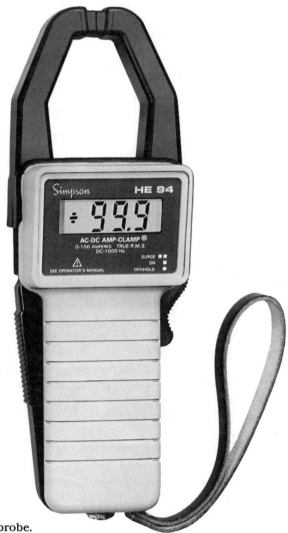

Fig. 2.50. Amp-clamp probe.
(Courtesy Simpson Electric Co.)

where φ is the phase angle. If voltage and current are in phase, the cosine will be 1.

There are several variations in wattmeter design principle. One design called the electrodynamic voltmeter and ammeter can be used to measure watts; however, its power consumption is rather high. It is connected as shown in Fig. 2.51 and the formula for compensation of the instrument power draw must be employed.

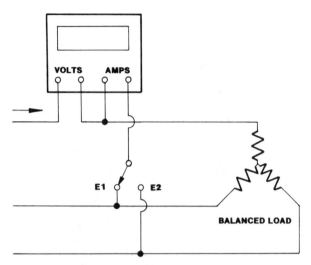

Fig. 2.51. Polyphase power measurement using single-phase wattmeter and voltage switching circuit.

A four-terminal electrodynamic wattmeter uses two sets of isolated coils to measure current and display watts simultaneously. An important consideration here is to match the meter with the power factor of the load to be measured. Unless properly matched, the meter could be overloaded causing serious damage to the instrument.

A four-terminal thermocouple wattmeter uses thermocouples as sensing elements. Thermocouples follow a square-law response, with the output indicating true rms value of ac waveforms, regardless of harmonic content.

Hall-effect wattmeter transducers contain transformers and resistive dividers, which are used to provide a highly accurate measurement of one side. An isolating transformer is attached to the power source, while the other side drives a Hall element. A second transformer attached to the load creates a magnetic field. This combination allows the Hall element to produce an ac and dc quantity proportional to the power draw.

Inductance wattmeters use a four-terminal wattmeter technique combined with a moving coil/permanent magnet movement. A direct curent in milliamperes is produced that is directly proportional to watts input of the electrodynamic circuit. This direct current, applied to a precision resistor, produces a dc millivoltage proportional to the power applied to the input.

A microjoule wattmeter works on the time-division-multiplier principle, using pulse width and pulse amplitude modulation of a

Fig. 2.52. True rms digital volt-ampere-wattmeter. *(Courtesy Dowty RFL Industries, Inc.)*

series of rectangular pulses. Input voltage and current transformers isolate the signals and provide usable signal levels. Current is switched by the pulse-width-modulated voltage signal, creating pulses whose average value is directly proportional to input power.

A true rms digital volt-ampere-wattmeter (Fig. 2.52) uses many of the techniques previously described. It samples volts and amps and compensates for any meter loads, thus providing accurate display of watts when potential is sampled on the load side of the instrument. If power is sampled on the source side, the user must manually compensate for power dissipated in the current input. This unit has good accuracy characteristics through a wide frequency range (40–440 Hz). Power factor can easily be measured and displayed using the formula

$$\text{power factor} = \frac{\text{true power (W)}}{\text{volts} \times \text{amperes}}$$

2.12 Data Acquisition Equipment

Thus far, we have discussed instruments used to detect certain system parameters. Many work on similar principles, while others are unique. Frequently, test data must be stored for engineering evaluation or furnished to a customer to verify product performance. In this case, the instrument must be capable of producing a signal usable by a remote device. When studying instrumentation literature, you will find information describing output features of a signal conditioner. "Signal conditioner" is a term for the electronics used to supply excitation voltage to an instrument or to translate the output from the instrument into a usable output or both. The conditioner will either display the value measured or provide an output proportional to the full-scale value of the instrument.

The typical outputs used include 0–5 Vdc, 0–10 Vdc, 4–20 mA or BCD. These outputs are usually a standard feature of any instrument. If not standard, output selection is usually available as an option. These outputs are generally sufficient to drive the electronics of data capturing equipment. It is recommended that when matching a conditioner output to a recording device, the characteristics of each be studied carefully. An output signal may not have enough power to provide accurate results.

2.12.1 Signal Conditioner Options

It is worth mentioning two features found on a majority of signal conditioners. These are peak hold and alarm set points. During certain production tests, only one or two data points need to be monitored to pass or fail a component. In this case, a very inexpensive method to provide a go/no go signal is to use optional features of the transducer signal conditioner or display (Fig. 2.53).

Peak hold is very useful when trying to monitor pressure spikes. A digital display is generally not fast enough to display a spike, although its internal electronics are fast enough to process the signal. The display is generally adjusted to negate fast data transients so that the display is readable. A sensitivity adjustment is often included to fine tune the display if the numbers are "bouncing" too fast. Using the peak hold feature stores the highest measured value and the display will lock in at that value until the display is switched back to its normal function. This stored value can then be used to compare desired high limits of test requirements. This often eliminates the high cost of an oscilloscope or waveform analyzer.

Fig. 2.53. Signal conditioner with limit alarm. *(Courtesy Viatran Corp.)*

Many electronic conditioners are also available with one or two set points. These setpoints are triggered when the transducer output matches the setpoint. Internal relays are available to provide an on–off signal to an external device. This feature has obvious advantages when testing a component for proper deadband characteristics or any single or dual data point. For example, if testing a pilot-operated device for pilot pressure required, this feature allows setting of the lower limit (point of minimum pilot actuation) and the high limit (point of maximum pilot actuation). If the device shifts between both set points, it is operating properly.

Another excellent test where this feature is useful is during leak testing. One set point can sense starting pressure, while the other can sense pressure decay point. In the case of high-resolution pressure decay leak testing, a pressure transducer and differential pressure transducer are used, each with their own display with dual setpoints. (See Chapter 3, Leak Testing.) This provides an exceptional test setup for high accuracy, high repeatability, and minimal cost.

2.12.2 *X-Y* Recorders

This device incorporates separate conditioning circuits to provide a graph hardcopy of two variable components. As shown in Fig. 2.54 an arm moves back and forth along the length of a piece of graph paper. Its movement is proportional to the full-scale set value of the transducer output. Attached to the arm is a mechanism that moves up and down along the length of the arm. Its movement is also proportional to the full-scale set value of a second transducer output. This mechanism contains a pen that marks the graph paper. Most

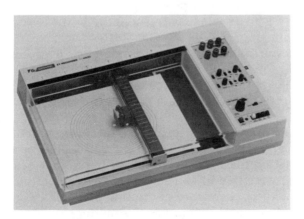

Fig. 2.54. *X-Y* recorder. *(Courtesy Western Graphtec, Inc.)*

X-Y recorders have a feature that lifts the pen off the paper when recording is not desired.

This device is especially useful when monitoring flow versus pressure, position versus torque, or any two parameters that can be expressed on an *X-Y* graph. Some recorders contain signal conditioning modules for various transducer types, while others accept only the common voltage or current analog outputs.

The graph paper is divided into equally spaced grids that represent a scale of the transducer output. The full-scale output of the transducer does not have to be equal to the full-scale travel of the graph paper. A section of the transducer output can be adjusted to provide full-pen-scale movement if necessary. For example, a pressure transducer with a range of 0–5000 psi is used to monitor system pressure. During a particular test a pressure variation of 2400–2800 psi is expected. Using a full scale of 0–5000 psi would produce a line with little movement and accurate reading of the trace may be difficult. By scaling the pen travel from 2000 to 3000 psi, a magnified trace of this range will allow more precise trace reading. However, note that this does not improve system accuracy.

These recorders are available as single *X-Y* graph units, or multiple traces can be provided along one axis. This feature allows an analysis of several affects caused by one variable.

2.12.3 Oscillographs

Similar to *X-Y* recorders is the oscillograph. The difference is that the oscillograph is a time-based unit. Graph paper is driven at a speed proportional to time. Several speed selections usually allow sufficient spacing between pen output traces. The graph paper is provided with grid divisions corresponding to the speed selections, such as 10 cm/sec, 1 cm/sec, and so forth.

Testing requiring time-referenced data such as pump response, pressure rise, and cyclic shock are often recorded on this type of device. Oscillographs can be purchased with one or several channels (Fig. 2.55) to record related data occurring at the same time reference.

A nice feature offered as an option on some units is an event marker. This provides an additional trace to depict the actuation of a device affecting the transducer output traces. For example, if the event marker were tied-in to a solenoid valve coil on–off switch, the marker would provide a trace of the time the coil was energized or deenergized. If one of the channels was monitoring flow through the solenoid valve, valve response time from energization to full flow, or

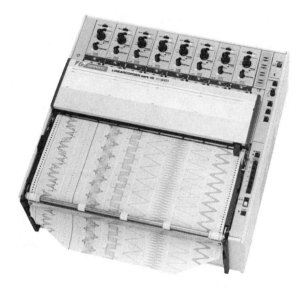

Fig. 2.55. Oscillograph. *(Courtesy Western Graphtec, Inc.)*

time from deenergization to full shut-off could be determined. Without the event marker to compare grid time between on to full flow, the data would be useless. (Note that this is just an example to describe the potential for event marker use. Flowmeter response time is generally not fast enough to provide accurate response time tests.)

2.12.4 Oscilloscopes

An oscilloscope (Fig. 2.56) uses a cathode-ray tube (CRT) to depict an X-Y graph function. One function is usually a time base reference. An electron beam is controlled via beam-deflection plates, and the resultant trace is produced on the phosphor-coated CRT viewing surface. The principle is similar to that described in the X-Y recorder operation. Input voltage or current is converted to a signal that proportionately deflects the electron beam in the X or Y axis of the curve. Depending on the complexity and, of course, cost of the scope, the more features are available.

Fluidpower testing does not require extreme high-speed conditioning as do other fields, so most oscilloscopes fill the need required for any test. Scopes with 50–100 MHz bandwidths are relatively inexpensive and serve most purposes. These scopes usually use what is called a postaccelerator CRT. The CRT contains an extra mesh screen that accelerates the electron beam to provide a much

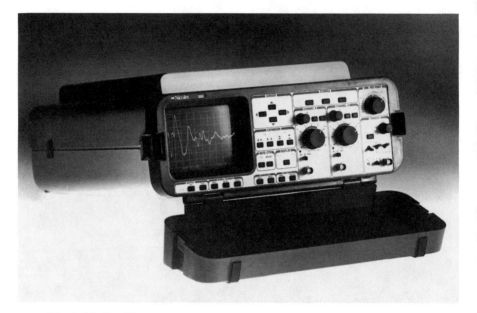

Fig. 2.56. Oscilloscope. *(Courtesy Nicolet Test Instruments Division.)*

brighter trace than lower cost oscilloscopes using single-accelera-
tion CRT tubes. This prevents problems encountered with less expensive units, which may not illuminate the screen sufficiently to
allow easy reading of a trace.

Oscilloscopes allow easy manipulation of input signals. Once the
X and Y full-scale values are selected, the scale can be tailored to
increase or decrease the wavelength, trace width, and intensity.
Triggering of the trace is adjusted so that the scope begins to display
a trace at the setpoint desired. This is especially useful when only a
small amplified section of a transducer output reading is desired.
Each time the transducer output reaches the setpoint, a new waveform will be displayed.

Once the trace is set properly and it corresponds to grid scales, a
photograph is usually taken to document the test results.

2.12.5 Analog Storage Oscilloscopes

Analog storage scopes work as described previously with an added
feature. A fine wire mesh or special phosphor coating on the screen
is used as a storage medium. Once the original electron beam has

created a trace on the screen, a secondary low-power electron beam source is used to maintain the trace brightness.

2.12.6 Digital Storage Oscilloscopes

Digital scopes (Fig. 2.57) work on an entirely different basis than the analog designs. The unit stores the input signal into digital memory after it amplifies the signal and converts it through an analog-to-digital (A/D) converter. This sample now has respective bit locations on an X-Y grid corresponding to the sampled waveform.

To reproduce the waveform, the scope must feed the stored information back into analog form through a digital-to-analog (D/A) converter. The analog signal is then amplified and used to drive the deflector plates of the CRT. The obvious advantage is that the sampled waveform can be stored indefinitely.

Digital scopes at present are not as fast responding as are analog units. In many cases the analog signal must be read, converted to digital memory, retrived from digital memory and converted to ana-

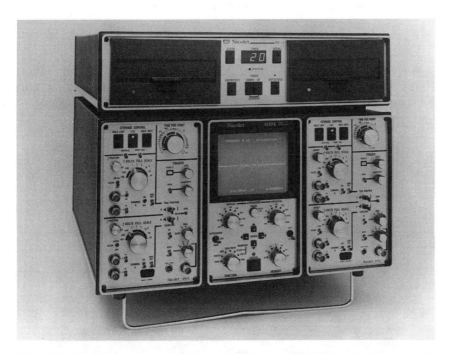

Fig. 2.57. Digital storage oscilloscope. *(Courtesy Nicolet Test Instruments Division.)*

log to drive the electron-beam-deflection plates. The signal conditioning rate of the A/D or D/A converters is presently not fast enough for high-resolution electronic testing. However, it is sufficient for the fluidpower industry. Some fully digital scopes will take a reading, go through the conversion process and display the sample when they are good and ready. This cuts down on the overall cost of the scope, as less expensive, slower response CRT's can be used. Again, this is usually sufficient for our purposes.

A benefit derived from digital technology is that once the sample is stored into digital memory, it can be manipulated as desired. For example, in analog scopes, the triggering setpoint only allows a trace of a section of the full picture. If full transducer output was displayed, the information desired at the small end of the spectrum would be impossible to read. However, the digital system would allow a full-scale trace to be stored and sections of the trace could be blown up and reviewed part by part. Also, the scope could take samples of a recuring waveform part by part and reconstruct the full picture at a later date. This technique allows a scope with slower response than the measured trace to be used.

2.12.7 Waveform Digitizers/Recorders

Waveform digitizers/recorders (Fig. 2.58) are very similar to a digital storage scope with two exceptions. The digitizer generally can be supplied with up to eight channels for simultaneous storage of

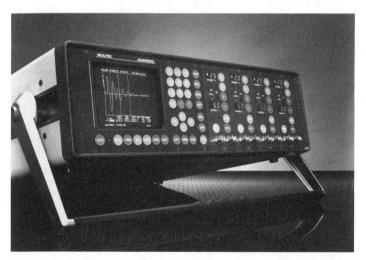

Fig. 2.58. Waveform digitizer/recorder. *(Courtesy Soltec Corp.)*

transducer data. These instruments can be purchased with transient capture bandwidths as high as 150 MHz. They are not supplied with a display, but are designed for use with computer-controlled data acquisition or waveform analysis programs.

2.12.8 Data Loggers

Production testing of a component determines whether or not all operating parameters are within specified limits. This testing is accomplished at the highest speed possible without sacrificing accuracy. The faster the test, the lower the cost applied to the manufacture of the component or the higher the profit. In some companies test speed is more important than complete testing. This usually leads to a higher warranty repair return because incomplete testing may not show up problems occurring during assembly or from faulty parts. By the time the problem has been uncovered, hundreds or thousands of faulty components could be out in the field.

An increasing trend to monitor these potential problem areas has been the application of data loggers or automatic data acquisition systems to production test stands. The term data logger encompasses a wide variety of devices that can monitor transducer data and maintain a statistical analysis file of desired test results. Some of the devices used include personal computers, special microprocessors, any variation of industrial computers, or dedicated data loggers.

The data desired is sometimes the most difficult area to be defined. It is important at the outset of a test system design that all parties understand the purpose and scope of a data collection system. A system that is underdesigned is normally quite expensive to modify for increased capacity. When in doubt, allow sufficient expansion by purchasing a system with higher memory or add-on module capability.

A common application for a data logger in the fluidpower industry is tracking a component test history. The program is designed to track the total number of units tested and the number of rejects, and to maintain a listing of the causes for rejection. At certain intervals the data can be reviewed to determine if corrective action must be taken to compensate for a high rejection rate due to a particular problem. The system can usually be programmed with limits to signal if immediate corrective action is required.

Data loggers can be purchased with one or more channels for simultaneous analysis of various system parameters (Fig. 2.59). The transducer inputs are generally in analog form and require internal

Fig. 2.59. Multichannel data logger. *(Courtesy Acurex Corp.)*

A/D conversion to digitize the signal. The A/D converter is the main element of the system, which dictates system resolution. This translates to accuracy and speed. Once the signal is in digital form, input channels can be scanned for data analysis and storage. The scan rate or synchronization of the scan can be another hinderance in high-speed data collection. The scan can normally be programmed to monitor various channels at different scan frequencies or monitor several channels simultaneously, reacting to the highest priority output. More complex systems allow a data acquisition system to monitor inputs, make optimizing decisions, and react to provide total system control.

Sampling rate (or scan rate) is an important feature when sampling at high speeds. The Nyquist sampling theory dictates that sampling rate must be at least twice the speed of an input analog signal. This allows the system to recover the signal without distorting it. It is also recommended that when scanning waveforms, such as a sine wave input, that the curve be scanned at least 10 times for reproducible results.

Data acquisition software and peripheral devices now allow the personal computer to be used for data collection. This is accomplished by optional I/O cards installed in the PC or in a remote card cage. Also, front end systems can be interfaced to a host computer. This flexibility allows an existing in-plant PC to also provide a data collection function. The cost is relatively inexpensive; thus, this method is gaining increasing acceptance.

Test Procedures

This chapter describes the more common tests applied to fluid power components. There are many other tests conducted to evaluate a component for a particular application in specific systems not described here. These tests are extremely specialized, and for general discussion would only provide filler material for the chapter.

Most tests described are based on years of development by many companies, both in the mainstream fluid power market and in totally unrelated markets. Very few test procedures are standardized, as you will see, although some progress have been made to overcome this problem. The consolidated material covered in this chapter hopefully will provide a useful tabulation of the various test methods used in our industry.

Figure 3.1 depicts a hydraulic schematic that incorporates a majority of the circuits required for fluid power testing. It consists of a power supply, control valving, pump/motor circuits, return line loading, and flow measurement. In addition, auxiliary circuits are depicted such as hydrostatic pressurization and a scavenge system for collecting sink drain oil.

3.1 General Flow and Pressure Tests

This section outlines tests that generally apply to most fluid power components. These tests mainly deal with the performance limits of a component rather than a specific function.

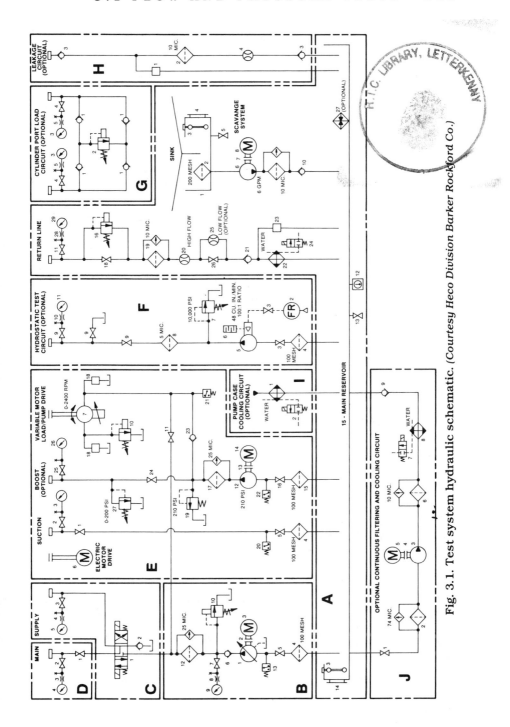

Fig. 3.1. Test system hydraulic schematic. (Courtesy Heco Division Barker Rockford Co.)

3.1.1 Pressure Drop versus Flow

Pressure drop versus flow is probably the most basic test conducted. Viscosity plays an important role in this test, since increased viscosity results in increased pressure drop. Some component manufacturers will use a low-viscosity fluid when developing basic performance data; this is misleading in certain cases. If the rated component pressure drop is not corrected for a higher-viscosity system, several problems could occur, including heat generation, insufficient pressure, response change, and decreased efficiency. In short, pay attention to fluid viscosity when testing for pressure drop or when reviewing pressure drop data.

An important consideration when conducting pressure drop tests is to eliminate (as much as possible) any pressure drop from fittings, hose, or other components. This is sometimes a difficult task, since the pressure gage or transducer used may already be plumbed into the system in a less than ideal location. The gage should be connected at the inlet port of the component or at the point where pressure drop data are needed. This can be accomplished with a small ID hose, as this hose will not contribute any error to pressure readings. This is because pressure within the hose is static (no flow); therefore, no pressure drop is created.

The next important step is to eliminate pressure drop on the valve outlet (or the downstream side of the test point). Any additive pressure will result in a higher pressure reading at the component inlet. The ideal situation would be to let the oil escape to atmosphere, but this could get rather messy. Instead, a pressure tap is usually installed at the component outlet and the return line pressure is monitored. Return line pressure is then subtracted from inlet pressure to produce the exact pressure drop value.

Collecting pressure drop data in this manner can produce errors. If two separate gages are used (one at the inlet, one at the outlet), then the gage accuracy of each, as well as the linearity of each, could open the error deadband. One way to eliminate this possibility is to use a selector valve connected to one gage and both test points. The same gage can then be used for both readings. However, if the pressure drop across the test component is too large, and return line pressure is very low, this may not help either.

Pressure gages are generally rated with an accuracy of percentage of full scale. The higher the pressure, the more accurate it becomes in terms of actual psi. Therefore the low end of the scale becomes rather inaccurate. It is important in pressure drop testing that this potential cause of error is reviewed.

Differential pressure gages can be used to eliminate the problems already discussed. The high-pressure side is connected to the valve inlet and the low-pressure side is connected to the valve outlet. Actual pressure drop can then be read. This device compensates for variable return line pressures due to changes in flow rate.

Whether standard gages or differential gages are used, it is important to use the lowest range gage possible; this increases accuracy. For example, a 5000 psi gage with ± 1% full scale accuracy equates to ± 50 psi, while a 100 psi gage with the same accuracy equates to ± 1 psi. In pressure drop tests, only enough pressure to create the desired flow rate is necessary. Even components designed for 10,000 psi service need only 100 psi supply to measure a 100 psi drop. However, if the system is to be used at higher pressures, the low-range gages must be removed or protected from pressures in excess of their rating. Some differential gages are designed for high-pressure service, so this does not always apply. Check your gage's ratings to verify maximum allowable operating pressure.

Pressure drop is a function of pressure at a certain flow. Therefore, accurate flow measurement is also necessary. Figure 3.2 depicts a typical pressure drop versus flow curve. Flow should be measured at the component outlet unless there is the possibility of internal leakage preceding the flowmeter. It is then recommended that inlet flow is measured, paying attention to pressure variations that cause flowmeter error.

Some components use the C_v factor to define pressure drop versus flow characteristics. Figure 3.3 shows a butterfly valve being tested with air to develop C_v rating. There are basically two C_v rating types: one used for gas as a medium, the other used for water applications.

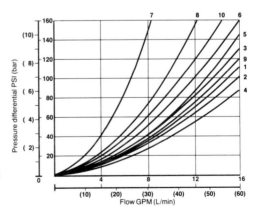

Fig. 3.2. Pressure drop versus flow curve. *(Courtesy Rexroth Corporation.)*

Fig. 3.3. C_v testing of a butterfly valve. *(Courtesy Rockwell International.)*

For air test formula, see Chapter 1, which provides information concerning the first C_v type.

When rating a component with water or liquids, the following formula applies:

$$C_v = \frac{Q}{\sqrt{\Delta P / G}}$$

where

C_v = flow capacity coefficient

Q = flow (gpm)

ΔP = pressure drop across the valve (psi)

G = specific gravity of the fluid

ANSI B93.49M-1980 (NFPA/T3.5.28-1977), Hydraulic Fluid Power Valves—Method of Measuring and Reporting Pressure Differential Flow Characteristics, defines a procedure for measuring pressure drop. Specific inlet and outlet piping considerations, location of measurement devices, as well as methods for evaluating tare pressure (test system pressure drop) are outlined. (Copies of this procedure are available from the NFPA.)

SAE J747a, Hydraulic Equipment Test Code—Directional Control Valves, defines a procedure used to measure pressure drop across single-acting and double-acting directional valves.

3.1.2 Minimum and Maximum Flows

Minimum flow is often required to overcome internal leakage or to ensure proper response time of a component. Response time is related directly to flow acting on moving parts within a component. This test is conducted with a system that can accurately measure flow at the low end of the test spectrum. Generally, flow is increased or decreased in successive steps, while all performance characteristics of the component are tested. In many cases this rating is also pressure oriented, since increased supply pressure may overcome losses due to leakage paths.

Maximum flow ratings can be applied because of excessive pressure drop, too fast of response, insufficient component force against flow forces, etc. Figure 3.4 depicts the maximum flow ratings (pressure-dependent) of a solenoid valve. In certain cases such as this, flow forces caused by a combination of flow and pressure are cause for reduced maximum flow limits at higher pressures. Again, a component is subjected to successive increasing steps in flow until malfunction or unsafe limits occur.

Figure 3.5 depicts a butterfly valve undergoing a maximum flow test. The term used here is "blowdown test," which is used to evaluate valve durability when subjected to high flow rates at high differential pressure.

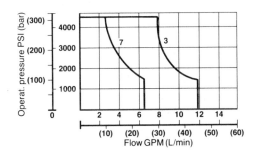

Fig. 3.4. Maximum flow versus pressure curve. (Courtesy Rexroth Corporation.)

Fig. 3.5. Maximum flow and pressure test. *(Courtesy Rockwell International.)*

3.1.3 Minimum and Maximum Pressures

Minimum and maximum pressures define malfunction points as well as safe operating limits. Minimum pressure is similar to minimum flow, in that a component must have sufficient supply to overcome the minimum forces required due to the basic design of the component. For example, a relief valve may have a range of 100–3000 psi (Fig. 3.6). In this case, the minimum operating pressure before the valve works is obviously 100 psi, because of spring rate, surface area of the poppet, etc., which must be overcome. If too low a rating were applied, the valve would become unstable at low pressure.

Pressure must be sufficient to operate or pilot spools across flow forces. Pumps require a minimum pressure at their inlet. Cylinders, motors, and actuators require minimum pressure for "breakaway" to overcome mechanical friction and other effects.

The component is subjected to increasing steps in pressure until proper operation occurs.

Maximum pressure, on the other hand, defines either the point at which a component will cease to function or a point of operation where the component safety factor of operation is reached. There are generally three terms used to define the maximum pressure a component may withstand:

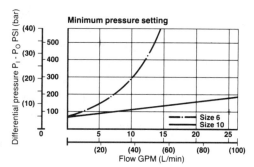

Fig. 3.6. Minimum relief valve operating pressure. *(Courtesy Rexroth Corporation.)*

Maximum Continuous Pressure means just that. A component with this type of rating can be used 100% of the time at maximum rated pressure without abnormal degradation of performance.

Maximum Intermittent Pressure is usually described as a percentage of total operating time a component can be used at a specified maximum pressure. For example, a specification may read: 3000 psi continuous pressure, and 4000 psi intermittent pressure—intermittent defined as not more than 5 min operation during a 1-hr period.

Maximum Surge Resistance concerns the component rating with occasional pressure spikes encountered in normal operation. This has some bearing on burst pressure and impulse cycle ratings. However, many components can easily withstand an occasional spike, while continuous impulse will cause premature failure.

Maximum pressure tests are conducted with the pressure increased step by step. Generally, the ratings described are a function of proof and burst pressures, described in the next section. Component performance is evaluated to determine any variance in proper operation.

NFPA/T2.6.1, Method for Verifying the Fatigue and Static Pressure Ratings of the Pressure Containing Envelope of a Metal Fluid Power Component, outlines static test procedures. The formula used in this instance is

$$STP = K_v \times RSP$$

where

STP = static test pressure

K_v = variability factor

RSP = rated static pressure, which is selected by a manufacturer as a component rating.

Both procedures should be reviewed carefully prior to use to ensure adherence to the intent of the specification. Copies of this standard may be obtained from the NFPA. The standard also suggests a possible method of using the RSP and RFP (rated fatigue pressure) ratings on a pressure versus cycles graph to determine finite life rating of a component. This would provide information necessary to predict a component's life expectancy based on working pressure versus pressure cycles.

3.1.4 Minimum and Maximum Return Pressures

It should not be forgotten that some components have different pressure ratings for the return ports than for the inlet ports. These pressure ratings are lower and are usually due to the housing design or other internal components affected by excessive pressure. Because of this, most component ratings will define maximum return pressure.

Conversely, some components require a minimum pressure at the return port to operate properly. This could be due to a minimum pilot pressure required or just to accomplish a stable condition within the component.

Tests of this nature should be conducted with a great deal of caution. Some components have very low pressure ratings in this case. Care must be taken not to allow pressure above recommended maximum limits.

3.1.5 Proof Pressure/Burst Pressure

These terms identify two important characteristics of a component. Proof pressure tests are conducted to determine the maximum pressure a component can be subjected to (over and above maximum operating pressure) without failure or change in operating characteristics. Most components are rated with proof pressure at 1.5–2 times maximum operating pressure. Pressure is brought up slowly, held for a short duration, and then decreased slowly. The part is visually inspected during and after pressurization to detect any mechanical changes, external leaks, jamming of moving parts, etc. During the development stages of a component, it may be totally disassembled, where all parts may be dimensionally checked for distortion or elongation.

Burst pressure means just that. Pressure is slowly increased until the component literally cracks and PSO's (pressure squirting out) are noted. This test determines the weak point of the mechanical design. This test is extremely dangerous, since loose particles of metal or high velocity oil may cause injury or death.

When this test is conducted, all precautions should be taken to prevent debris from exiting the test chamber. The component should be housed within a heavy-metal-wall enclosure. If viewing is necessary, use bulletproof Lexan. If the component is tested with a gas, allow a large path for compressed air to expand and escape without allowing debris to exit the chamber. If a large enough air path does not exist, the test chamber itself must be tested for burst pressure. If it fails, it too can cause death or injury.

When conducting burst tests, it is best to use a gage or transducer that will store maximum pressure seen by the component. At times this test is distracting, so that the component is viewed more often than the gage—during all the excitment, the engineer forgot to look at the gage! A pressure gage can be purchased with a follower pointer and a transducer display can be provided with a "peak hold" feature.

3.1.6 Minimum and Maximum Temperatures

There are several factors involved with temperature extremes. Seal material is the first consideration. Most components are available with Buna N seals for moderate service and with Viton for service in excess of 275°F. A difference in the function of the seal is also a consideration. Static seals will generally withstand higher temperatures than dynamic seals. A dynamic seal that softens or expands at elevated temperature must also overcome friction and uneven loading from linear or rotating movement. Often, severe applications require that seals be used at less than their maximum limits. Extremely low temperatures will reduce a seal's resiliency, and make it useless.

Temperature can also play a role in choice of metals within a component. Dissimilar metals expand and contract at different rates. When subjected to temperature extremes, a component may jam owing to decreased clearances. On the other hand, increased clearances can occur, causing excessive leakage and a decrease in efficiency.

Figure 3.7 depicts a butterfly valve being subjected to a temperature of 308°F. The test engineer is testing the valve for functional operation while electrically heated water passes through the valve and test loop. In this case, seat leakage, operating torque, and stem leakage are checked.

During temperature extreme testing, the component is checked

Fig. 3.7. High-temperature
valve test. *(Courtesy Rockwell
International.)*

to determine any performance degradation including leakage (Fig.
3.8). After the test, internal parts are checked for fatigue or dimensional change. This test is conducted with a constant test temperature or an incrementally increased or decreased temperature over a
period of time. This procedure differs significantly from the temperature cycle tests discussed subsequently in this chapter.

3.1.7 Minimum and Maximum Viscosities

You will note that most fluid power components are rated over an
optimum fluid viscosity range. In addition, the types of oil recommended define the best selection for that particular component. Performance ratings, such as pressure drop, response time, etc., are all
related to a specific fluid viscosity.

Minimum viscosity limits define the point where internal leakage
may become excessive, reducing the overall efficiency of a component. In particular, pumps and motors are affected by lower viscosity. A pump that is 95% efficient at 150 SSU may drop to 50% effi-

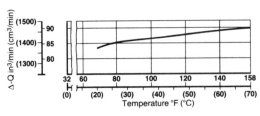

Fig. 3.8. Temperature versus
leakage curve *(Courtesy Rexroth Corporation.)*

ciency at 40 SSU. If this pump is run by an electric motor sized at 85%, the motor will become overloaded, or insufficient flow and pressure will result. In many applications, the minimum viscosity limit of a component is overlooked during the design stage and the problem is discovered after the fact.

Maximum viscosity limits define the points at which pumps can no longer pull a suction or require higher supercharge. Response time increases until a component is too slow. Pressure drop increases considerably until insufficient pressure supply causes a system to stall. This is particularly common with mobile equipment, which sits outside in cold weather.

Yes, viscosity is dependent on temperature (and pressure). But, this rating does not account for temperature. When the test is performed, fluid temperature is one method used to modify viscosity to set the proper parameters for the test. However, the component is never subjected to any temperature extremes that may shadow the effect of viscosity. In fact, for proper test data different fluids may be used to achieve various viscosity points at a common temperature range. When a component is selected, viscosity ratings have no bearing on operating temperature. However, when a component is selected, the system temperature swing should be reviewed to determine that fluids' viscosity at both ends of the temperature spectrum. If viscosity of that fluid is within the recommended range, then seal selection should be reviewed to determine proper elastomer material.

3.1.8 Cycle Tests

This type of test is conducted to determine how well a component withstands multiple cycles of normal operation. Solenoid valves, for example, are sometimes rated for 10^6 or 10^7 cycles capability under normal operating conditions. This test is conducted under conditions in which the component is rated. Temperature shock, pressure shock, etc., do not apply here.

The component is connected to a system that supplies flow, pressure, and temperature within the component's range. Backloading is sometimes used to simulate a particular field condition. The test system automatically cycles the component over a timed period, with amount of cycles recorded on a counter. This type of test can run for hours or weeks depending on the severity of the test and the time it takes for one complete cycle.

A typical cycle would be for a solenoid valve rated at 5 gpm and

3000 psi. The valve is energized, flow goes across a relief valve set at 3000 psi, pressure builds, a pressure switch signifies pressure, the valve deenergizes, and the counter counts one cycle.

Because this test may go on for some time, the system is normally designed to operate without any operator supervision. This requires that extra safety devices be included to monitor any situation which could cause damage to the test component or system. In addition, if the component fails, the system must also be designed to sense this, preventing tabulation of further test cycles after the component has failed.

After a component has passed a test like this, it is disassembled and examined for wear or any other changes in performance or dimension. If the component fails, the failed component may be redesigned to further increase the component's mechanical integrity.

3.1.9 Fluid Compatibility

With the various fluids used today and the development of newer fluids, a component may have to be tested to determine the potential changes of any characteristic due to the fluid. The first area of concern is seal material. Elastomers have many characteristics that must be considered; temperature range, resiliency, and expansion are just a few. There may be situations where a seal used in a static application may work fine, but in a dynamic application it would fail.

The concern is not only with 0-rings and the like, but also with diaphragms or plastic internal components. Some of the aerospace fluids in application today will melt various synthetic compounds.

Water-based fluids, used mainly in steel mill applications, are utilized in components designed for oil service. After testing, many of the components are approved for the application, but normal operating parameters are severly derated. Some components normally used at 3000 psi are derated to 1000 psi.

A fluid currently under developement is CTFE (chlorotrifluoroethylene). This fluid is being developed by the Air Force because of its increased flame resistance and lower viscosity at lower temperature. However, one of its main drawbacks is the fact that steel immediately begins oxidizing when exposed to air after being wetted by CTFE. It was initially thought that the fluid itself was corrosive by nature. It was learned that the fluid stripped any oil impregnated within the steel left during its manufacture. This caused the unprotected steel to immediately oxidize when exposed to air.

It is problems such as this that form the basis for fluid compatibility tests. Seals are a main, but not the only, consideration. Unfortunately, the only way to test a component completely for compatibility is under dynamic situations. This requires that the test system be flushed completely and then filled with the test fluid. This is a time-consuming process when various fluids are to be tested. It should be noted that many fluids cannot be mixed and contamination may be caused by a mix. For example, it was learned that one drop of MIL-H-83282 fluid in 1 gal of CTFE was sufficient enough to cause the fluid to be certified as contaminated.

3.2 Pump Tests

There are many pump designs available, each having its own specific set of operating parameters. In the fluid power industry the more common types of pumps include gear, vane, and piston, in open-loop and closed-loop systems. In the vane and piston category, we have fixed-displacement as well as variable-displacement pump types.

All pump tests are similar with particular attention paid to fluid type, viscosity, outlet pressure and flow, drive speed, and inlet conditions. There are a few pump designs that are exceptions. Examples are metering pumps, which provide pulse output, or centrifugal pumps, which have a highly dependent flow versus pressure curve relationship.

Pump testing can be fairly simple or may involve the use of complex equipment (Fig. 3.9) to measure all the parameters required. Depending on who performs the test, the test methods used, and the location and accuracy of the instrumentation used, pump test data may be highly accurate or in error by as much as 25%. This is particularly true when flowmeters are misapplied in a test circuit. (See Chapter 4.) In all cases, the pump test circuit and instrumentation must all be engineered into a single package to ensure accurate results.

This section discusses the more common tests applied to all pumps, particularly fixed-displacement designs. Variable-volume pumps are subjected to these tests as well as the tests defined in Section 3.4.

A general test procedure for hydraulic pumps has been developed by the SAE. Test procedure J 745c outlines various procedures used to define pump performance. These procedures also include test fluid type and temperature, instrumentation accuracy, recom-

Fig. 3.9. Pump test facility. *(Courtesy Rexroth Corporation.)*

mended test circuitry, and general definitions. Specific test procedures as outlined by the NFPA are also detailed throughout various
sections of the pump and motor test procedures.

ANSI/B93.27M-1973 (R1979) (NFPA/T3.9.17-1971), Method of
Testing and Presenting Basic Performance Data for Positive Displacement Hydraulic Fluid Power Pumps and Motors, is also a procedure developed to standardize various tests used during pump evaluation.

3.2.1 Pump Flow versus Speed Test

The graph depicted in Fig. 3.10 shows the typical flow versus
speed relationship found in most pump manufacturer's literature.
On this graph, several pump sizes (4–22) are represented at two different pressures. Note that the curves start at 500 rpm, which represents a minimum speed rating for most models. Flow versus speed
tests are, for the most part, a linear relationship, with the exception
of efficiency effects. Note the difference of flow output of size 22 at
290 and 3000 psi. The increased pressure causes decreased efficiency (internal leakage), thus providing less flow at the same speed.
This test is also performed with a specific fluid at a specific temperature in order to define the viscosity effect on pump performance. It
is very important to evaluate published data versus actual condi-

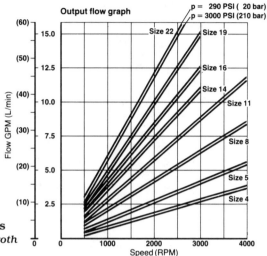

Fig. 3.10. Pump flow versus speed curve. (Courtesy Rexroth Corporation.)

tions in order to select the proper pump. Lower operating speeds and lower viscosity can combine to affect pump output drastically.

When conducting this test, it is important to have sufficient pump inlet pressure, accurate flow and pressure instrumentation, back pressure regulation, and a variable speed drive with sufficient torque and tachometer output.

3.2.2 Input Horsepower versus Speed Test

Continuing the evaluation of pump size 22, the curve shown in Fig. 3.11 depicts the amount of horsepower required at a specific drive speed and operating pressure. Insufficient HP will cause a system to stall completely or possibly to damage other system components. Excessive HP is never a problem, provided that pump output and pressure are limited. The pump will only draw as much HP as is required by the system. As you can see, HP requirement increases as speed increases owing to the nature of the HP equation, which is

$$HP = \frac{\text{torque (in.-lbf)} \times \text{rpm}}{63025}$$

This test is also affected by pump efficiency. Many pumps have been undersized owing to low viscosity or low speed. An accurate test would require a good torque transducer, tachometer, and pressure-measuring equipment.

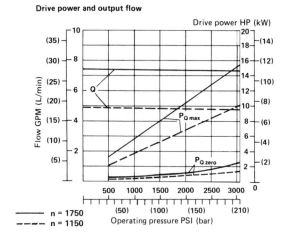

Fig. 3.11. Pump HP, speed, and pressure curve. *(Courtesy Rexroth Corporation.)*

If the HP versus speed data lines were to be extended, the line would taper upward or downward above the maximum speed and below the minimum speed. Exactly what occurs depends mainly on the pump design. Some pump efficiencies would increase to a point, while others reach such a level of inefficiency as to not produce any measurable output.

3.2.3 Flow versus Pressure Test

The curve shown in Fig. 3.12 depicts the flow output of pump at various degrees of output pressure. Specifically, a size 16 pump shows an output just under 10 gpm at 0 psi output pressure. As pressure increases, the flow line drops off owing to increased internal leakage caused by higher pressure (decreased efficiency). This line continues downward until the pump output drops to about 8.5 gpm at about 900 psi. The seven vertical lines shown below the flow

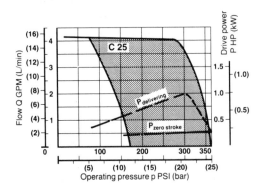

Fig. 3.12. Pump flow versus pressure curve. *(Courtesy Rexroth Corporation.)*

line represent the characteristics of a pressure-compensated pump set at seven different compensator settings. (This will be discussed later.) The chart also depicts the HP versus output pressure relationship of the same pump, to offer an overview of several simultaneous characteristics.

Flow versus pressure tests are also very viscosity sensitive. Not only can the test results vary because of fluid temperature, but also fluid pressure. Both can completely change a flowmeter calibration to the point of producing useless data. It is strongly suggested that Chapter 4, Flowmeter Accuracy, be reviewed prior to conducting this test.

3.2.4 Minimum/Maximum Flow and Pressure

The minimum/maximum flow and pressure test determines if a pump is either operating over the specified range of flow and pressure or is used to rate these ranges. Pump flow is directly related to displacement, drive speed, and efficiency. In variable-volume designs, flow is also related to the function of the controller. Pressure ranges relate to the minimum output possible as well as the maximum design pressure of the pump or the maximum controller setting.

It is advisable when testing pumps to collect data at both extremes of flow and pressure ratings. In addition, several flow and pressure points within the rated ranges should be verified. This will prevent a faulty pump from passing a test. Often, problems only show up at specific flow or pressure combinations.

3.2.5 Inlet Pressure/Suction

A pump is always rated with a required pressure condition at its inlet. Some pump designs require a minimal full flow supercharge to prevent cavitation, while others pull a suction. Open-loop pumps that require a pressurized inlet either are provided with a boost pump rated at a flow above the maximum flow output, or the reservoir is pressurized with compressed air or nitrogen. Closed-loop pumps generally have a small boost pump at the inlet sized at a flow above system leakage. Return flow from the circuit is diverted back to the pump inlet at 50–200 psi. Leakage returns to a small reservoir, where the boost pump draws fluid to maintain proper inlet pressure and provide cooling and filtration flow.

In all cases, pumps should be tested to be certain that they operate at all extremes of inlet pressure/suction. In instances where open-

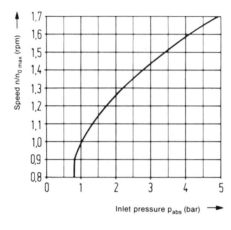

Fig. 3.13. Pump inlet pressure versus speed. *(Courtesy Rexroth Corporation.)*

loop pumps with suction capability are used at high altitudes, particular attention should be paid during this test, since increased altitude equates to lower barometric pressure, thus providing more demanding lift requirements on the pump. An inlet restrictor valve should be used to restrict pump inlet and pressure or vacuum should be read on a gage with a very small range to maintain the accuracy of pressure reading. Be careful not to exceed pump rating, since a starved pump will cavitate and possibly self-destruct.

The main factor involved in a pump's inlet rating is speed as depicted in Fig. 3.13. A pump with suction capability eventually requires boost pressure at some drive speed point. Tests should be conducted at the actual speed of the system in which the pump is applied, with attention paid to fluid viscosity and type.

3.2.6 Pump Efficiency

Pump efficiency can be related in three different ways: volumetric efficiency, mechanical efficiency, and overall efficiency.

Volumetric efficiency is the percentage of actual flow versus theoretical flow. All pumps have some amount of leakage due to clearances between moving components or through control valves or orifices. The amount of leakage is proportional to the operating pressure of the pump outlet. The higher the pressure, the larger the leakage. Testing volumetric pump efficiency is easy if the pump has a case drain. In this case, output flow versus case drain flow are monitored at various pressures and the ratio of these two values are calculated as

$$\frac{\text{output flow} - \text{case flow}}{\text{output flow}} \times 100$$

Where no case drain is provided, the actual output flow at various pressures versus calculated mechanical displacement must be used. Pump manufacturers have the machining drawings necessary to derive this information. Field testing procedures must rely on published theoretical flow data in manufacturers' literature. In this case, the formula used is

$$\frac{\text{actual pump flow}}{\text{theoretical pump flow}} \times 100$$

Another formula widely used by pump manufacturers (when actual pump displacement is not known) to determine pump flow with adequate accuracy is

$$\text{pump displacement (in.}^3\text{/rev)} = \frac{\text{gpm} \times 231}{\text{rpm}}$$

This test is conducted by running the pump at rated rpm at 100 psi outlet pressure.

It is generally important to know efficiency data when applying a pump to a system. Often this is overlooked and a system operates slower than desired due to either unexpected leakage or operation at a viscosity lower than used when deriving efficiency ratings. Pump volumetric efficiency is very dependent on viscosity and drive speed extremes. Temperature greatly affects fluid viscosity characteristics, so pumps should be tested not only with the proper fluid, but also at the highest temperature at which the pump must operate.

Mechanical efficiency describes the relationship of theoretical input torque versus input torque required. Enough input power must be provided for required output plus the mechanical losses present. In this case, input torque is monitored with a torque transducer, while output flow and pressure are unrestricted. To achieve accurate test data, the pump should be run at zero output pressure and output flow totally unloaded without fittings and tubing to provide even minimal restriction.

This efficiency rating is used to define the amount of power necessary just to turn the pump at a given speed. Input torque is recorded at various speeds and then compared to input torque readings at the same speeds but at various pressures. This allows the calculation

$$\text{mechanical efficiency} = \frac{\text{theoretical torque}}{\text{actual torque}} \times 100$$

Where theoretical torque is determined by

$$\frac{\text{pressure (psi)} \times \text{displacement (in.}^3\text{/rev)}}{2\pi}$$

Overall efficiency takes into account the effects of volumetric and mechanical losses. Again, input torque is monitored and the actual output horsepower is calculated from output flow and pressure. This is recorded at various flows, pressures, and speeds, and output horsepower versus input horsepower (torque versus speed) are compared by

$$\frac{\text{output horsepower}}{\text{input horsepower}} \times 100$$

Again, as with any pump test, knowledge of fluid type, viscosity, and temperature are important to collect accurate test data.

3.2.7 Case Leakage

Some pumps are supplied with a case drain port that allows leakage from around piston clearances and control valves to vent at low pressure. The drain port generally is located at the top of the pump in order to maintain the pump housing full of oil. A certain amount of oil is expected from this design as shown in Fig. 3.14. The curve depicts the expected leakage of four pump sizes at various pressures. Leakage increases the larger the pump due to increased piston diameter. Increased diameter means more annular opening area caused by the difference between piston OD and piston bore ID.

This test is critical because it can detect a pump that is about to fail. Excessive flow from the case drain port indicates a condition of piston wear or improper control operation. In addition, excessive heat from the case drain is a warning sign that mechanical friction is present due to swashplate wear or rotary group wear.

When conducting this test, all that is necessary is some method to backload the pump outlet and an accurate method of monitoring case flow. It is very important to make certain that you know and adhere to the maximum case pressure rating of the pump. New pump housings are not cheap. The flowmeter design and return

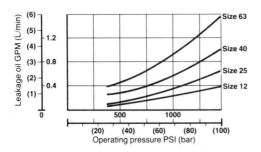

Fig. 3.14. Pump leakage flow versus pressure. (*Courtesy Rexroth Corporation.*)

plumbing should be reviewed to determine the total pressure drop possible at case flow. Remember, if the pump is defective, case flow can increase drastically, thus increasing pressure drop.

3.2.8 Case Drain Pressure (Maximum)

The case drain pressure (maximum) test is conducted by a manufacturer in the rating of the pump. This test should not be conducted in any other situation, since pump damage would most likely occur. Most often, the shaft seals will leak before the housing is cracked.

Figure 3.15 depicts the maximum case pressure for four pump sizes at various speed ratings. These ratings are based on a specific shaft seal design.

3.2.9 Pump Sound Level

The most common source of system noise is from a hydraulic pump. Many steps have been taken by pump manufacturers and system designers to reduce noise. These steps were necessitated by the standards set by OSHA.

Figure 3.16 depicts the sound level output of a variable volume pump at full stroke and zero stroke at various pressures. The curve was taken at a specific viscosity and at 1750 rpm. This curve depicts the sound level output only at one speed. In most cases, sound level increases with increased drive speed. As you can see, the output fluctuates up and down when the pump is at various speeds at full stroke. This is generally caused by the resonant frequencies generated by the internal rotating group. This is a function of specific pump design and nearly all pumps follow unique curves.

The important factor here is the accuracy of the sound meter, its location, and the surrounding ambient noise.

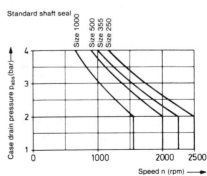

Fig. 3.15. Maximum pump case pressure. (Courtesy Rexroth Corporation.)

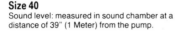

Size 40
Sound level: measured in sound chamber at a
distance of 39" (1 Meter) from the pump.

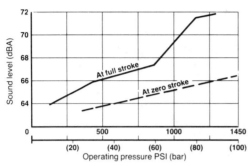

Fig. 3.16. Pump sound level output. *(Courtesy Rexroth Corporation.)*

ANSI/B93.71M-1986 (ISO 4412/2) (NFPA/T2.7.4M-1983), Hydraulic Fluid Power Pumps—Test Code for the Determination of Airborne Noise Levels, establishes a defined test procedure for monitoring and recording pump noise levels. Prior to the issuance of this standard it was difficult to determine pump noise based on available data. While many pump manufacturers used proper care in gathering test data, others did not.

This standard provides detailed data concerning measurement uncertainty, test environment, instrumentation, pump installation conditions, operating conditions, and sound instrument location. Instructions for conducting the test procedure, recording of data, and calculations for pump mean sound pressure levels and sound power levels are discussed.

Pump characteristics to be recorded during the test include description of fluid, viscosity of fluid, pump inlet fluid temperature, shaft speed, inlet pressure, outlet pressure, and pump flow.

Copies of the procedure can be obtained from the NFPA.

3.2.10 Shaft Loading

Pumps are often applied within systems that do not provide in-line drive to the pump shaft. Pump manufacturers usually recommend that a pump shaft and drive be interconnected with a flexible coupling to prevent any shaft loading due to eccentric or nonparallel shaft positions.

Some pumps are used with a belt drive, which is generally the worst case situation and the reason for this test. As force is applied to the belt or chain the pump turns, but forces generated by the rotational input also convert to side loads, attempting to pull the pump toward the drive shaft. This side load acts against the bearing surfaces of the pump. Some pumps are designed for this situation, since oversized bearings are installed on the shaft. Others will fail due to insufficient bearing surface design.

This is usually a destructive test conducted by a manufacturer to compile bearing life rating data or by a user to predict pump life in specific field application.

3.2.11 Flow Degradation from Contamination

The flow degradation from contamination test is used to determine the effects on a pump that is continually subjected to a particular level of contamination. This problem decreases pump efficiency owing to the abrasive action on moving part clearances.

NFPA T3.9.18M R1-1978, Method of Establishing the Flow Degradation of Fixed Displacement Hydraulic Fluid Power Pumps when Exposed to Partculate Contaminant, defines a procedure used to standardize this test. The claim is made that this test may not exactly simulate field conditions because additive effects from "water and the action of oil chemistry and water on pump wear surfaces" is not considered during this procedure.

Basically, the test pump is installed in a test loop that consists of a back loading valve, clean-up filtration loop, heat exchanger, and flowmeter. An injection chamber is used to add AC Fine Test Dust to the test loop so that specified contamination levels can be attained. The test pump is run within the manufacturer's limits of speed, temperature, outlet pressure, and inlet pressure during the test cycle. Flow output is recorded during specified time intervals to determine the potential reduction of flow at various pressure levels. Flow degradation at the end of the test is expressed as flow degradation ratio (Q_r), which is the ratio of the stabilized flow rate after test versus flow rate prior to the test.

Copies of the exact procedure may be obtained from the NFPA.

3.2.12 Pump Ripple

Pump ripple is caused by the frequency of the rotating components, which produce output flow and pressure. In slow motion, pump gears mesh or vanes and pistons provide cyclic pressurization of small internal chambers. The combined output of each of these pressurization cycles produces what appears to be smooth flow output.

In essence, this is true when flow is monitored by a slow-response flowmeter. In addition, this is not a problem in most systems, since high-frequency variations cause no apparant degradation of system performance.

However, in servosystems where a servovalve is required to maintain a constant output, this could be a problem. A high-response component may have a difficult time in a closed-loop control circuit if pressure variations are sensed, when in actuality the variations are due to pump ripple.

Pumps are designed so that several chambers are alternately pressurizing fluid, while others are drawing in fluid for the impending pressurization. The term used to define this synchronization is internal port timing. It is the overlap or lack of overlap between pressurized chambers exiting these timed ports that creates ripple.

Ripple frequency is exactly proportional to the input drive speed to the pump and the number of pressurized cavities.

To determine pump ripple, the pump outlet must be loaded across a relief valve or needle valve as long as outlet pressure remains constant. Flowmeters do not have the response necessary to detect ripple, so a pressure transducer is used. Ripple is defined as the frequency (Hz) of pulses versus the amplitude of the pulse. Amplitude is the lowest and highest pressure generated by the pump along a mean pressure line. Pressure transducers and signal conditioners with fast response should be used. An output of pressure versus time should be recorded on an oscillograph or similar instrument.

3.3 Motor Tests

Hydraulic motors are supplied in two variations: standard duty and low-speed high-torque (LSHT). The major difference in testing the two types is the speed range of the motor and the motor loading circuitry. Often LSHT motors cannot be tested on conventional pump test stands because of their extreme low speed. Of these two types there are several common designs. However, most tests described are used on any style or variation of motor.

ANSI B93.27M-1973 (R1979) (NFPA/T3.9.17-1971), Method of Testing and Presenting Basic Performance Data for Positive Displacement Hydraulic Fluid Power Motors, was established to standardize the method of representing basic performance characteristics of fixed-displacement motors.

The procedure depicts test schematics and performance graphs derived from motor tests. It defines general requirements for the following tests: volumetric displacement, structural integrity, input

flow, power output, overall efficiency, torque, no-load torque and reduced displacement tests.

Copies of this procedure are available through the NFPA.

3.3.1 Torque versus Pressure

This test determines the amount of rotary force (torque) a motor
produces at a specific pressure. To perform this test, a motor must
be mounted to a back-loading system such as a dynamometer or a
pump driven across a relief valve. The relief valve is used to control
torque loading. The loading device must provide sufficient load to
produce the torque desired plus enough load to overcome any mechanical and volumetric efficiencies of both the test motor and itself.

Output torque should be tested at minimum and maximum rated
motor speed, as well as minimum and maximum pressure. In addition, intermediate speed and pressure points will further define the
complete output characteristics of the motor.

Figure 3.17 depicts the output torque of several motor sizes. The
curve depicts that minimum pressure is approximately 400 psi.
This highlights the fact that mechanical forces and efficiency must
first be overcome before motion and torque can be produced. This
also depicts the reason why LSHT motors cannot always be tested

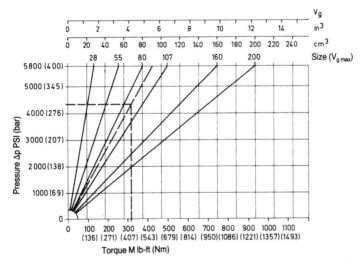

Fig. 3.17. Motor output torque versus pressure. *(Courtesy Rexroth Corporation.)*

on standard test stands. Hydraulic pumps with similar characteristics are generally used to test motors in the range of 200 to 3600 rpm. Speeds below 200 rpm typically freewheel the pump, because inefficiency has not yet been overcome.

The best way to measure torque is with a rotary torque transducer mounted between the test pump and loading device. This instrument can be calibrated in exact engineering units with a high degree of accuracy. An inexpensive method to closely approximate torque when using hydraulic pumps to load the motor is to install a pressure gage at the pump outlet. The pressure developed by the pump across the relief valve is directly proportional to the pump displacement minus efficiency effects. In fact, some pressure gage manufacturers will print a secondary range on the gage dial for a nominal fee. If the torque versus pressure relationship is known, the torque scale can be printed on the dial. The procedure is not recommended for high-accuracy testing owing to the inconsistency of efficiency, but this is a reliable method to use to verify that nothing drastic is wrong with a rebuilt motor.

One word of caution when using a hydraulic pump for back-loading. The cooling system must be designed to dissipate the total amount of input HP created by the flow and pressure across the relief valve. Without appropriate cooling, system heat will increase rapidly during continuous testing.

When testing LSHT motors, accurate control of load at low speeds is important. Some units achieve 1 rpm or less, which makes loading difficult, particularly for larger units with higher torque capacity. Stall torque may even be necessary to achieve complete test results.

Output torque is expressed by

$$\text{torque (in.-lbf)} = \frac{\text{gpm} \times \text{psig}}{27.2}$$

3.3.2 Torque Efficiency

This test defines the ratio of hydraulic input power to output torque and is expressed as

$$\text{efficiency} (\%) = \frac{\text{power output}}{\text{theoretical hydraulic power}} \times 100$$

or

$$\text{efficiency} (\%) = \frac{\text{output torque}}{\text{theoretical torque}} \times 100$$

This test is accomplished the same as was described for torque versus pressure, except careful attention is paid to the input flow and pressure at the motor inlet. The flow and pressure relationship is converted to torque (or power) and compared to the torque read from the torque transducer.

3.3.3 Torque Ripple

The torque ripple test is conducted as described previously, except that the output of the torque transducer is recorded on a fast-response oscillograph or similar device. Hydraulic motors do not achieve smooth torque characteristics owing to the multiple pistons, gears, or vanes that accept the inlet pressure through timed passages. During rotation, high-pressure oil is diverted to the proper piston, while oil is vented elsewhere. It is this crossover period that causes torque to rise and fall. Multiple pistons attempt to overcome this situation, but the effect is almost always present. Whether it creates a problem within the system is questionable during actual operation. Typically, high-response servovalve systems are the major application where ripple may be objectionable.

3.3.4 Volumetric Efficiency

The internal fluid leakage in a hydraulic motor defines volumetric efficiency and is expressed:

$$\text{volumetric efficiency} (\%) = \frac{\text{output flow}}{\text{input flow}} \times 100$$

During various loads, pressures and, most important, speeds, inlet flow and outlet flow must be recorded to define leakage. Motors that have case drains make this test simple, since all leakage exits the motor through this port. In this case output flow = input flow − case flow.

Motor speed is important as detailed in Fig. 3.18. A low-speed motor is tested at various pressures from 0 to 270 rpm. Note that volumetric efficiency drastically drops between 5 and 10 rpm. This reduction in efficiency often gets fluid power designers into trouble. Insufficient torque at lower speeds is sometimes important, but sometimes overlooked.

Note that as pressure increases, so does inefficiency. Higher pressure creates higher leakage flow through constant area leak paths.

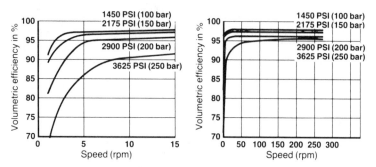

Fig. 3.18. Motor volumetric efficiency. *(Courtesy Rexroth Corporation.)*

3.3.5 Mechanical Efficiency

Figure 3.19 depicts the mechanical efficiency of a low-speed hydraulic motor. Note that mechanical efficiency in the 0–15 rpm graph is relatively constant except for the slight drop at minimum rpm. In the 350 rpm graph, efficiency drops as speed increases. Test results depict an optimum speed range of 5–250 rpm for this motor even though maximum speed is rated at 350 rpm.

Mechanical efficiency is expressed as:

$$\text{mechanical efficiency}\,(\%) = \frac{\text{actual torque output}}{\text{theoretical torque output}} \times 100$$

To extract these data, the motor is loaded at a specific torque (actual torque) and inlet pressure is recorded. Theoretical torque is then calculated using the formula:

$$\frac{\text{psi} \times \text{motor displacement (in.}^3/\text{rev)}}{2\pi}$$

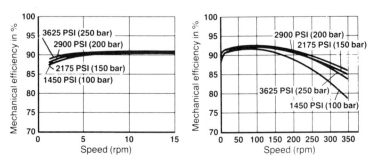

Fig. 3.19. Mechanical efficiency of a hydraulic motor. *(Courtesy Rexroth Corporation.)*

3.3.6 Overall Efficiency

This test for overall efficiency includes the effects of mechanical and volumetric losses and is expressed as

$$\text{overall efficiency} (\%) = \frac{\text{output horsepower}}{\text{input horsepower}} \times 100$$

Testing is conducted as described previously.

3.3.7 Speed versus Flow

A motor must be tested to determine the actual flow required to develop a specific speed at a specific load rating. Speed will vary with varying loads at constant flow owing to volumetric efficiency. Therefore, it is important to conduct this test at "worst case" conditions. Worst case in this instance is at highest rated motor pressure (or torque).

This test requires an accurate measurement of flow at the motor inlet or at the motor outlet and case drain. The pressure effect on viscosity is present during this test, so flowmeter accuracy on the motor inlet may not exist. (See Chapter 4.) It is preferable to use two flowmeters—one at the motor outlet and one on the case drain. This provides the advantage that all flow measurements are taken at fairly constant return line pressure. Now, the only concern for viscosity swing would be a temperature variation, which would also cause flowmeter inaccuracy.

If a motor is to be operated at either minimum or maximum speed extremes, it is mandatory that actual flow requirement be tested. This is where volumetric efficiency plays an important part in defining actual flow requirement.

3.3.8 Minimum and Maximum Speed/Flow

This test defines the upper and lower stability points of a motor. Flow is directly proportional to speed except when the factor of volumetric efficiency overrides if too much leakage occurs. Figure 3.20 depicts case drain flow of a motor from 0 to 4000 psi at 10 and 100 rpm. This more or less defines the minimum flow required to start the motor turning. However, some motors may still be unstable after this flow is provided. This can be caused by internal timing of ports, the number of pistons, the method of converting linear to rotary motion, etc. Motors may exhibit a pulsating movement at low speeds until all design areas have been satisfied with sufficient flow. This

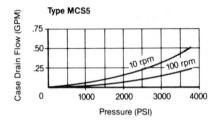

Fig. 3.20. Motor case drain flow. *(Courtesy Rexroth Corporation.)*

problem is also viscosity-sensitive, since more viscous fluids increase efficiency.

Maximum flow relates to the mechanical design of the motor, and includes bearings, timing passage area, mechanical stress factors, etc. A manufacturer will test a motor to determine acceptable life cycles at the upper speed range.

Both these tests are conducted at loaded and unloaded conditions to determine the full operating characteristics of the motor.

3.3.9 Motor Breakout/No Load Pressure

The pressure required to start a motor turning is termed the breakout pressure. Typically it takes more pressure to start a motor than it does to run at minimum speed. Figure 3.21 depicts breakout and no load pressure curves for several motor sizes. Note that the motor requires 160–170 psi to start turning and only 29 psi at 70 rpm.

When conducting this test, the motor shaft should be disconnected from anything that will cause any loading. Slowly increase

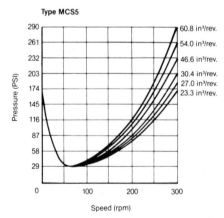

Fig. 3.21. Motor breakout and no load curve. *(Courtesy Rexroth Corporation.)*

pressure until shaft movement is noted. This will be the breakout pressure. For no load running tests, a tachometer should be connected to sense rpm, but remember, no loading on the shaft.

3.3.10 Acceleration/Deceleration

The acceleration/deceleration test may be destructive and is usually performed only by a motor manufacturer. This test allows the manufacturer to derive data that will help predict the life of a motor used in high-cyclic applications. Usually the motor is run at no load and loaded conditions for a predetermined amount of up/down cycles. At preselected intervals, the motor is visually and dimensionally checked for fatigue-related problems.

3.3.11 Degradation due to Contamination

NFPA Standard T3.9.25M-1977 (R1982), Methods of Establishing the Speed Degradation of Hydraulic Fluid Power Motors when Exposed to Particulate Contaminant, was established to provide a "uniform repeatable method for determining and reporting motor speed degradation due to contaminate wear."

The procedure describes the speed and pressure cycles a motor should be subjected to while various prescribed contaminant particles are injected into the motor test circuit. Motor speed at the start of the test is recorded, and speed is monitored throughout the test with inlet flow, pressure, and temperature remaining constant. The contaminant used during the test is AC Fine Test Dust.

Copies of the exact procedure may be obtained from the NFPA.

3.3.12 Airborne Noise Levels

ANSI B93.72M-1986 (NFPA/T2.7.5M-1985) (similar to ISO 4412/2), Hydraulic Fluid Power Motors—Test Code for the Determination of Airborne Noise Levels, describes a standard method of obtaining sound power levels under controlled test conditions.

The method describes procedures for the installation and operating conditions required as well as instrumentation requirements. The test procedure details step-by-step instructions for eliminating background noise error and the method of sound level measurement. Equations are also provided to calculate the motor sound pressure rating used to define motor characteristics.

Copies of the procedure are available through the NFPA.

3.3.13 Bearing Life/Side Loads

Because hydraulic motors may be used in applications where "perfect installation" conditions do not exist, they must be tested to determine resistance to undesirable loading of the shaft. These conditions include belt-driven nonparallel and eccentric shaft mountings and vibration. Motors are subjected to side loads and subjected to visual examination for bearing life and internal component wear.

Figure 3.22 is a graph that details the maximum radial force allowable at various distances from the bearing. The farther from the bearing, the less side load permissible.

This is a destructive test and is performed either by the manufacturer or a company that is aware of an existing side loading problem. In some cases, correcting the cause of side loading is extremely expensive, so several motors may be tested to see which model lasts longer under known adverse conditions.

3.4 Pump/Motor Controller Tests

Pump and motor controllers are responsible for varying the output characteristics of a variable-displacement device. Controllers can be tested for their particular characteristics or pump output can be tested as a response to a controller command. In most cases (with the exception of direct-acting mechanical displacement) a variable-volume pump/motor is piloted hydraulically to provide variable displacement. The input signals, whether they be mechanical, electrical, or hydraulic, are all converted to the proper hydraulic signal re-

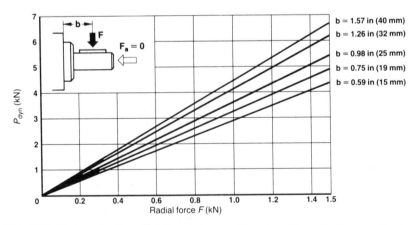

Fig. 3.22. Motor shaft radial force curve. *(Courtesy Rexroth Corporation.)*

quired to stroke the pump/motor. The flow and pressure applied control the response characteristics of the device.

While there are many design variations available, the basic concepts always remain the same. This holds true for open- or closed-loop systems as well as unidirectional or bidirectional pumps and motors (one or both sides over center).

Generally, the controller is first tested as a separate item to verify that it performs within specified limits. This is true during manufacturing, but in the field, the controller is tested as part of the overall pump/motor package. This is due to the many design variations available. Testing pump output to a given situation is much easier than performing numerous tests on a controller assembly.

The controller can consist of a single valve subassembly or be composed of many spools and poppets, depending on the pump design and application. The exact function and operating principle of the controller valve assembly may not be known or understood in the field. The manufacturer of the assembly generally performs tests based on the type of valve design used in the assembly. Specific test requirements for each control valve should be reviewed and understood prior to testing in the field.

Actual tests performed on pumps with variable-displacement controllers are conducted with several critical pieces of instrumentation. The word critical is used because all these functions are interdependent. Instrumentation used includes flowmeters, pressure transducers or gages, tachometer, timing device, torque transducer, and also a pump drive or motor loading circuit. Characteristics are generally tested at various operating speeds, temperatures, viscosities, and loads. This allows a thorough overview of different operating conditions. Data are usually taken point-by-point owing to the slow response of most flowmeters. Inaccurate test data are frequently acquired owing to flowmeter response and viscosity-effect problems. (See Chapter 4.)

The following subsections describe the more common pump/motor tests performed by manufacturers as well as rebuild facilities.

3.4.1 Response Time

The response time test defines the amount of time (in milliseconds) that a pump or motor takes to go from full to zero or zero to full displacement. In some cases, the response of the input signal is also considered an additive factor to the pump controller response. In other cases, input response is treated as a separate test.

Response time is generally tested as time versus pressure. This allows a pump output to be monitored at both a flowing condition and at deadhead. A back-pressure device and fast-acting solenoid valve are placed at the pump outlet and a pressure transducer, is hooked to an oscillograph (or similar device), to record output pressure. Actual test parameters are generated by a manufacturer or by a specific application. Actual flow during response cannot be monitored owing to the slow response of most flowmeters.

With the pump at a stable flow and pressure, the solenoid valve is energized and the time between solenoid energization (minus valve response time) and stable deadhead pressure is recorded as response time. Conversely, the solenoid valve is deenergized, or a redundant manual valve is opened, and deadhead pressure to stable flowing pressure is recorded as response time.

Other response tests may require response to flow, which is monitored by swashplate position. Additional problems encountered are response of other components, instrumentation, and the compressibility of the fluid.

Figure 3.23 depicts the response characteristics of several pump sizes with a specific control option. As shown, response time is dependent on system (deadhead) pressure. Response to deadhead in this design decreases slightly with increased deadhead pressure. Response to full flow depicts a more defined variation with changing

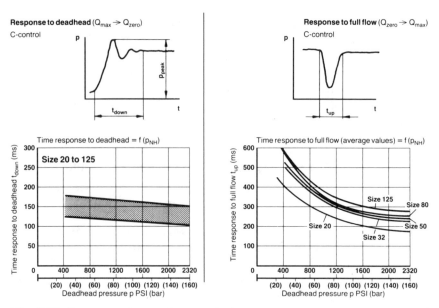

Fig. 3.23. **Pump response characteristics.** *(Courtesy Rexroth Corporation.)*

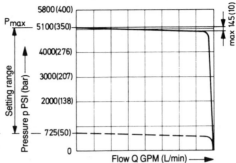

Fig. 3.24. Pressure-compen-
sation curve. *(Courtesy Rexroth Corporation.)*

deadhead pressure. In addition to flow, pressure is monitored to define the potential peak pressures caused by rapid response. Response to deadhead tests shows an overshoot in pressure with a following droop and stabilization time prior to constant pressure
output.

3.4.2 Pressure/Flow Compensation

The pressure/flow compensation test is probably the most common control test applied to variable-volume pumps and motors, because the pressure compensator is the most frequently used control
type. Pressure and flow compensations are similar and often identical. Once set pressure is reached on a pressure-compensated pump,
the pump destrokes to the flow that maintains the set pressure. Figure 3.24 depicts this situation with the compensator set at 5100 psi
(high end) and 725 psi (low end). A flow compensator will maintain
a set flow regardless of changes in load or speed. Figure 3.25 depicts
a flow-compensated control with the function of flow and pressure
reversed on the X-Y scales.

These controls are tested for the flow versus pressure relationship

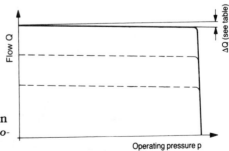

Fig. 3.25. Flow-compensation
curve. *(Courtesy Rexroth Corporation.)*

in both static and dynamic situations. Also, response time is a critical function of any control variation. As you can see, flow or pressure tends to drop off slightly as set point is reached. A sharp drop is then seen at close proximity to set point, until the variable drops to zero.

3.4.3 Horsepower Limiting

Horsepower-limiting control is very similar to the pressure/flow compensator, with the exception as seen in Fig. 3.26. The curve depicts an area of maximum flow or pressure unattainable because of the controller destrokes prior to the maximum of both functions. This control is used where the combination of both high flow and high pressure exceeds the limits of the input drive device, but where high flow or high pressure singularly are required. The control allows higher flow or higher pressure than could normally be attained with a standard flow or pressure compensator.

The tests as described in Section 3.4.2 are used.

3.4.4 Load Sensing

Load-sensing control senses the variables of pressure, flow, and input HP and maintains a maximum set point for any combination of these variables. The curve is similar to the HP limiting control, except that maximum flow moves up and down the vertical axis to further limit maximum power draw.

Testing of this design is more complicated because of the additional variable. Generally, each variable is adjusted to the minimum and maximum limits to determine its relationship to the overall set point. Input drive torque and speed, as well as accurate flow and pressure measurements, are necessary.

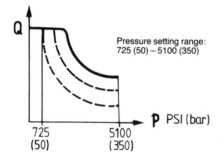

Fig. 3.26. Horsepower-limiting curve. *(Courtesy Rexroth Corporation.)*

3.4.5 Pump Output versus Mechanical Input

This test defines the mechanical position as well as torque or force required versus pump/motor displacement. Generally, torque or force values are derived based on maximum requirements in a worst case condition.

Pump/motor displacement versus mechanical input is generally expressed as shown in Fig. 3.27. The curve shows an overcenter pump with a 31°–0°–31° lever arm travel. At 31° the pump is at 100% displacement, with linear position versus flow output until zero delivery is attained at about 3°. The 3° deadband on each side of center assures zero pump null output.

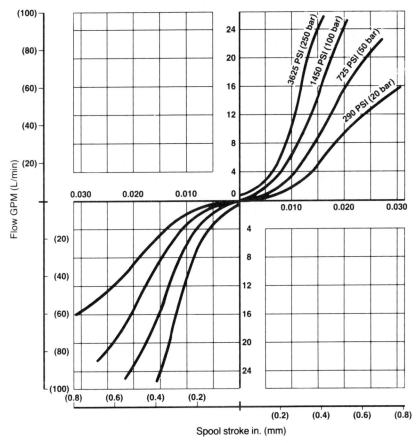

Fig. 3.27. Displacement versus manual lever position. *(Courtesy Rexroth Corporation.)*

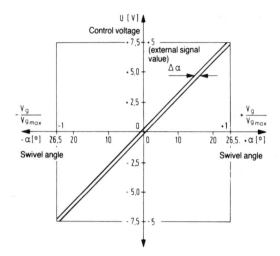

Fig. 3.28. Flow versus input voltage curve. *(Courtesy Rexroth Corporation.)*

3.4.6 Pump Output versus Electric Input Signal

In Fig. 3.28, pump flow (swivel angle) versus control voltage is depicted. This defines the linear relationship of an input voltage to a control device, versus the resultant displacement from that signal. Again, an overcenter pump is depicted with 26.5°–0°–26.5° travel. The curve also represents a hysteresis deadband that is labeled as ≤ + 1.5% of full scale. This test requires that accurate voltage (low current levels) and flow readings be taken.

As an example of the electric to hydraulic signal conversion required to stroke a pump, Fig. 3.29 depicts the mechanism. A spring-

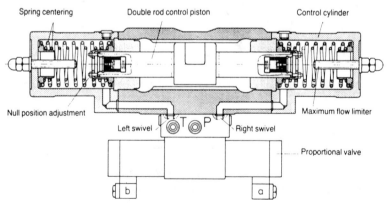

Fig. 3.29. Electric to hydraulic signal conversion. *(Courtesy Rexroth Corporation.)*

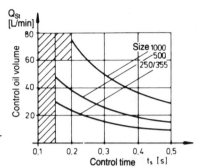

Fig. 3.30. Control oil volume versus control time. *(Courtesy Rexroth Corporation.)*

centered double rod cylinder is positioned by a proportional valve. The control piston is mechanically linked to the swiveling cylinder block assembly. An electric signal to the proportional valve provides a hydraulic pilot to the control piston that controls pump displacement.

Another test required on this control type is control oil volume versus control time as depicted in Fig. 3.30. The faster control oil is fed to the control piston, the faster the response.

3.4.7 Pump Output versus Hydraulic Input Signal

Pump output versus hydraulic input signal control design directly converts a remote pressure source to appropriate displacement. Figure 3.31 depicts an overcenter pump with a required pilot pressure of 85–260 psi from 0 to 100% displacement. Again, a deadband below 85 psi exists to eliminate false signals and ensure pump destroking.

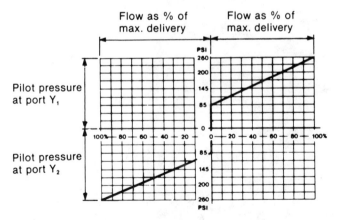

Fig. 3.31. Pump output versus hydraulic pilot. *(Courtesy Rexroth Corporation.)*

In some pump designs, control signal flow would be sized high enough to prevent slow response. In this design, however, control signal flow required is negligible because the hydraulic signal is supplied by a control piston piloted from the remote hydraulic signal.

3.4.8 Repeatability (Hysteresis)

A repeatability test is conducted on any of the control types to determine their operational deadband tolerance. This tolerance is necessary to define the proper operating requirements possible for any given application.

Exact parameters are maintained, such as fluid temperature, viscosity, flows, and pressures, to measure the repeatibility of the control design. For example, a pump with a pressure compensation control will be cycled back and forth to simulate a flow versus pressure curve. If an X-Y plotter (Fig. 3.32) is used, the pen will not exactly follow the same path. Over an extended cycle period the plot line will be wider than a single trace. The width of the line will define the hysteresis value.

Fig. 3.32. *X-Y* plotter used to record test data. *(Courtesy Rexroth Corporation.)*

After this, a control may be subjected to the same test with variable temperatures, flows, etc., to see the effect on hysteresis. Typically, controllers are rated by manufacturers using a specific fluid at a specific viscosity. Changes of fluid type may greatly affect all control parameters published.

3.4.9 Control Valve Leakage

Like any other component in a system, displacement control valves also allow a certain amount of leakage to pass through spool openings and control orifices. These leak paths are generally necessary for proper operation of the control package. It is nearly impossible to measure this leakage with the control mounted on the pump or motor. Most control packages vent back through the pump/motor housing.

To define the leakage characteristics of a particular control package, the manufacturer tests the controller separately. This leakage value is then added to the pump/motor volumetric efficiency rating in order to account for this additional loss. Actual control valve pressures, flows, and viscosities must be exactly simulated to provide accurate leakage ratings.

3.4.10 Minimum and Maximum Characteristics

Each control type has a range of operating limits for proper operation. At the low end of the spectrum, erratic pump/motor operation may occur owing to the lack of pilot pressure needed. Too low a current signal in an electrically operated control may reduce voltage, thus providing insufficient voltage to begin control. A high-pressure signal may surpass the recommended operating pressure of the component.

All these limits are tested for, to define the proper range of operation. Many are a function of temperature or viscosity or whatever effects the intended design function of the test component.

When testing a rebuilt component, it should be subjected to the full range of rated parameters to reveal any potential problem areas prior to operation.

3.5 Valve Tests

This section describes common tests applied to various valve types. In addition to this section, Section 3.1, General Flow and Pressure Tests, also applies.

SAE J747a, Hydraulic Equipment Test Code—Directional Control Valves, outlines several procedures for testing pressure drop leakage rate, operating effort, relief valve characteristics, and metering characteristics.

ANSI B93.49M-1980 (NFPA/T3.5.28-1977), Hydraulic Fluid Power Valves, Pressure Differential–Flow Characteristics, Method of Measuring and Reporting, defines a standard procedure used in gathering test data on valve pressure drop.

3.5.1 Valve Response Time

A valve response time test is used to predict the time it takes for a valve to respond to an input signal. The signal can be electric, pneumatic, hydraulic, or mechanical. The valve can be a directional control valve or a pressure or flow control valve.

Directional control valves are supplied in a wide variety of operator types and designs. Spool valves, for example, can be direct-acting, pilot operated, or manual design. When spool valves are tested for response time, the main spool is usually connected to a position-sensing instrument such as an LVDT. A timer is used to measure the time between input signal to the valve and the point where the main spool is fully shifted. Again, the response is checked for the spool to return to center from the time the input signal is released.

In actual evaluation, there are a series of individual events, each of which have their own response time. The additive responses of each event are combined as a rating for the valve. For example, let us take the case of a pilot-operated solenoid valve as pictured in Fig. 3.33. The initial input from where the time starts is when voltage is applied to the coil. the coil has a rating called force versus air gap, which is used to determine the force exerted by the coil along the full length of coil plunger travel. The pilot spool has a certain mass on which this force is exerted to create pilot spool shift. There are several forces in addition to mass that must also be overcome. Mechanical friction caused by spool/sleeve fit and spool side loading from pressure forces must be overcome. Once the spool begins to shift, passages open and close, causing flow that must be sheared by the spool lands. There is a relationship between shear forces and fluid type as well as effects of viscous dampening caused from spool displacement.

Once these factors have been overcome, the pilot spool begins to supply pilot fluid to one side of the main spool, while the opposite side of the main spool is vented to tank. In order to avoid excessive main spool speed, some valves are designed with orifices either re-

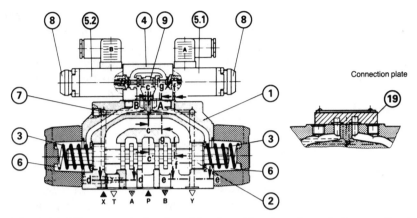

Fig. 3.33. Pilot-operated solenoid valve. *(Courtesy Rexroth Corporation.)*

stricting pilot flow to the spool or metering fluid out the vented side of the main spool. In these cases, small check valves are sometimes used to bypass these orifices in the opposite direction of controlled flow. The main spool response therefore depends on orifice size, applied pressure, and fluid viscosity. The rate at which pilot flow is supplied into the displacement volume of the shifted main spool governs the response time. Other additive effects are back pressure in the return line, return spring force, spool surface area, and radial jet pressure and axial jet pressure (causing increased spool friction).

Tests should be conducted with instrumentation having high response characteristics so that test results do not include instrumentation response. It is also imperative to know oil viscosity, pressure, and flow information.

In some devices response time is measured by flow or pressure output rather than just spool travel. Spool travel can define mechanical response, but system response is also dependent on fluid compressibility. Compressibility is dependent on system fluid volume, rate of oil flow, and pressure desired. Compressibility is also governed by fluid characteristics. Most flowmeters do not have the response necessary to provide accurate data on events occurring in less than 1–5 sec, so flow is usually monitored as a function of pressure. Pressure instrumentation is generally fast enough for most tests.

3.5.2 Seat Leakage/Spool Leakage

Valve internal leakage is sometimes a critical factor in system design. Load holding capability of a check valve used to maintain position of a manlift platform is a good example. A leaky check valve seat will cause a maintenance man to change that lightbulb real fast. On the serious side, manlift platform drift can cause catastrophic results. Valve spool leakage causes system inefficiency and heat.

Based on the size of the leak you are trying to detect, various methods are available. See Section 3.9 Leakage Tests for a description of common methods used.

3.5.3 Cross-Port Flow

Another type of internal leakage deals with the events occurring during the valve shift. Directional control valves come in a variety of spool configurations, with center position having the widest variety. If a closed-center, three-position valve opens the pressure port to tank for a short portion of the shift cycle, can this cause a problem? Yes.

Picture an accumulator connected to the pressure port and a cylinder operated by the valve. As the cylinder bottoms out, pressure rises and the accumulator stores all this potential energy. The valve is now shifted to retract the cylinder. For several milliseconds the pressure port is connected to return, allowing the accumulator to dump at extreme velocity. This short shot of fluid can damage components downstream by subjecting them to a flow rate higher than they can handle. Thus, the importance of cross-port flow.

The test is generally conducted by mechanically moving the valve spool through the shift and applying flow and pressure at a known viscosity. Flow is recorded from port to port whether it is drops/minute or several gpm. Pressure is applied not only to the pressure port, but also to each cylinder port. In all cases, flow is usually recorded at the tank port, since this is the path of least resistance for leakage flow.

The valve leakage is recorded as leakage versus spool position so that leakage can be analyzed on an *X-Y* chart. This leakage can often be tailored to a specific application by changing valve land timing (increasing or decreasing each spool land width) or sleeve opening dimensions.

3.5.4 Spool Force versus Spool Position

The spool force versus spool position test can be applied to two different situations. First, this can be applied during the development stage of valve design. Often it is not possible to predict internal

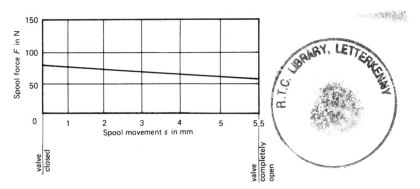

Fig. 3.34. Manual spool shift force. *(Courtesy Rexroth Corporation.)*

forces by calculation. It is then necessary to measure the force required to shift a spool through its full travel at specific dynamic conditions.

Once the proper flow, pressure, and viscosity are supplied, spool force is monitored by a load cell and spool travel is monitored by a position device. Force versus position is then plotted so that variable forces encountered through full travel are defined.

The second situation concerns the force required to shift a manually operated valve. (See Fig. 3.34.) This force is determined by the ratio of actual spool force as discussed previously versus lever arm design of the manual handle. Manual valves are rated with a maximum operating force at worst case conditions. Production manual valves are often tested with a force gage to detect any jamming caused by improper machining or assembly.

3.5.5 Metering Spool Flow versus Position

Directional control spools are often provided with a means to control flow to some degree. The spool can be machined with tapers, notches, or grooves, or the sleeve can be modified. Instead of pure on–off control, this allows an operator to throttle an actuator.

The center section of spool travel is where this metering takes place (Fig. 3.35). The spool must still provide positive shutoff and full-open conditions, so metering does not occur through the full travel of the spool. Some land material must be left to isolate these controlled openings.

The openings are based on the "pressure drop through an orifice" principle; therefore, in order to provide a near linear flow versus position curve, precise machining is required. This requires that test-

s · Q · curve (transition characteristics)
s · Δp · curve
Connection between spool movement "s", flow "Q" and pressure drop "Δp"

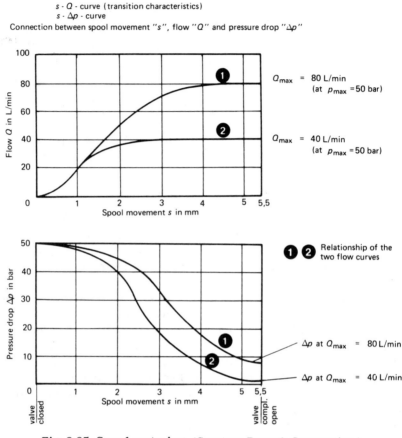

Fig. 3.35. Spool metering. *(Courtesy Rexroth Corporation.)*

ing be accomplished to determine flow regulation under field conditions.

Testing is performed with a flowmeter at the regulated port outlet, while pressure at the inlet and outlet are monitored. Spool position is monitored and flow versus position curves are generated.

3.5.6 Minimum and Maximum Coil Voltages

In this test, a solenoid valve is subjected to its maximum flow or pressure rating or both while voltage to the coil is monitored. Voltage is rarely available at its nominal value such as 120 Vac or 12 Vdc. Voltage fluctuates due to demand of other circuits or the availability of power. Solenoid valves are generally rated with a voltage rating of ± 10% of nominal. So, a solenoid valve with a 120 Vac rating can operate as low as 108 Vac or as high as 132 Vac. Too low a voltage will not generate enough force to the solenoid plunger. The plunger then stalls against the forces subjected to the spool or poppet from flowing fluid or static pressure. The result is a valve that won't shift or that chatters. High voltage causes a coil to overheat due to increased resistance. If this occurs, coil life can be drastically shortened or total failure can occur.

When conducting this test, it is important to have an accurate voltmeter and an adjustable power source. Operate the valve at its worst case condition and adjust the voltage to the lowest rating. Operate the valve two or three cycles to make certain the valve shifts and that it does not hum or chatter. Do not leave the valve energized for any length of time if the valve does not shift properly at low voltage. This could cause the coil to burn out.

In most cases, if the valve shifts at minimum voltage, it will shift at maximum voltage. This is due to an increase in coil force. Maximum coil ratings are based on insulation rating and maximum resistance values of the coil.

3.5.7 Pilot Pressure Requirement

Many valves are designed to accept a hydraulic pilot signal to accomplish valve operation, an example being a pilot-operated check valve. The check portion of the valve is rated to hold maximum system pressure. In order to release the trapped pressure, a pilot signal is used to push the poppet off the seat. If the valve were subjected to maximum available pressure, then the pilot piston would have to have a larger area than the seat area in order to overcome the force present. This is the case with pilot-operated check valves. The pilot piston is anywhere from 1.5 to 10 times the surface area of the seat. In two-stage designs the pilot piston initially unseats a smaller check to vent trapped pressure (decompression). This design can then be rated up to and over 40 : 1 ratio between operating and pilot pressure.

This test requires two pressure sources, one at the main check cylinder port and one at the pilot port. With zero pressure at the pilot port, apply a maximum pressure to the pilot-operated check cylinder port. Then slowly increase pilot pressure until the check valve opens. Note pilot pressure and determine from valve rating whether the valve is operating at the specified ratio or minimum pilot pressure (Fig. 3.36).

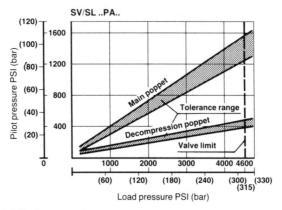

Fig. 3.36. Minimum pilot pressure. *(Courtesy Rexroth Corporation.)*

3.5.8 Valve Operating Force/Torque

Manually operated valves are supplied with maximum ratings for operation at worst case conditions. Spool valves generally are supplied with a lever or plunger that must be shifted to control the valve. Some shutoff valves are opened and closed with a 90° or 180° turn of a handle. Such is the case with ball valves, butterfly valves, etc. Other valves such as needle valves, globe valves, etc., require multiple turns from full open to full closed.

In all cases, at one time or another, the valve should be tested for required operating force. This detects assembly or machining problems, or excessive seal friction. Rotary operated valves are often used with an actuator for remote control capability. Operating torque value is required to properly size a rotary actuator. Linear valves are often controlled remotely with a cable or lever control. Again, resultant force required is necessary for proper design of the actuation system.

This test is conducted with the valve at worst case conditions (flow and pressure) and a linear or rotary force transducer mounted between the operating handle and the valve itself. Shift the valve slowly to avoid any force peaks from excessive input force. The required force may not be equal through the full valve travel. In this case it may be helpful to use a force transducer and display with "peak hold" feature. This would store the maximum force required during the test.

3.5.9 Flow Regulation

The flow regulation test defines the regulating characteristics of a flow control device. The flow control is generally tested to determine how far off the desired flow setting will deviate under specific conditions. Flow controls are available as fixed or adjustable, pressure compensated or nonpressure compensated. For the most part, nonpressure compensated valves are tested for pressure drop versus flow rate at a known viscosity. This discussion will focus on the pressure-compensated design and fixed flow rate. Adjustable flow controls are tested as fixed flow controls except that tests are conducted at several points throughout the adjustable range, particularly at the highest and lowest ratings.

Flow hysteresis and repeatibility tests are conducted with the valve subjected through a cyclic increase and decrease of system demand. For example, an on–off valve can be installed upstream or downstream of the flow control and cycled. With pump flow higher than the flow control setting, record the actual flow during each on cycle. This should be done with a flowmeter on the downstream side of the flow control to avoid pressure/viscosity errors from the flowmeter. The variance in flow readings will be the repeatibility characteristics of the valve.

With the valve at a steady-state flowing condition and unchanging loads, record the variance of flow regulation at several points over a period of time. This will provide steady-state hysteresis data.

Flow regulation with variable inlet and outlet pressure tests provide additional data under variable system conditions. Even the best pressure-compensated regulators are affected by variable conditions. Figure 3.37 depicts the flow regulation of several valve models

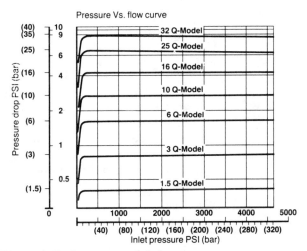

Fig. 3.37. Flow regulation versus inlet pressure. *(Courtesy Rexroth Corporation.)*

under varying inlet pressure. In some cases flow increases with increasing pressure, while in other cases, flow decreases. This is dependent on valve design and operating principles.

This test is accomplished with a relief valve and pressure gage upstream of the flow control valve. As the curve depicts, the regulation does not occur until minimum operating pressure is attained. With pressure adjusted at several points above minimum pressure, record flow and inlet pressure. Make sure that sufficient time is allowed for stabilization of the flowmeter after each adjustment.

Return pressure loading is accomplished by installing a relief valve between the valve outlet and flowmeter. Be certain that the relief valve has no external drain. Also, the use of a vent valve to control the main relief will result in a flow loss, causing error in the test. Adjust return pressure through the minimum to maximum pressure ratings of the valve at various inlet pressures to measure the effect of this load. Be sure that the minimum differential across the test valve is sufficient to overcome the minimum operating pressure of the valve.

Flow response tests are conducted to evaluate performance under rapidly changing system conditions. Again, remember the infamous forklift example. A forklift mast cylinder is fully extended, with a load on the pallet. Our flow control regulates the speed at which the mast descends. The directional valve is opened. The pallet drops, bounces, and then stabilizes to steady descent. What caused the bounce? Flow control response. Most flow controls are normally open devices. This means that the valve is fully open to allow unrestricted flow to flow rates less than set point. Once flow set point is reached, the valve throttles to maintain desired flow by increasing pressure drop with increased flow.

The valve is tested to determine how long it takes to regulate flow from an instantaneous flow condition. Again, because flowmeters are too slow, pressure is used versus time measured in milliseconds. An on–off valve is opened, and this event is recorded on the X-Y graph. A restriction in the valve outlet creates a certain pressure drop at a certain flow. Once this preset pressure is stable, so is the flow. This is the end time of the response curve. Be certain that additive response of test circuit valves, instruments, or variable pumps are not included as part of the test results. Careful system analysis is required for this test.

Flow regulation versus temperature defines valve performance throughout the rated operating temperature range. Temperature affects viscosity; viscosity affects pressure drop. Since the valve operates on a pressure drop principle, this is important. Many flow con-

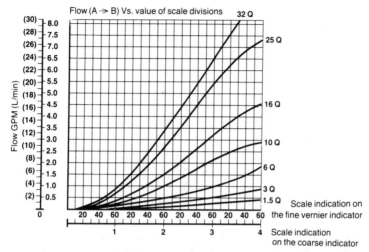

Fig. 3.38. Flow output versus handle position. *(Courtesy Rexroth Corporation.)*

trol manufacturers use sharp-edged orifices within the valve. This type of orifice helps negate the viscosity effect from both temperature and pressure variations. These valves are generally referred to as temperature- and pressure-compensated designs. All tests previously described are redone at various operating temperatures.

Secondary flow capability tests are done on priority flow controls and flow divider valves. Both sides are subjected to the same tests as previously described to determine the affect one side has on the characteristics of the other. Some valve designs can provide regulated flow at any pressure to both outlets, while other designs cannot operate properly with higher pressure on the secondary port.

Flow output versus handle position tests are used to define the (hopefully) linear relation between adjusting handle or screw position to actual output. This is useful information when a flow control is installed in a system without a flowmeter. Figure 3.38 depicts this relationship on several sized valves which have a coarse and fine division vernier indicator. Actual flow is measured at a particular viscosity and temperature.

3.5.10 Pressure Regulation

There are many types of pressure regulation devices: for example, relief, sequence, pressure reducing, counterbalance. Each has a specific application, but all have a common purpose—to regulate pressure.

Pressure testing is probably one of the easiest tests performed because the instrumentation used is highly responsive. There are several common tests applied to this type of control valve.

Pressure hysteresis/repeatibility tests are conducted to see how stable a pressure control valve is during a fixed situation. A preset flow is passed across the valve at a known viscosity and temperature. Over a period of time pressure is recorded to monitor pressure variation in a steady-state system. This variation is known as hysteresis.

To determine valve set point repeatibility, an on–off valve at the test valve inlet is used to cycle a steady flow at a stable viscosity and temperature. The resultant pressure is recorded each time the valve is open and the variation provides repeatibility data.

Pressure variation with flow is basically a pressure drop test. Once the regulator has been set at a pressure at minimum flow, the flow is increased. Pressure regulation is monitored to determine the effect of additive pressure drop through the valve.

Crack/full flow/reseat pressure tests are generally applied to relief and sequence valves to measure the rise in flow setting between crack and full flow. Crack pressure is the set pressure point at which a valve will begin to pass flow. This is usually defined as a certain amount of drops per minute. As pressure increases, so does flow through the valve (Fig. 3.39). This increases pressure drop and, therefore, the valve regulated pressure. Now pressure is decreased to monitor the pressure at which the valve stops all flow, thus reseat pressure. Reseat pressure is sometimes lower than crack pressure, because in addition to overcoming pressure on surface area and spring force factors, the flowing fluid through the valve must be sheared.

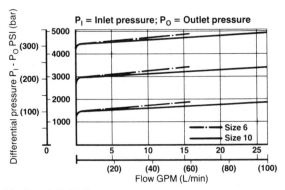

Fig. 3.39. Crack/full flow pressure. *(Courtesy Rexroth Corporation.)*

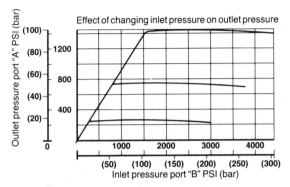

Fig. 3.40. Variable inlet and outlet pressure effect. *(Courtesy Rexroth Corporation.)*

Reduced pressure regulation applies to pressure-reducing valves. In addition to the previously mentioned tests, this valve is subjected to variable inlet and outlet pressures (Fig. 3.40). With these pressures varied, the valve must still regulate outlet pressure within the specified limits. Some valves are venting and some nonventing. This translates to regulation in a static condition. At no flow a venting valve should still maintain downstream pressure at setpoint. This is accomplished by a small amount of bleed flow out the vent port. Nonventing valves may overshoot set pressure by trapping a compressed volume of fluid during sudden shutoff of flow. Be certain that vent ports are unrestricted during testing. Usually added vent pressure affects set pressure accordingly.

Test the valve in a static and dynamic situation at variable inlet and outlet pressures. Always provide sufficient flow and pressure to overcome recommended minimum values. Figure 3.41 depicts a

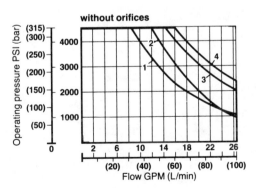

Fig. 3.41. Reduced flow versus pressure curve. *(Courtesy Rexroth Corporation.)*

typical reduced pressure versus flow graph. Note that pressure setting decreases with flow in this case.

Pressure setting versus handle position tests are used to define the pressure output of a device when a gage is not present. The valve can be supplied with a screw, handknob, or calibrated handle for adjusting pressure. The regulated pressure versus position of the adjusting device is recorded at normal system operating conditions. For example, a relief valve with a screw adjustment may be rated at 120 psi change in pressure per one full revolution of the set screw.

3.6 Proportional Valve and Servovalve Tests

The tests described in this section are probably the most critical owing to the complexity of the components involved. It is important to understand how these components function and how the test instrumentation functions prior to conducting these tests.

While proportional valves and servovalves differ in design and operation, the tests required for both are quite similar. In order to avoid redundancy, the tests most common to servovalves will be presented.

Servovalve testing (Fig. 3.42) requires special equipment not used in any other form of fluid power components. While many qualified servovalve repair companies exist, there are also many companies

Fig. 3.42. Typical servovalve test setup. *(Courtesy Moog, Inc.)*

that perform incomplete repairs due to lack of familiarity. Moog, Inc. suggests that both static tests and dynamic tests are important if a repaired servovalve is to provide the same performance characteristics as when new. The recommended static tests include null test to determine null bias current, valve polarity, a plot of pressure gain (average slope of load pressure versus input current), quiescent flow, no-load flow characteristics, loaded flow characteristics, hysteresis, threshold (resolution), proof pressure, and null/shift characteristics.

These tests are normally performed on a point to point basis. The use of continuous data plotting equipment for recording static performance information has generally replaced point to point procedures. This is the method used at Moog. Recommended dynamic tests include frequency response, change in frequency response with supply pressure, and valve transient response.

The following is a reprint (courtesy of Moog, Inc.) of the Recommended Test Procedure section of Moog Technical Bulletin 117.

Schematics are given in conjunction with the test descriptions which follow in order to illustrate the equipment requirements. Some equipment especially suited to the purpose is available from various manufacturers and a partial listing of this equipment has been included.

When selecting individual test instruments, gages, etc. to be used in special test set-ups for measuring servovalve performance, the accuracies of the equipment chosen must be compatible with the precision of the tests to be performed. Most hydraulic and electrical test equipment requires frequent calibration to maintain suitable accuracy. In addition, most hydraulic flowmeters require reading corrections to account for fluid viscosity variations. These variations can be expected with fluid use or changing fluid temperature, or between different fluid types.

In designing special equipment set-ups, care should be exercised to minimize unnecessary line lengths and hydraulic fittings so as to reduce line pressure drops. Likewise, tubing sizes should be selected for compatibility with peak flows to be encountered. Location of static pressure gages at the valve manifold will avoid inaccurate indications of actual valve pressures.

The input to electrohydraulic servovalves is defined as the *current* in the servovalve coils. In order to provide for unambiguous testing of servovalves, the electrical equipment used should be capable of providing undistorted current waveforms, irrespective of the valve coil inductance and back emf's due to valve action.

Hence, a high output impedance driver, such as a current feedback dc amplifier, is especially well suited.

Valves supplied with three lead coil connections are generally intended for use with differential electrical inputs. Accurate measurement of true differential current with most driving amplifiers is usually inconvenient, so the general practice is to drive the coils in series with a single-ended (two wire) connection. A single-ended amplifier may be connected to pins B and C of a three pin valve connector to achieve full coil operation.

Measured currents will correspond to the series coil connection, so will be one-half corresponding differential currents. Valve performance measured in this manner will accurately represent differential coil operation.

Several manufacturers and users are now utilizing semiautomatic, continuous data plotting equipment to obtain static performance characteristics for flow control servovalves. Either continuous or point-by-point test equipment may be employed for static valve testing. The recommended test procedures that follow recognize either technique as acceptable.

Unless specified otherwise, flow control servovalves should be tested under the following standard conditions:

Air temperature	70°F ± 20°F
Fluid temperature	100°F ± 10°F
Altitude	normal ground
Vibration	none
Humidity	10–90% relative
Hydraulic fluid	that for which the valve has been designed
Supply pressure	rated pressure ±2%
Return pressure	0 to +3% supply
Fluid filtration	10 microns nominal

3.6.1. Electrical Tests

COIL DC RESISTANCE

The dc resistance of the coil, or coils, should be measured with an accuracy of ±1% for 70°F ±2°F ambient temperature. If the valve utilizes a three or four lead coil connection, the resistance of each side of the total coil should be measured separately. A conventional Wheatstone bridge type electrical test instrument is acceptable for measuring coil resistance (Industrial Instruments,

Inc., Type RN). The valve need not be supplied with pressurized oil during measurement of coil resistance.

COIL INDUCTANCE

The recommended procedure for measuring valve coil inductance establishes the apparent inductance to signal frequency variations. This inductance value is suitable for transfer function representation in system dynamic studies.

The *total* coil inductance should be measured with the valve operating under conditions specified. The recommended test set-up is shown in Fig. 3.43.

A suitable audio frequency oscillator is connected to drive the total valve coil, connected in series with a precision resistor. The oscillator should have sufficient power output to supply undistorted input currency of at least one-half rated value. The oscillator frequency is set at 50 Hz and the amplitude adjusted to give a peak-to-peak current amplitude of one-quarter valve rated current.

The vectorial relationship of the voltages is indicated by a sketch in Fig. 3.43. The voltage drop, e_L, across the apparent valve

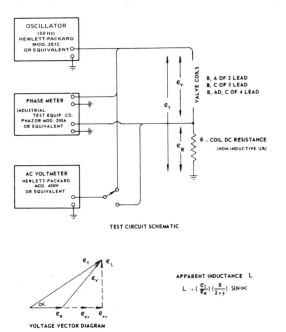

Fig. 3.43. Coil inductance test setup. *(Courtesy Moog, Inc.)*

inductance is determined by indirect measurement. From this, the total valve coil inductance can be determined.

The procedure is

Step 1 Measure the ac voltages e_R and e_T.

Step 2 Measure the angle, α, between e_R and e_T.

Step 3 Determine L from the following:

$$L = \left(\frac{e_T}{e_R}\right)\left(\frac{R}{2\pi f}\right)\sin \alpha \qquad \text{henrys}$$

If a suitable phase angle meter is not available, actual construction of the voltage vector diagram from measurement of e_I, e_R, and e_V can establish the magnitude of e_L with sufficient accuracy for determination of valve coil inductance.

INSULATION RESISTANCE

The insulation resistance of the valve coil and connections can be measured by the following steps. Valves need not be pressurized for the insulation test; however, wet coil and stale coil valves should be filled with hydraulic fluid during the test.

Step 1 Apply a dc voltage of five times the maximum anticipated coil voltage, or 200 volts (whichever is greater) between the combined coil connections (connected together) and the valve body (using Industrial Instruments, Inc., Type P6 Voltage Tester, or equivalent).

Step 2 Maintain the test voltage for 60 seconds.

Step 3 With the test voltage still applied, measure the current flow.

Step 4 The applied voltage level divided by the measured current value gives the insulation resistance. (Equivalent electrical instruments may give the insulation resistance directly.) No evidence of insulation breakdown should be apparent and the insulation resistance should exceed 100 megohms.

3.6.2. Static Performance Tests— Point-by-Point Data

The valve should be securely mounted to a suitable static test stand. Figure 3.44 gives the hydraulic schematic for a static test stand utilizing return line flowmeters.

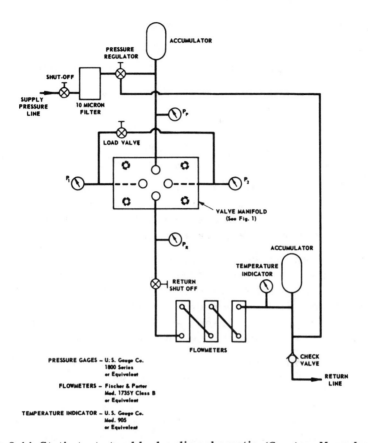

Fig. 3.44. Static test stand hydraulic schematic. *(Courtesy Moog, Inc.)*

NULL

The procedure to determine valve null follows.

Step 1 Open the return line shut-off valve.

Step 2 Close the load valve to block the servovalve control ports.

Step 3 Apply rated pressure to the valve under test by opening the supply line valve and adjusting the pressure regulating valve. The supply pressure to the valve should be adjusted until

$$P_p - P_R = \text{rated pressure}$$

where

P_p = pressure at port P, psi

P_R = pressure at port R, psi

Step 4 Apply full positive rated current.

Step 5 Decrease the input current slowly to zero, then increase current in a negative direction to full negative rated current.

Step 6 To eliminate electrical hysteresis, continue cycling the input current slowly between positive and negative values while gradually decreasing the maximum current levels.

When the current is decreased to zero by this method, note the readings of the control port pressure gages, P_1 and P_2.

Step 7 If the load pressure drop, $P_1 - P_2$, is not zero, the valve is off-null. To null the valve, slowly increase the input current in the proper direction to equalize the control port pressures. A positive input current will tend to increase the pressure at port 1.

Step 8 When the load pressure drop is zero, note the value of input current. Then, slowly increase the input current in the same direction until the load pressure gage readings change. Stop and reverse the direction of applying current input until the load pressure drop is again zero. Note the value of input current. The two values of current will differ due to valve threshold.

Step 9 The null bias current (in mA) is the average of the two current values noted in the preceding step.

POLARITY

Correct valve polarity can be determined from the control port pressure indications following step 4 of Null Test. Pressure at control port 1 should be equal to P_P and pressure at control port 2 should equal P_R.

PRESSURE GAIN

Data for a plot of the pressure gain characteristic (see Fig. 3.45) may be obtained by the following procedure.

Step 1 Null the valve, as in Null Test.

Step 2 Slowly apply increasing negative input currents until $P_2 = P_p$. Record the input current value and the pressures P_1 and P_2.

Step 3 Slowly apply positive increments of input current, stopping to read the pressures P_1 and P_2 following each increment.

Step 4 Repeat step 3 until $P_2 = P_R$.

Step 5 The load pressure drop (P_1 to P_2) may be calculated from each set of data. Approximately 10 data points, representing nearly equal increments of load pressure drop should be obtained.

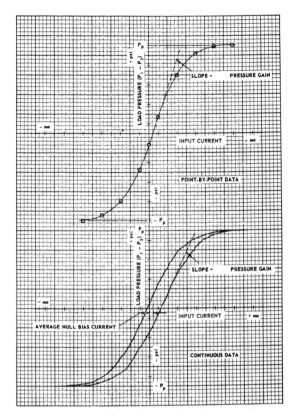

Fig. 3.45. Pressure gain characteristics. *(Courtesy Moog, Inc.)*

Step 6 The average slope of the load pressure versus input current plot in psi/mA in the region between $\pm 40\%$ P_P is the valve pressure gain.

INTERNAL LEAKAGE

Data for a plot of valve leakage flow versus input current (see Fig. 3.46) may be obtained by the following procedure. Often a statement of maximum internal leakage (which exists at the null point) may suffice in lieu of a plot.

Step 1 Follow steps 1 through 4 of Null Test.

Step 2 Slowly decrease the input current, stopping at suitable points to record the flowmeter readings. Care should be taken to obtain the maximum flowmeter reading that will occur at the valve null. Note that the null bias current obtained here will be different from that of Null Test by approximately one-half the valve hysteresis.

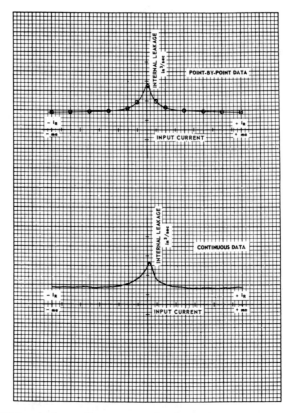

Fig. 3.46. Valve leakage flow versus input current. *(Courtesy Moog, Inc.)*

Step 3 Continuation of step 2 until full negative input current is obtained will provide data for a complete internal leakage plot. A minimum of eleven data points is recommended. These data are also necessary to obtain the no-load flow plot of the following section.

NO-LOAD FLOW CHARACTERISTICS

Data for a no-load flow plot (see Fig. 3.47) may be obtained by the following procedure.

Step 1 Open the return line shut-off valve.

Step 2 Open the load valve.

Step 3 Apply rated pressure to the valve, as in step 3 of Null Test. Under conditions of maximum control flow, special care should be exercised to maintain full valve rated pressure drop.

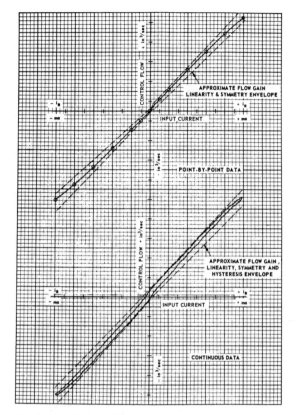

Fig. 3.47. No load flow plot. *(Courtesy Moog, Inc.)*

Step 4 Apply full positive rated current.

Step 5 Decrease the input current in the same increments utilized in steps 2 and 3 of Internal Leakage Test. At each value of input current, record the flowmeter readings.

Step 6 Calculate the equivalent control flow by subtracting the internal leakage flow values from the corresponding flow readings obtained in step 5. It should be noted that this method of determining control flow (i.e., subtracting internal leakage from total return line flow) is subject to error, since many valves are influenced by flow reaction forces. Since there is no control flow from the valve during the internal leakage test, the correspondence of like data points is upset. These errors will be particularly evident through the null region where the control flow is determined by differences of nearly equal flow readings.

Step 7 A plot of the data obtained in step 6 will provide valve linearity, symmetry, flow gain, and flow saturation information. Minor flow gain discontinuities will not necessarily be uncovered by this point-by-point test method.

H Y S T E R E S I S

Maximum valve hysteresis can be measured quite accurately by the following procedure.

Step 1 Perform steps 1 through 4 of the preceding section (No Load Flow Characteristics).

Step 2 Decrease the input current slowly to full negative rated current.

Step 3 Increase the input current slowly to full positive rated current. Slowly decrease input current to approximately 10% of positive rated current. Note the current value (i_0) and the equivalent flowmeter reading.

Step 4 Decrease the input current to full negative rated, then increase the current in a positive direction, continuing through zero and the value i_0 until the same flowmeter reading is obtained. Record the input current (i_1) at this point.

Step 5 The valve hysteresis is given by

$$\% \text{ hysteresis} = \left(\frac{i_1 - i_0}{i_R} \right) 100$$

where

i_0 and i_1 = input current readings obtained in steps 3 and 4, mA

i_R = rated current, mA

T H R E S H O L D

Valve threshold may be obtained by the following procedure:

Step 1 Perform steps 1 through 3 of No Load Flow Characteristics.

Step 2 Slowly apply input current of one polarity, stopping near 10% rated input, Note the current (i_0) and equivalent flowmeter reading.

Step 3 Slowly decrease the input current until a change in the flowmeter reading is noted. At this point, note the input current indication (i_1).

Step 4 The incremental current $\triangle i = i_0 - i_1$ is a measure of valve threshold.

Step 5 Repeat steps 2 through 4, stopping at other values of input current. Readings taken near the null point will be most accurate owing to increased flowmeter sensitivity.

Step 6 The largest increment of input current obtained should be recorded as the valve threshold, and this can be stated in percent of i_R.

LOADED FLOW CHARACTERISTICS

Step 1 Perform steps 1 through 4 of No Load Flow Characteristics.

Step 2 Without changing the input current from the full positive rated value, commence closing the load valve.

Step 3 At suitable points, stop and obtain readings of flow, P_1 and P_2. About five sets of data should be obtained at nearly equal increments of load pressure drop.

Step 4 Repeat step 3 at other constant values of input current, as appropriate.

Step 5 For each set of data from steps 3 and 4, calculate the load pressure drop $(P_1 - P_2)$. Also, correct the return line flow data for the corresponding internal leakage flows obtained at like input current values (see Internal Leakage).

Step 6 From the calculations of step 5, the valve loaded flow characteristics may be plotted (see Fig. 3.48).

PROOF PRESSURE

The valve may be subjected to rated proof pressures by the following procedure:

Step 1 Close the return line valve.

Step 2 Close the load valve.

Step 3 Apply rated pressure to the valve supply port by opening the supply line valve and adjusting the pressure regulating valve.

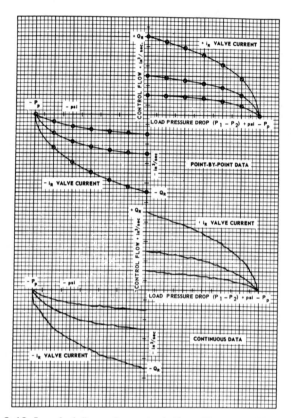

Fig. 3.48. Loaded flow characteristics. *(Courtesy Moog, Inc.)*

Step 4 Apply full positive rated current, $+i_R$, then apply $-i_R$. No external leakage should be present.

Step 5 Open the return line valve.

Step 6 Increase the valve supply pressure to $1\frac{1}{2} P_P$ by adjusting the pressure regulating valve.

Step 7 Repeat step 4.

N U L L   S H I F T   W I T H   S U P P L Y   P R E S S U R E

Data for a plot of valve null shift with supply pressure variations may be obtained by the following procedure:

Step 1 Null the valve as in Null Test.

Step 2 Decrease the supply pressure in suitable increments by adjusting the pressure regulating valve. At each value of supply

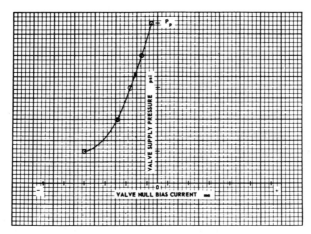

Fig. 3.49. Null shift with supply pressure. *(Courtesy Moog, Inc.)*

pressure, repeat steps 7, 8, and 9 of the valve null procedure to obtain the corresponding null bias currents.

Step 3 Data of step 2 may be plotted as in Fig. 3.49.

NULL SHIFT WITH RETURN PRESSURE

Valve null shift with changing return line pressure may be presented as in Fig. 3.50. Data for such a plot may be obtained by the following procedure:

Step 1 Null the valve as in Null Test.

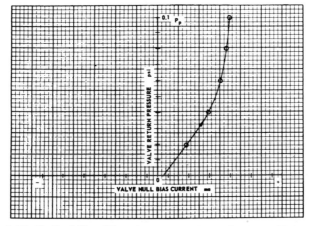

Fig. 3.50. Null shift with return pressure. *(Courtesy Moog, Inc.)*

Step 2 Slowly close the return line valve to establish suitable increments of back pressure. At each back pressure value, repeat steps 7, 8, and 9 of the valve null procedure to obtain corresponding null bias currents.

Step 3 Data from step 2 may be plotted as in Fig. 3.50.

NULL SHIFT WITH QUIESCENT CURRENT

Data for a plot of valve null shift with quiescent current may be obtained by the following procedure.

Step 1 Null the valve as in Null Test.

Step 2 Reconnect the valve coils to give a differential coil connection (if necessary). See sketch in Fig. 3.51.

Step 3 Increase the coil quiescent current in suitable increments, and at each current value repeat steps 7, 8, and 9 of the

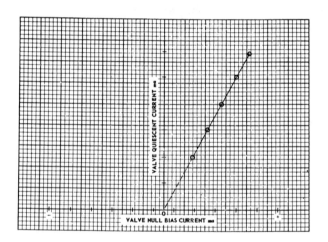

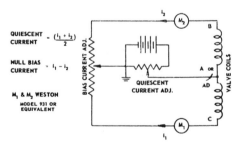

Fig. 3.51. Null shift with quiescent current. *(Courtesy Moog, Inc.)*

valve nulling procedure to obtain corresponding null bias currents.

Step 4 Data from step 3 may be plotted as in Fig. 3.51.

NULL SHIFT WITH TEMPERATURE

The recommended test equipment for this test utilizes a simple position servoloop to provide continuous null bias currents. See Fig. 3.52.

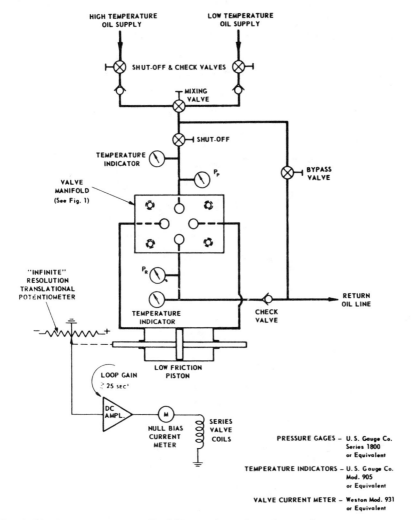

Fig. 3.52. Temperature null shift equipment schematic. (*Courtesy Moog, Inc.*)

Step 1 Open the return line valve.

Step 2 Set both the hot and cold oil supply pressures to the rated supply pressure.

Step 3 Adjust the mixing valve to give 100°F oil supply temperature to the valve under test. Record the input current.

Step 4 Readjust the mixing valve to establish suitable increments of increasing oil supply temperature. At each successive temperature allow the valve to stabilize for one minute preceding the input current reading. It may be desirable to introduce some dither signal into the system to avoid piston stiction.

Step 5 Continue to obtain null bias current readings for temperature increments to the rated maximum temperature; then de-

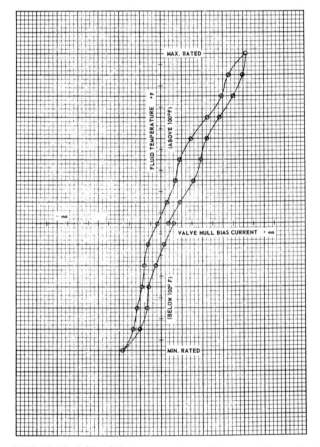

Fig. 3.53. Null shift with temperature. *(Courtesy Moog, Inc.)*

crease oil temperature in like increments to the minimum oil temperature. Similarly, increase the temperature and obtain null bias readings to 100°F.

Step 6 The data may be presented as in Fig. 3.53.

NULL SHIFT WITH ACCELERATION

The recommended equipment for this test utilizes electrical pressure transducers at the control ports as indicated in Fig. 3.54. The transducer outputs may be connected to give a differential load pressure indication following appropriate transducer cal-

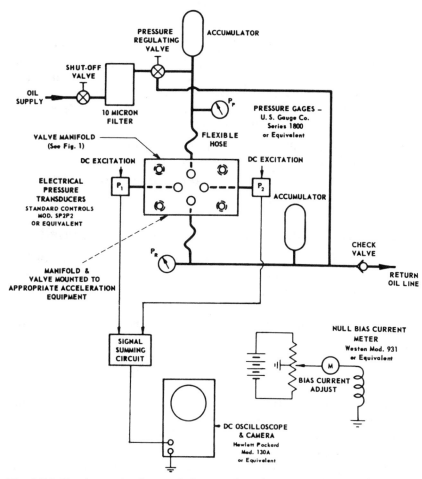

Fig. 3.54. Equipment schematic for acceleration null shift. *(Courtesy Moog, Inc.)*

ibration. The following procedure may then be utilized to measure the valve null shift with acceleration inputs.

Step 1 Mount the valve and manifold on a suitable acceleration input device (centrifuge for constant acceleration, shaker for vibration inputs, or drop equipment for shock).

Step 2 Open the return line valve.

Step 3 Apply rated supply pressure to the valve.

Step 4 For constant acceleration inputs, the null bias current necessary to obtain zero load pressure drop may be determined for appropriate acceleration values.

Step 5 For vibration inputs, an oscillating load pressure signal will exist. The peak-to-peak amplitude of the null shift may be determined by monitoring the load pressure signal on a dc oscilloscope, while slowly varying the bias input current. The algebraic difference of bias current values necessary to produce zero load pressure drop, first at one peak and then at the other peak, is the peak-to-peak null bias current.

Step 6 For shock inputs, a dynamic recording of the differential load pressure signal is required. The equivalent null bias currents can then be determined from the pressure gain data of Pressure Gain Test.

3.6.3 Static Performance Tests—Continuous Data

The use of continuous data plotting equipment for recording servovalve static performance information is rapidly displacing point-by-point testing procedures. In general, more complete data are obtained more expeditiously through the use of semiautomatic plotting equipment. Figure 3.55 gives a simplified schematic representative of this equipment.

Valve output flow is supplied to a large capacity piston equipped with motion transducers. Piston velocity is equivalent to the valve metric flow from the valve and an electrical signal proportional to the piston velocity can be recorded. A programmed current sweep input is supplied to the valve. Usually the sweep signal is a triangular waveform as sketched in Fig. 3.56.

The piston position prior to the sweep cycle is near one end, so the piston traverses the cylinder and back during the cycle. The steep duration is programmed such that the piston does not bottom during the cycle. A low gain position feedback signal with

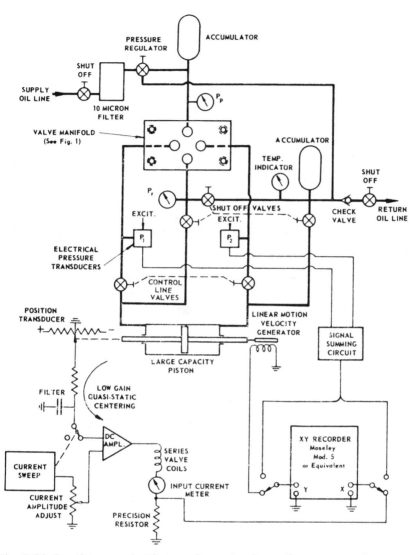

Fig. 3.55. Continuous plotting equipment schematic. *(Courtesy Moog, Inc.)*

bias adjustment is used to maintain the piston position prior to a sweep input. During a current sweep, the valve is operating open loop.

Most continuous data plotting equipment for flow servovalves is supplied with control port pressure instrumentation. By closing off the control line shut-off valves, plots of valve pressure gain characteristics can be made. For these plots, the amplitude of the

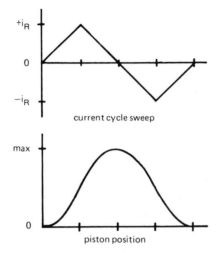

Fig. 3.56. Current sweep input.
(Courtesy Moog, Inc.)

current sweep can be adjusted for a suitable cycle about the valve null.

Some plotters can produce valve loaded flow characteristics. Different techniques for obtaining loads are utilized. The technique illustrated in Fig. 3.55 uses a manually adjusted load line valve to vary the load pressure drop. Some plotter pistons can be loaded with dead weights to furnish flow versus input current plots at constant values of load pressure drop.

It is usually possible to obtain plots of valve internal leakage flow versus input currently by blocking the control lines and diverting return line flow to the plotter piston. A half-sweep input commencing at $+i_R$ can be utilized to produce the internal leakage plot.

At the present time, two manufacturers are known to have complete data plotting equipment commercially available for testing flow controls servovalves. These manufacturers are

1. Industrial Measurements Corporation (Model 11 Flow Plotter), Pomona, California.
2. Moog Inc. (Model TE 1004 Flow Plotter), East Aurora, New York.

Detailed operation and calibration procedures are available from these manufacturers to explain the correct use of their equipment. Frequent calibration of plotting equipment is especially important for accurate data presentation. Electrical circuits within the plotters should be arranged to produce positive plotter displacements with the recommended standard positive current, flow and pressure specifications.

Prior to each test the zero axes should be established by actually plotting zero Y input versus X sweep and zero X input versus Y sweep. This can be accomplished by manually rotating the corresponding axis centering control.

BLOCKED LOAD CHARACTERISTICS

Pressure gain, polarity and null characteristics can be ascertained from the following procedure:

Step 1 Connect the valve, both hydraulically and electrically, to the plotting equipment.

Step 2 Close the control line valves.

Step 3 Open the return line valves.

Step 4 Set the valve supply pressure to the rated value by opening the supply valve and adjusting the pressure regulating valve. The adjustment should compensate for any return line back pressure.

Step 5 With the plotting pen raised, automatically cycle the input current sweep several times between plus and minus rated current values, stopping at zero input current.

Step 6 Establish a plotter configuration to permit recording load pressure drop along the Y axis and input current along the X axis. Use scale factors to give $\triangle P_{max} = \pm P_P$ and $i_{max} \simeq \pm 0.2 i_R$.

Step 7 Set the automatic current sweep to approximately $\pm 0.2 i_R$ maximum (or, as appropriate for the valve under test).

Step 8 With the plotting pen still raised, automatically cycle the input current between values set in step 7. During cycling, ascertain that pen motion is unrestricted and that a sufficiently slow sweep is in use to ensure a smooth plot, free of recorder dynamic effects.

Step 9 Lower the plotter pen and record the pressure gain characteristic for one complete current sweep, as in Fig. 3.45.

Step 10 The slope of the plot in the region between $\pm 40\% P_P$ is the valve pressure gain.

Step 11 Correct valve polarity is indicated by positive differential pressures ($\triangle P = P_1 - P_2$) corresponding to positive input current.

Step 12 The valve null bias is assumed to be midway between the abcissa intercepts of the pressure gain plot.

I N T E R N A L   L E A K A G E

A plot of the valve internal leakage may be obtained by the following procedure:

Step 1 Repeat steps 1 and 2 of the preceding section.

Step 2 Set the return line shut-off valves to allow return flow to pass to the flow metering piston.

Step 3 Establish a plotter configuration to permit recording return line flow along the Y axis and input current along the X axis. Use scale factors to give reasonable plotter deflections for the maximum leakage flow value and to allow $i_{max} = \pm i_R$.

Step 4 Set the input current sweep for a half cycle plot starting at $+i_R$. Some plotters are not equipped for half cycle plots, so will require a manually controlled sweep current input.

Step 5 Apply rated supply pressure to the valve. The plotter piston will commence moving, so immediately lower the plotter pen and introduce the input current described in step 4. Note that the sweep speed is sufficiently slow, especially through the null region, to give a smooth, accurate plot. See Fig. 3.46.

Step 6 Should the piston bottom prior to the completion of the previous step, or should another plot be required and insufficient piston stroke remain, the piston must be moved back to the starting end. This will require a temporary change in the shut-off valve configuration of the return and control lines, followed by a repetition of steps 2 and 5 above.

N O - L O A D   F L O W   C H A R A C T E R I S T I C S

Flow gain, linearity, symmetry, hysteresis, and saturation may be obtained by the following procedure:

Step 1 Repeat steps 1 through 5 of Blocked Load Characteristics.

Step 2 Open the control line valves.

Step 3 Establish a plotter configuration to permit recording control flow on the Y axis and input current along the X axis. Use scale factors to permit recording $Q_{max} = \pm Q_R$, $i_{max} = \pm i_R$.

Step 4 With the plotter pen raised, automatically cycle the input current throughout the entire range. During the sweep note that the flow piston does not bottom, that the supply and return pressure values remain sufficiently stable, and that the sweep speed is sufficiently slow to avoid excessive plotter dynamic inaccuracies.

Step 5 Lower the plotter pen and obtain a no-load flow plot as in Fig. 3.47.

Step 6 Acceptable flow gain, linearity, symmetry, and saturation are best ascertained by use of a transparent overlay upon which are drawn allowable envelope limits.

Step 7 Valve hysteresis can be measured as the maximum width (i_w) of the no-load flow plot. Percent hysteresis is given by

$$\% \text{ hysteresis} = \left(\frac{i_w}{i_R}\right) 100$$

THRESHOLD

Valve threshold can be determined on continuous data plotting equipment by the following procedure:

Step 1 Following the no-load flow plot obtained by the procedure of no load flow characteristics, increase the sensitivities of both the flow current recording channels so that changes corresponding to the valve threshold will be discernable.

Step 2 With the plotter pen lowered, slowly adjust (manually) the input current until a flow change is recorded. At this point, slowly reverse the direction of application of current until a corresponding change of flow is noted. Repeating this procedure will establish a threshold plot for the valve.

Step 3 Step 2 may be repeated at various input current levels near the valve null to determine the consistency of the threshold indication. Observation of plotter piston displacement during this test will be necessary to avoid bottoming of the piston.

Step 4 Threshold may be determined by the maximum width of the threshold plot (i_w). Percent threshold is given by

$$\% \text{ threshold} = \left(\frac{i_w}{i_R}\right) 100$$

LOADED FLOW CHARACTERISTICS

The change in control flow with load pressure may be plotted by the following procedure:

Step 1 Repeat steps 1 through 5 of Blocked Load Characteristics.

Step 2 Establish a plotter configuration to permit recording control flow on the Y axis and load pressure drop on the X axis. Use scale factors to permit $Q_{max} = \pm Q_R$ and $\Delta P_{max} = \pm P_P$.

Step 3 Set the input current to a constant value of $+i_R$.

Step 4 Open one control line shut-off valve.

Step 5 With the plotter pen raised, slowly open the other control line valve. By trial and observation, establish a rate for opening the control line valve such that maximum flow can be obtained prior to bottoming of the plotter piston. Between trials, apply negative input currents to move the piston back to the starting point; then close the control line valve.

Step 6 Lower the plotter pen and repeat step 5 to obtain one plot of the family shown in Fig. 3.48.

Step 7 Repeat this procedure to obtain plots at other appropriate values of input current.

3.6.4 Dynamic Tests

The valve dynamic response while operating into no-load with sinusoidal current inputs can be determined by equipment equivalent to that illustrated in Fig. 3.57. In this equipment, valve control flow is supplied to a low mass, low friction actuator. Motion of the actuator is detected by electrical transducers. The design of the actuator should be such that negligible dynamic differential pressure variations exist in the frequency range of interest (usually to several hundred cycles per second). In this way the valve dynamic characteristics in the presence of a true no-load can be ascertained.

A low gain, quasistatic position feedback signal is utilized for nominal centering of the response actuator. A linear motion velocity generator is recommended for deriving dynamic flow signals. The servoamplifier must be designed to supply sinusoidal input currents to the servovalve throughout the test frequency range. Input current signals may be obtained across a resistor in series with the valve coils.

The amplitude and phase relationship of flow to current may be determined directly from an oscilloscope Lissajous pattern, or by use of suitable transfer function measuring instrument.

Considerable caution must be exercised when setting up a servovalve dynamic test installation to achieve adequate pressure regulation. In any new installation the dynamic pressure variations in both the supply and return lines should be observed during a typical test. Excessive pressure variations should be elimi-

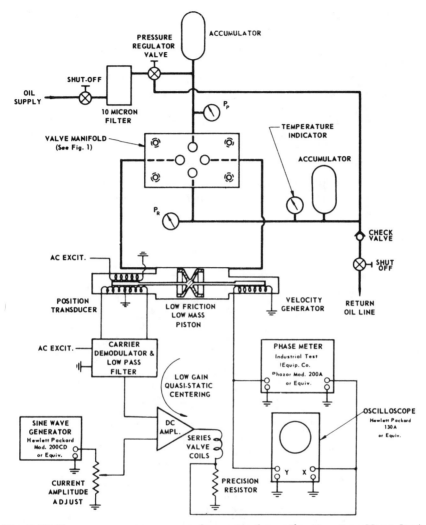

Fig. 3.57. Frequency response equipment schematic. *(Courtesy Moog, Inc.)*

nated by careful choice of regulator, accumulators, and plumbing configuration.

A significant source of ambiguity and data inaccuracy can arise if the flow waveform is nonsinusoidal. Distortions of the flow waveform are often introduced by valve nonlinearities caused by either too small or too large input current amplitudes. With commercial dynamic measuring equipment, the results obtained with distorted waveforms depend upon the specific design of the equipment. Since wide differences in data arise under these conditions, careful monitoring of both the flow and current waveforms is essential for accurate determination of servovalve dynamic characteristics.

Equipment especially designed for servovalve dynamic testing is available from at least two manufacturers.

1. Industrial Measurements Corporation Model 601.
2. Moog Inc. Model TE 1006.

AMPLITUDE RATIO

Data for a plot of valve amplitude ratio (see Fig. 3.58) may be obtained by the following procedure:

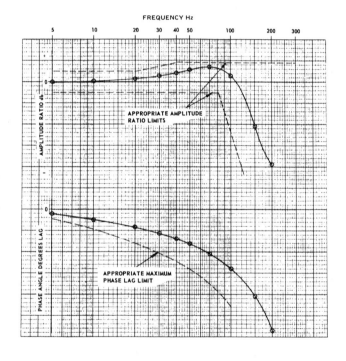

FREQUENCY RESPONSE DATA FOR VALVE
NO-LOAD FLOW OUTPUT TO CURRENT INPUT

SUPPLY PRESSURE	$-P_p$
CURRENT INPUT	$-1/2\ I_R\ (pp)$
OIL TEMPERATURE	$-100°$ F.

Fig. 3.58. Valve dynamic performance. *(Courtesy Moog, Inc.).*

Step 1 Mount the valve to be tested on the response actuator. Connect the valve coils to the servoamplifier.

Step 2 Open the return line shut-off valve.

Step 3 Pressurize the valve to rated pressure by opening the supply line and adjusting the pressure regulating valve. This adjustment should compensate for any return line back pressure. With proper polarity of the valve and test equipment, the position feedback should tend to center the actuator.

Step 4 Set the oscillator frequency to a prescribed reference frequency, usually 5 Hz.

Step 5 Increase the oscillator output amplitude until the desired input current amplitude is achieved. A peak-to-peak input current of one-half rated is recommended. Observe the actuator velocity signal on the oscilloscope to verify that the actuator does not bottom near the zero velocity points. Slight adjustment of the valve bias current may be necessary to center the response piston.

Step 6 Adjust the oscilloscope controls to produce a reference height (peak-to-peak piston velocity) and width (peak-to-peak current) for the output/input Lissajous pattern.

Step 7 Increase the oscillator frequency to the next test point.

Step 8 If necessary, readjust the oscillator amplitude to maintain the reference peak-to-peak input current.

Step 9 Measure and record the velocity amplitude with respect to the reference amplitude in the manner prescribed by the equipment manufacturer. Some equipment furnishes direct readings in decibels. With other equipment, db may be calculated from the measured amplitude ratio, if desired.

Step 10 Repeat steps 7, 8, and 9 for the remainder of the test points. Approximately ten data points in the frequency region to the 90 degree phase lag point are desirable.

PHASE ANGLE

Data for valve phase angle may be obtained concurrently with the amplitude data of the previous section.

Step 1 At each frequency setting use a suitable phase meter to ascertain the phase difference between the velocity and current signals. If a suitable phase meter is not available, the phase difference may be determined from the oscilloscope Lissajous pattern. The following may be used

$$\theta = \sin^{-1}\left(\frac{\Delta y}{y_{pp}}\right)$$

where

     θ = phase angle (degrees)

     Δy = vertical separation of the Lissajous pattern at zero current

     y_{pp} = peak-to-peak height of the Lissajous pattern

3.7 Filter Tests

Filter testing is probably the most standardized and researched area in the field of fluid power testing. As components developed into finely tuned precision control devices, they became more susceptible to contamination. This created a problem in the filter industry, as no standard procedures were available. One manufacturer's 10 micron filter was not even close to the performance of another brand with the same rating. Early component failure, wear, noise, all belonged to a group of symptoms directly caused by poor filtration. The industry decided that enough was enough.

This section has been developed from information provided by Purolator Technologies, Inc. and Mr. Alvin Lieberman of Hiac/Royco. Also included is a brief description of test standards developed by the National Fluid Power Association relating to this section.

3.7.1 Bubble Point Test

The bubble point test is the differential gas pressure at which the first steady stream of gas bubbles is emitted from a wetted filter element under specified test conditions.

Bubble point tests are conducted by immersing the filter element in a liquid that thoroughly "wets" and saturates the filter medium. A measured air pressure is applied to the outlet of the element in such a manner that it is applied to the porous wall of the element from the inside to the outside. While the element is held horizontally approximately 1/2 in. beneath the surface of bubble test liquid and is rotated, the air pressure is slowly increased until a steady stream of air bubbles appears on the outside of the filter medium and breaks away toward the liquid surface. The air pressure required to blow the first stream through a pore is inversely proportional to the size of the largest pore in that element.

The bubble point test data are related to the maximum pore opening of the filter element by the capillary tube equation (Poiseuille's law):

$$P = \frac{K4y \cos \theta}{d}$$

where

P = Bubble point pressure

y = surface tension of the wetting fluid (i.e., interfacial surface tension between filter medium and test liquid)

θ = angle of contact of the wetting fluid to the filter medium

d = maximum pore diameter

K = shape correction factor

The shape correction factor K is necessary because, in actuality, virtually no pores are shaped like capillary tubes but are of random shapes providing a tortuous path through the filtration medium. The factor remains constant for a given medium.

Since it has the advantage of being nondestructive, the bubble point test is an excellent means that can be used by both manufacturer and user to determine the integrity of the element and whether it has been damaged in processing or handling.

In the production testing of filter elements, the bubble point test is used to determine the element's absolute rating. Maximum particle passed tests are used in laboratory development to measure the absolute rating, and these transmission tests are tedious and expensive; they also "destroy" the elements, since the glass beads used in the tests contaminate them.

Bubble point tests may be correlated to glass bead transmission test data for a specific filter medium; thus, they may be used in production on a 100% basis to measure the element's absolute rating.

ANSI B93.22 (ISO 2942) (NFPA/T3.10.8.4M-1972), Method for Determining the Fabrication Integrity of a Hydraulic Fluid Power Filter Element, outlines a procedure using the bubble point test to determine the acceptability of an element for further testing and use. The element is mounted with its centerline horizontal to the fluid surface and below the surface 1/2 in. The element is submerged for 5 min at room temperature. Next, air pressure (pressure value optional) is applied, and the element is rotated 360° while evidence of a persistent stream of bubbles is monitored.

Copies of the exact procedure may be obtained from the NFPA.

3.7.2 Clean Pressure Drop Test

This is a nondestructive test used to evaluate the flow capabilities of a unit in the new, or clean, condition. The unit is installed in a system similar to that shown in Fig. 3.59.

Clean fluid is pumped through the system at varying measured flow rates, at standard controlled temperature, typically 100°F. At five to seven approximately equally spaced flow rates, the differential pressure, or "pressure drop," across the test unit is measured and recorded.

Normally tests are performed on an empty test housing and then on the housing with element installed. The data from these tests can be combined to yield the pressure drop across the element alone if desired. (Also see Section 3.7.14, Multipass Test.)

3.7.3 Collapse Test

The collapse test determines the integrity of the filter element at high differential pressure by subjecting the element to contaminant

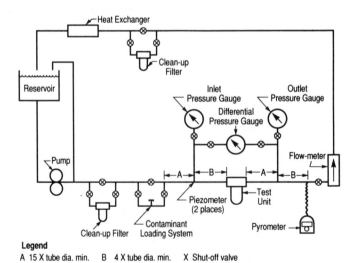

Legend
A 15 X tube dia. min.   B   4 X tube dia. min.   X Shut-off valve

Fig. 3.59. Clean pressure drop and dirt capacity schematic. (*Courtesy Purolator Technologies, Inc.*)

under rated flow until the pressure drop exceeds the required value, or until the element fails or collapses. The collapse test is a destructive test.

ANSI B93.25 (ISO 2941) (NFPA/T3.10.8.5M-1972), Hydraulic Fluid Power—Filter Elements—Verification of Collapse/Burst Resistance, is used to define the capability of an element to withstand a designated differential pressure at normal flow. The element is subjected to the fabrication integrity test prior to start of this test. After successful completion, the element is mounted in a housing connected to a contaminant injection test stand. The element is subjected to the nominal flow rate and temperature in a clean condition and element pressure drop is recorded.

While maintaining the same flow rate and temperature, a controlled amount of contaminant is injected upstream of the element. Element pressure drop versus contaminant (in grams) is recorded until the collapse/burst rating of the element is reached.

After completion of this test, the element is then subjected to another fabrication integrity test, and inspected for structural, seal, or filter medium failure.

Copies of this standard are available from the NFPA.

3.7.4 Dirt Capacity Test

The dirt capacity test determines the weight of a specified artificial contaminant that must be added to the influent to produce a given differential pressure across a filter at specified conditions. It is used as a indication of relative service life.

This is a destructive test which is performed in a test system shown in Fig. 3.59.

Under steady-flow conditions, test contaminant, usually AC Fine test dust, is added to the test fluid upstream of the test unit, at regular time intervals. As the filter retains the contaminant, the differential pressure across it increases; this differential is recorded, so that a graph of contaminant added versus differential pressure may be plotted. (See Section 3.7.14. Multipass Test.)

3.7.5 Degree of Filtration Test

Degree of filtration test is a measure of the efficiency of the filter element, expressed as a percentage, in removing a specified artificial contaminant (AC Fine or Coarse Test Dust and/or glass beads) at a given concentration under specified test conditions.

The test system and method are similar to those used for the maximum particle passed test, except that the weight of test contaminant that passes through the test unit, and is subsequently retained by the analytical membrane filter, is determined by gravimetric analysis.

Then the degree of filtration, or

$$\text{nominal rating} = \frac{A - M}{A} \times 100\%$$

where

A = weight of test contaminant added

M = weight of test contaminant retained by membrane filter

3.7.6 Maximum Particle Passed Test

Maximum particle passed test measures the largest hard spherical particle that will pass through a filter under specified test conditions. This is an indication of the size of the largest pore in the filter element and thus of its absolute rating. The test is destructive in that it contaminates the element.

The fluid in the system is circulated through the "clean up" filter until a sample of the fluid yields a cleanliness level of 0.0004 g/100 ml, by gravimetric analysis. (See Fig. 3.60.)

A measured quantity of artificial contaminant of graded and known particle size range—typically spherical glass beads in the range 2–80 μm diameter—is injected into the contaminant mixing chamber. After thorough agitation in the test fluid, a measured volume of the resulting mixture is passed through the test unit and caught in a clean beaker. That effluent is then passed through a very fine membrane filter, typically of 0.45 μm pore size. The membrane filter is then examined under a high-powered microscope fitted with a graduated scale (ocular reticle). By visual observation, the diameter of the largest glass bead on the membrane is determined.

The diameter of that largest bead, expressed in microns, is deemed to be the absolute rating of the element.

If the bubble point of the element is measured prior to the maximum particle passed test, the value of the shape correction factor (K) can be determined for any other filter element using the same filter medium by bubble point testing the element and using the value of K calculated from the maximum particle passed test in the capillary tube equation. It is wise, however, to correlate maximum particle passed test results to the bubble point test results for each type of filter element.

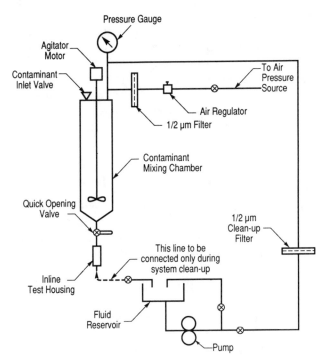

Fig. 3.60. Maximum particle passed test. *(Courtesy Purolator Technologies, Inc.)*

3.7.7 Cold Start Test

This determines the integrity of the filter element under conditions that simulate the high differential pressures generated at the start up of a hydraulic system, especially under cold environmental conditions.

The element is bubble tested, then subjected to a high differential pressure generated by the pumping of oil cooled typically to −65°F.

After the application of this high differential, the element is again bubble tested to ensure that the integrity of the media and construction has not been reduced by the loading of the differential pressure.

3.7.8 Media Migration Test

The media migration test is a nondestructive test used to determine the amount and nature of material that a filter releases from its own medium, under controlled test conditions of flow and, for certain specialized applications, vibration. Typically, such material is comprised of fibers or small lumps of binder resin or both. The release may be due to particles of medium loosened or built in during manufacture or to failure of the medium to retain its integrity under load. Such released material, of course, becomes a contaminant in a system. This test is, thus, of importance in that it provides a means of ensuring that a filter will not in fact contribute to system contamination. The test comprises microscopic and gravimetric analysis of effluent from a test filter passed through an analytical membrane.

3.7.9 Flow Fatigue Test

The flow fatigue test determines the integrity of the element to withstand pulsating flow by subjecting the filter element to a flow cycle that ranges from zero to the rated flow. The integrity is determined by bubble testing the element before and after the test.

Acceptable levels of cycle life range from 10,000 for steady-flow applications to 100,000–1,250,000 cycles for pulsating-flow applications.

The flow fatigue test is a destructive test.

ANSI B93.24M-1972 (R1980) (NFPA/T3.10.87M-1972), Method for Verifying the Flow Fatigue Characteristics of a Hydraulic Fluid Power Filter Element, is used to test the ability of a filter to withstand flexing caused by cyclic pressure shock without affecting its collapse/burst rating.

After successfully completing a fabrication integrity test, the element is mounted in a flow cycle test stand in a housing. Contaminant is added to the system at 25–100% of the element flow rate until the maximum pressure drop rating of the element is reached. Next, the element is subjected to pressure pulses (90–100% of rated maximum pressure drop) by adjusting flow. The amount of cycles is optional.

The element is then subjected to visual inspection, collapse/burst resistance test, or fabrication integrity test, or all three

Copies of the exact procedure can be obtained from the NFPA.

3.7.10 Material Compatibility Test

This test determines the integrity of the filter element to withstand the environment in which it is to be used by artificially aging it by soaking the filter element in the service fluid, for not less than 72 hr, at an elevated temperature. The integrity is determined by running a number of the preceding tests on the artificially aged element, depending on the service for which the element is intended.

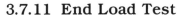

3.7.11 End Load Test

ANSI B93.21M-1972 (R1980) (ISO 3723) (NFPA/T3.10.8.2M-1972), End Load Test Method for a Hydraulic Fluid Power Filter Element, outlines the purpose and sequence for testing an element's capability to withstand the axial loading imposed by installation.

Basically, the element is initially subjected to the fabrication integrity and hot soak tests (72 hr—hot soak). After the element has cooled to room temperature, the element is subject to the rated load with weights or in a loading fixture for 5 min. After this, the element is inspected for structural, seal, or filter medium failure.

The element must then successfully complete the recommended collapse/burst test.

Copies of the exact procedure may be obtained from the NFPA.

3.7.12 Slosh Test

This test is used as a simple method to transfer contaminates from a component to a clean test fluid, while in a static condition.

SAE J1227, Assessing Cleanliness of Hydraulic Fluid Power Components and Systems, discusses a method of filling a test component one-third to one-half full with clean test fluid. The component is sealed and mechanically agitated, vibrated, and shocked. The fluid is removed and analyzed. Specific descriptions for the vibration geometry, sealing method, and fluid sample removal are presented.

Copies of the exact procedure can be obtained from the SAE.

3.7.13 Element Pore Size Test

ANSI B93.46M-1978 (NFPA/T3.10.8.12M-1976), Method of Determining the Pore Size of a Cleanable Surface Type Hydraulic Fluid Power Filter Element, describes two procedures for determining pore size. The first recommends a visual examination with a microscope (or other equipment) to measure the openings of 10 random areas and record the largest opening diameter in each area. The second procedure recommends using SAE/ARP 901-1968 Bubble Point Test Method as a procedure.

Copies of the exact standard can be obtained from the NFPA.

3.7.14 Multipass Test

The multipass test is used to determine the B-ratio of a filter element and is a destructive test. (See Fig. 3.61.)

Fig. 3.61. Multipass test stand. *(Courtesy Pall Industrial Hydraulics Corp.)*

This test was introduced by Dr. Ernest Fitch and fellow scientists of Oklahoma State University. The test is an attempt to develop a reproducible realistic laboratory test that simulates actual use conditions by the recirculation, within the test system, of contaminant that is not trapped by the filter under test. Fresh contaminant is also continuously fed into the test system, so that a constant contamination level of the test fluid is maintained. The test contaminant is AC Fine Test Dust.

Samples of the test fluid are withdrawn simultaneously upstream and downstream of the element under test, at predetermined levels of differential pressure across the test filter.

The samples of fluid are then analyzed with an automatic particle counter calibrated per ISO Standard 4402. The cumulative particle size distribution per milliliter of fluid is determined, usually at particle sizes of 5, 10, 20, 30, and 40 μm.

From these data, the beta filtration rating may be calculated, as described in Chapter 1.

The basic circuit used in the multipass test is depicted in Fig. 3.62, and consists of an injection circuit and a filter test circuit. The injection circuit is continually circulated by a low-pressure pump through a heat exchanger to maintain constant test temperature. The system is designed to ensure that contaminant distribution also remains constant. The filter shown in the injection loop is used during injection loop "clean-up" cycles. A shutoff valve is adjusted to add contaminant from this loop into the filter test circuit.

The filter test circuit is similar to the injection circuit, in that a

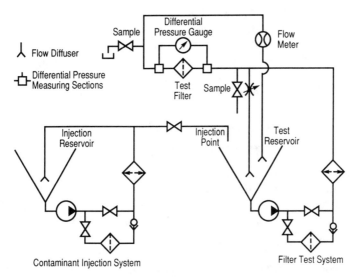

Fig. 3.62. Multipass filter test circuit. *(Courtesy Purolator Technologies, Inc.)*

low-pressure pump circulates oil in open-loop system, passing fluid through a heat exchanger. The test filter is mounted at the outlet of the heat exchanger. A differential pressure transducer is used to monitor increasing pressure drop as contaminant is added. The inlet and outlet of the test filter are provided with sampling valves so that fluid samples can be analyzed. These are used to determine fluid condition and make certain that the system is calibrated properly. Many manufacturers now use in-line particle counters rather than taking fluid bottle samples. A flowmeter is used to determine flow rate at the filter outlet. Flow rate is adjusted for turbulent flow to ensure continuous mixing of the test contaminant.

This test procedure is used to determine beta value/separation efficiency, dirt-holding capacity, and element pressure drop characteristics.

ANSI B93.31M-1973 (R1981) (NFPA/T3.10.8.8M-1973), Multi-Pass Method for Evaluating the Filtration Performance of a Fine Hydraulic Fluid Power Filter Element, is an extremely detailed procedure developed to provide reproducible test data for appraising the filtration performance of a fine hydraulic fluid power filter element.

The procedure is based on the use of the multipass test method using AC Fine Test Dust and an automatic particle counter (or any ISO-approved counting method). It describes the system design requirements, test filter preparation, contaminant injection procedures, test result extraction, and other related topics.

NFPA T3.10.8.18M-1972 (R1982), Multi-Pass Method for Evaluating the Filtration Performance of a Coarse Hydraulic Fluid Power Filter Element, is similar to the test described above, with the exception that AC Coarse Test Dust is used for the evaluation of filters with a filtration (B_{10}) ratio of less than 2.0.

Copies of both standards may be obtained from the NFPA.

3.7.15 Using Particle Counters— Preliminary Preparation

This section is a reprint of an article originally published in the *BFPR Journal* and written by Alvin Lieberman of Hiac/Royco. It is reprinted with the permission of the author and the Fluid Power Research Center, Oklahoma State University, Stillwater, OK.

SAMPLE CONTAINER CLEANING

Analyses of fluid contamination are frequently made on samples that have been collected and transported to a laboratory in a suitable container. If new containers are used, then the responsible agency has the option of choosing the container type and closure and of specifying the necessary processing to ensure reasonable sample conditions.

Container size for contamination samples is usually in the range of 100–500 ml. A popular intermediate level is approximately 250 ml. Containers can be procured that are made of glass, polypropylene, polyethylene, or metal. Glass is the best material for sample containers because it can be easily cleaned and, with suitable packaging, can be transported even under hostile conditions of temperature and vibration. Glass can resist mild corrosive conditions and will not add debris to the sample that may be confused with original contamination. Polypropylene is also a useful material but it can be easily confused with polyethylene which is much less suitable. Polyethylene is subject to shredding under mildly abrasive conditions. Polyethylene is much less resistant to acid corrosion than glass or polypropylene. Metal containers should usually be avoided. Stainless steel containers are suitable but are extremely expensive. Typical drawn aluminum or lined mild steel containers corrode rapidly and add debris to the sample.

The typical shape of a container is cylindrical with a height-to-diameter ratio of three or four. Either wide or narrow mouth containers are suitable. Narrow mouth containers can be sealed more easily and securely. Wide mouth containers can be cleaned more easily.

A convenient closure and seal system for sample containers consists of a one-mil nylon or Mylar film that is wrapped around the container and is held loosely in place with an elastic band. The seal for the container is typically a screw cap. With this arrangement, it is possible to use a sample delivery port that can penetrate the film and deliver sample fluid directly into the container with no further handling. Alternately, one may use a polypropylene cap of the flexdome type or a screw cap with a polypropylene or Teflon insert.

When a supply of new containers and caps is received, it is usually wise to ensure that these items are clean. The containers should be cleaned either by scrubbing with a detergent solution or in a commercial dishwasher. Following adequate rinsing with tap water, a rinse with deionized water is recommended. The deionized water should be removed from the interior of the container and the interstices of the closure by rinsing with filtered isopropanol or methanol. The alcohol can be dried by a stream of filtered dry air. Alternately, the alcohol can be removed by rinsing with a suitable solvent mutually miscible in the alcohol and in the hydraulic oil that will be later placed in the container. Because such solvents have appreciable vapor pressure, a small amount can be left in the covered container. During shipment, no breathing will occur that may result in ingestion of atmospheric particulate debris.

If large quantities of sample containers are required, then a mass cleaning procedure should be used. This cleaning procedure should include scrubbing, rinsing, a fluid exchange process, and cleanliness verification using procedures similar to those described in the American National Standards Institute Procedure, B93.20-1972, entitled: "Procedure for Qualifying and Controlling Cleaning Methods for Hydraulic Fluid Power Fluid Sample Containers." The clean containers should then be stored in a clean location until they are to be shipped for sample acquisition. Shipping containers can be fabricated by molding or boring polyurethane foam blocks to fit the bottles and by using sturdy fiberboard shipping containers. It is recommended that the usual "popcorn" packing material not be used. This material deposits excessive quantities of plastic particulate debris all over the exterior of the container which can easily be brought into the sample. Popcorn packing cannot be dedusted like the foam block. Never use sawdust.

In many cases, sample containers are recycled and used many times. Before a container is to be recycled, it should be inspected to ensure that the container is cleanable and undamaged. Glass containers should be examined to make sure that the rims of the containers are not chipped and that the closures are reusable. New nylon or Mylar film is always required if the container is to be reused. If metal containers are required, then they should be inspected to ensure that no corrosion or leaks have occurred and that no deformation around the closure area exists. Following this inspection, the containers for reuse can be cleaned, verified, and stored as if they were new containers.

PARTICLE DISPERSION

In analyzing samples that have been stored for any period of time, particulate material in that sample may agglomerate, precipitate, or become attached to the container walls. If this occurs, the particulate material must be redispersed or reentrained. In this way, the correct size distribution will be recorded, and the concentration measured will not be affected by recording a number of particles which have agglomerated as a single larger particle.

The forces that cause particles to agglomerate in hydraulic oil during storage are primarily gravitational forces resulting in sedimentation. Because the contaminant particles are typically at least twice as dense as the hydraulic fluid, they will settle to the bottom of the container during storage. If the concentration is sufficiently high, one particle will deposit on top of another. Eventually, the intervening oil film is removed, and the two particles achieve intimate contact. A combination of forces such as mechanical linkage for irregular particles, van der Waal's forces and electrostatic forces will retain particles in intimate contact with one another once the intervening liquid film is removed. In general, the agglomeration forces vary inversely with particle size; that is, smaller particles are more difficult to redisperse than larger particles.

Consider the procedure for preparing calibration material indicated in ANSI B.93.28-1973 entitled, "Method for Calibration of Liquid Automatic Particle Counter Using AC Fine Test Dust." This common practice, employed in many laboratories, is adequate to redisperse particles in samples. The fluid sample can be placed in an ultrasonic cleaner and sonicated for a period of 2 to 5 min. Following this high intensity ultrasonic application, the sample is violently agitated by paint shaker or by vigorous manual shaking for a period of at least 5 min to ensure complete dispersion of particulate material.

Note, of course, that violent agitation and mixing result in air bubble formation. The air bubbles must be removed from the sample before any particulate analysis is made especially with an optical particle counter. Techniques for air bubble removal will be discussed later.

In general, samples that are stored for no more than one or two days will not require this intensive mixing action for adequate particle dispersion. Vigorous shaking for a period of no more than 1 or 2 min may be adequate. After the particles are well dispersed in the fluid, measurement must be made before larger particles settle from the zone where the particles are aspirated or allowed to flow into the particle counter inlet tube. Particles settling in the fluid reach their terminal velocity very quickly. In general, the time to achieve the terminal velocity for most particles of interest is based on Stokes law.

When considering the requirements for particle dispersion in stored samples, the parameters of concern include: storage time before measurement, sample transport from container to measuring system, sample container configuration, initial sample concentration, and sample/liquid relationships. If a sample is stored for a long period of time, gravitational settling results in the possible formation of coherent flocs, agglomerates, or both. These require violent agitation for adequate dispersal. Following dispersion, the particles in the sample will begin to settle to the bottom of the sample container again, even while fluid is being withdrawn. If the flow rate to the particle measuring system is very low, a series of successive samples will differ in recorded concentration and particle size distribution because larger particles will continue to settle out of the sample acquisition zone. Therefore, it may be necessary to continue some mixing during sample removal, taking care to prevent possible air entrainment from vortex formation. The level of liquid in a sample container controls mixing effectiveness and prevents vortex formation at the point of sample withdrawal.

These problems are aggravated if the specific gravity difference between the particles and the liquid increases, and if the surface nature of the particulate material in a particular liquid can lead to the possibility of an electrical double layer formation. If the latter is the case, coagulation rates are affected. In a similar manner, coagulation is accelerated at higher concentrations. In this case, Brownian motion and thermally caused turbulence result in the increased probability of intra-particle contact.

DEAERATION

In the process of providing adequate dispersion, the liquid is often mixed to the point where air is entrained, and air bubbles are formed. In this process, the high energy shear forces needed to ensure dispersion also fracture large air bubbles into smaller bubbles that are incorporated within the body of the liquid and will remain suspended for relatively long periods of time. Some hydraulic oils, particularly those with a high viscosity, retain air bubbles from submicrometer to nearly 100 micrometers for up to an hour. In addition to entrained air, it is also possible that oil vapors may produce gas bubbles due to cavitation.

When large quantities of air are present in the oil sample, the quantity of entrained air can be determined by noting the volume of the sample by reducing the pressure above the sample to exhaust air from the liquid, and then noting the change in volume. If air is present in very small bubbles, only a relative measure of entrained air can be noted by observing the turbidity of the suspension.

When air bubbles are present within the oil, their effect on the measurement depends on the instrument in use. If an optical counter is used, air bubbles are detected as any other particulate material. The optical counter will report the air bubbles as particulate material, and erroneous data will be developed. The same effect occurs if an electrical sensing counter is used when measuring oil samples with added conductive liquid, or if a light-scattering instrument is used. If microscopic methods, gravimetric analysis, or Ferrography is used, then air bubbles will not interfere with the data produced because they are removed from the analytical zone before data are recorded.

If air bubbles are present and if it is necessary to eliminate them, three methods are generally used. The first is to expose the sample to sonic energy. Cavitation energy tends to agglomerate the smaller air bubbles into larger bubbles that escape relatively rapidly from the surface of the liquid. Following sonication, application of reduced pressure to the surface of liquid aids in removing air bubbles from the body of the sample. The bubbles rise to the surface as a foam and do not interfere with the samples taken from the interior body of the liquid. If the air bubbles are 10 to 15 micrometers or less in diameter, it is possible to cause the gases to go into solution in the liquid by increasing the pressure.

DEHYDRATION

Many samples are obtained from environments where excessive quantities of water are present. Water may be ingested into the system being analyzed. Water may enter a sample during the collection process, in inclement weather, or from careless storage where breathing or direct ingestion of water can occur directly from the atmosphere. As received MIL-H-5606 is required to contain no more than 100 parts per million of total water, both dissolved and free water. This concentration is equivalent to 0.01 percent by weight and is well within solubility limits even at relatively low temperatures. As temperatures increase, more and more water will dissolve in the fluid. Typical methods for determining the water content of oil depend on visual examination to determine the presence of the immiscible phase for gross contamination with water. For determination of dissolved and emulsified water at low concentrations, the Karl Fischer method may be used. Water content is determined by titration.

If undissolved water is present in the oil, emulsion droplets will usually be present. These droplets are indicated by optical and electrical particle counters in the same way as air bubbles; that is, as contaminant particles. Water in fluid samples can also cause oxidation of metal containers and generate secondary contamination.

To eliminate water from fluid samples, several dehydration methods can be employed. The first is mixing with miscible high vapor pressure fluids such as alcohol. The water-alcohol azeotrope can be evaporated. This method is time-consuming, and the mixing processes may result in dissolution of solid materials that would normally be retained in the fluid. A second method involves passing the sample through a porous tube where the water diffuses through the tube and is removed from the fluid sample. Problems can exist from the retention of particulate material on the walls of the tube. A third method for removing water from oil samples combines heating and evacuation. The water is evaporated from the surface of the oil and is distilled off as water vapor. At a low pressure, the particulate material remains within the oil. Note, however, that the evaporation time of up to an hour or more may be required to remove excessive quantities of water. Following evaporation, a particle measurement is often possible; for example, with optical particle counters at elevated temperatures, if the particle counter is not exposed to high temperatures for excessive periods of time.

DILUTION

Frequently, samples are obtained in which the particulate content is known to be high. In this case, any single particle counter can be affected by coincidence error and possible electronic saturation. Coincidence is defined as the simultaneous occurrence of more than one particle at a time in the sensitive volume. This problem occurs with any optical counter, electrical sensing counter, and with microscope measurements. Because groups of particles are detected as a single larger particle, coincidence error is not affected by flow rate. The number of particles present in the sensitive volume is a function only of the concentration and the size of the sensitive volume, and it does not increase or decrease with changes in flow rate.

The electronic system counting rate capability is affected by flow rate as well as by concentration. Eventually, either coincidence error or electronic saturation causes the system to miss so many particle counts that gross errors occur. In such a situation, it is necessary to dilute the sample to the point where the particle counting device is able to measure the concentration in the diluted sample. In some cases, dilution of a viscous oil with a lower viscosity solvent is necessary. This requirement arises if the particle analyzer contains small flow passages that will not permit the sample to flow at a rate that is satisfactory. For example, most particle counting instruments are calibrated for operation at a fixed flow rate. If the flow rate is changed, the particle counter sizes particles incorrectly. To ensure correct sizing, the flow rate must be maintained at or near that recommended by the manufacturer. Viscosity control may be required to permit flow rate control. Dilution is required if hydraulic fluids with viscosity appreciably greater than 200 cSt are used, or if the particle counting device cannot handle fluids at temperatures higher than 100 to 125° C.

When a fixed ratio dilution is required, a relatively simple procedure can be used. First, establish the desired dilution ratio. For dilution ratios larger than approximately 25:1, a sequential procedure is required because volumetric measuring accuracy for small quantities of liquid is extremely difficult. Next, fill a clean container to a preset marked level with clean diluent. A convenient diluent container size is 100 mL. Then, remove a quantity of clean diluent equal to the inverse of the dilution ratio. For example, if a dilution ratio of 25:1 is to be used, remove 4 mL of clean liquid from the 100-mL container. Next, the original sample should be well mixed, and a sufficient sample should be poured into the marked diluent container to bring the total quantity to the original level. In this way, 4 mL of sample are added to 96 mL of clean diluent to make a 25:1 dilution.

If the required dilution ratio is not known but the original sample is suspected of requiring dilution, the following procedure can be used. First, make a measurement of the particle concentration in at least five size ranges with the particle measuring device set to produce a cumulative size distribution. Realizing that the vast majority of contaminant samples have a logarithmic or power function distribution, examine the data. If the first two or three particle size ranges, beginning with the smallest range, do not follow the typical size distribution function, dilution is indicated. Note the particle size distribution data in all five size ranges for future reference. Next, make a 2:1, 5:1, or 10:1 dilution depending on the indicated size distribution using the procedures described in the preceding paragraph. Reexamine the particle size distribution that is obtained following the first dilution. If this first dilution is adequate, the particle concentration ratio for all the size ranges decreases in accordance with the dilution ratio. If the dilution ratio is not adequate, the particle concentration ratio in the larger size ranges may decrease. Repeat the dilution process as many times as necessary to ensure that the particle concentration in the smallest size range decreases in accordance with the dilution ratio last used. Record the particle concentrations for each of the ranges in accordance with the dilution ratio effective for that size range. Care is required. Sequential dilutions of the type described here will result in a decrease in particle concentration in the larger size ranges to the point where the data are statistically insignificant.

In performing any dilution, some precautions are necessary. Remember that some particles may be soluble in solvents other than the original hydraulic fluid sample. It is also possible that artifacts can be introduced in the dilution process from handling, from inadequately cleaned solvents, and from containers and measuring devices that are used in the dilution procedure. A sufficient concentration of particles in all size ranges must be measured so that the data are statistically significant. Excessive dilution to reduce concentration at the small particle size may reduce the large particle concentration to the point where handling and liquid measurement errors may be excessive in comparison with statistical accuracy of the particle concentration that is measured. Further, in the liquid handling process, good mixing is required throughout the procedure because large dense particles may easily be lost during the handling process. Finally, the requirement for extreme accuracy in handling fluids and for measuring small quantities of liquid cannot be overemphasized.

3.7.16 Additional Filter Test Standards

The following is a list of other standards related to filter testing:

Standard No.	ANSI Standard No.	Contamination ISO No.	Brief Title
1. NFPA/T2.9.1	B93.19-1972	ISO 4021 1977	Sampling Fluid from Pressurized Lines
2. NFPA/T2.9.2	B93.20-1972	ISO 3722 1976	Verification of Sample Bottle Cleanliness
3. NFPA/T2.9.6	B93.28-1973	DIS 4402	Calibration of Automatic Particle Counters
4. NFPA/T2.9.3	B93.30-1973	DIS 3938	Reporting Contamination Analysis Data
5. NFPA/T2.9.9	B93.44-1978	—	Sampling Fluid from a Reservoir
6. SAE/RP–J1165 1977	—	DIS 4406	ISO Solid Contaminant Code (Ref. Fitch paper OSU/P76-26)
7. NFPA/T2.9.8 19xx	—	—	End Item (Roll-Off) Cleanliness (in process) (Ref. Karhnak 1973 and 1974 NCFP papers)
8. SAE/ARP 598A	—	—[a]	Microscopic Particle Counting Method
9. SAE/ARP 785	—	—[a]	Gravimetric Method for Measuring Contamination
10. ASTM/D 95	—	—	Total Water in Oil
11. ASTM/D 1744	—	—	Total Water in Oil
12. ASTM/D 96	—	—	Free Water in Oil
13. SAE/RPXJ1277	—	—	Assessing Cleanliness of New Hydraulic Fluid[b]

		Sampling Plan
1. ANSI Z1.4-1971 (MIL-STD-105)		

[a]ISO is currently finalizing an international version.
[b]Proposed draft currently under review.

3.8 Cylinder and Rotary Actuator Tests

Cylinders and rotary actuators have been combined because many tests are applicable to both actuator types. Unlike motors, these devices have limited travel, so tests conducted are quite different than those used for continuous motion actuators.

3.8.1 Output Force Versus Pressure

The output force versus pressure test defines the force an actuator creates at a specific input pressure. For cylinders this is easily equated using the formula:

$$\text{force} = \text{pressure (psi)} \times \text{area (in.}^2)$$

This equation is always accurate, with the exception of running frictional forces caused by mechanical tolerances and seal forces.

Rotary actuator force, however, is more difficult to calculate. Initial force generated is equivalent to the preceding cylinder formula, but the linear motion must be converted to a rotary motion. There are several variations in rotary actuator designs as shown in Chapter 1. Each variation has its own specific method to convert linear motion to rotary. In most cases the standard lever arm formula can be used to determine actual output torque:

$$\text{torque} = \text{linear force} \times \text{moment arm}$$

In most rotary actuator designs the distance between the center line of the output shaft and the point at which the lever arm is pushed by the actuator piston remains constant. In these cases the torque versus pressure relationship remains relatively constant. However, in actuator designs, such as the scotch yoke method, the lever arm length varies. This causes a drop in torque in the center area of travel. The highest degree of torque is generated at both extremes of stroke travel. This should be accounted for when testing this type of actuator.

When conducting this test for cylinders or rotary actuators, they must either be stalled against an immovable fixture or stalled by an actuator of the same type while in motion. The fixture method is usually cumbersome when trying to take readings at more than one point along the total travel. However, by using a load actuator to stall the test actuator, actual output force can be measured throughout the full travel. There are certain precautions that must be considered when using this method. First, the load actuator's speed must be controlled to a speed slower than the test actuator. If the test actuator moves too quickly, supply pressure may drop, thus reducing torque. Also, fast load actuator movement may reduce the backload on the test actuator, thus reducing the required torque.

The best method for monitoring output force is a load cell or rotary torque transformer. These devices can display actual force calibrated to whatever engineering unit you desire. These devices should be mounted between the test unit and load device with precautions taken for proper alignment based on manufacturer's recommendations. Improper alignment can cause errors in readings and instrument damage.

When conducting this test (especially for rotary actuators), it is wise to test in both directions of travel, which will ensure that the actuator is mechanically sound in all operating conditions. Also, some actuators may be single acting with spring return. Many manufacturers recommend checking spring return force to detect spring failure or wear.

3.8.2 Breakaway Pressure

A breakaway pressure test determines the amount of force necessary to commence movement of an actuator. There are several factors involved here. Piston and rod seals have a tendency to bind after sitting for long periods of time with no lubrication. Even in normal use, seal preload is required to maintain a positive seal. The tolerances between moving parts also has an effect. Even though several thousandths of an inch clearance between parts is standard, unconcentric fit, wear, and perpendicularity problems can increase the amount of normal mechanical friction. Another consideration is the weight of the moving parts. Inlet pressure must also exert sufficient force to move internal weight. This weight can also change depending on the position the cylinder or rotary actuator is mounted. Picture a ram cylinder with a 12-in. rod and 36 in. stroke. With the cylinder mounted horizontally, the load is supported by the bearing surfaces of the piston and rod bearing. Horizontal force is a fraction of vertical force. Now try and lift that rod weight vertically. That's different.

When conducting this test, it is important that the actuator is cycled several times first, which will ensure that the seals are lubricated and that the unit exibits no unusual or erratic movement through the full travel. You will then require a good pressure regulator and gage to read a low pressure (in most cases) for breakaway pressure. Typically in small and medium size actuators, breakaway pressure is 20–100 psi, while operating pressure is 3000–5000 psi. If an accurate breakaway pressure is needed, a 3000 or 5000 psi gage is not very accurate or readable down in the low-pressure-range required.

After the actuator begins moving, the pressure will usually drop to running force pressure. The highest pressure detected prior to unloaded running pressure is considered to be breakaway pressure. SAE J214, Hydraulic Cylinder Test Procedure, defines a procedure for measuring packing drag (breakaway pressure).

3.8.3 Piston Seal Leakage

The piston seal leakage test is conducted to determine the amount (if any) of oil "blowby." This measures the effectiveness of the piston seal. Some cylinders are designed with positive seals, while other cylinders allow a small amount of leakage. Basically, there are three methods most frequently used to measure piston seal leakage.

END OF STROKE PRESSURIZATION TEST

In the end of stroke pressurization test, the cylinder is bottomed out and pressurized at one end. The opposite end is vented to atmosphere so that any fluid leaking past the piston seal can be seen exiting the vented port. In the event of leakage, the vented port can be plumbed to a flowmeter or into a beaker to monitor the relationship of leakage versus applied pressure.

The only drawback to this test is that leakage can only be measured at extreme ends of the stroke. In many cases, if leakage is to occur, it is usually at some point midstroke. Cylinder tube machining sometimes results in the tube "belling out" in the center owing to the difficulty in maintaining close tolerances farther away from the tool holder of the boring machine. Also, score marks can occur at any point along the travel during machining or from operation with contaminated fluid.

MIDSTROKE LOAD TEST

This test provides for a method to measure piston seal leakage at any point throughout the actuator stroke. It is similar to the test previously described, except that the test equipment necessary includes an additional slave cylinder and valve, and zero leak shutoff valves and pressure gages at each port of the test actuator.

Basically, the test cylinder and slave cylinder are mechanically linked at the rod ends. After several air purge cycles, the test cylinder is positioned at midstroke. With the test cylinder cap end shut-off valve closed, the slave cylinder cap end test pressure is obtained. The rod is now monitored to detect any evidence of drift.

The same test can then be conducted on the rod side of the cylinder by closing the shutoff valve on the rod end port of the test actuator, and pressurizing the rod side of the load actuator until desired rod pressure is attained. Once again, rod drift is monitored.

SAE J214, Hydraulic Cylinder Test Procedure, outlines a recommended practice of this test method where the cap end and rod end are each presurized for 15 min at 5%, 50%, and 100% of the designated test pressure.

Note that when conducting this test, the unpressurized side of the test cylinder must be vented to atmosphere to prevent excessive pressure intensification and false leakage readings.

M I D S T R O K E   P R E S S U R E   T E S T

The midstroke pressure test is used to measure piston seal leakage at any point of travel without the use of a load cylinder and fixturing required to resist the opposing forces of load and test actuators.

This is a fairly common test that is used often. In practice, pressure is applied to one side of the test cylinder while the opposite side of the cylinder is connected to a leak-tight shutoff valve. The force created by the piston against the oil trapped on the opposite side also pressurizes the opposite side. The actual force created is the pressure on the cap side against the area of the rod on the other side. This unbalanced force would force the cylinder to extend, just as would a regenerative circuit. In this case, the only way the rod would extend is if the oil from the rod side of the cylinder could leak around the piston back to the cap side. Rod drift, therefore, is an indication of piston seal leak. Actual leak rate can be calculated by converting the actual rod travel versus the cylinder internal volume.

It is important when conducting this test that rod side pressure be monitored at all times. Pressurizing the cap end with the rod side port blocked will create a pressure intensification. This could cause the test cylinder to be pressurized beyond its rated limits. It is suggested that the shutoff valve be installed on the cap side only for leak testing in both directions. The test results will prove to be the same.

The midstroke pressure test and the midstroke load tests will not provide the same rod movement at the same leak rate, owing to the different operating principle of each test. The midstroke load test allows full force to be generated on the trapped fluid because the opposite side of the piston is vented to atmosphere. Therefore, full surface area of the piston is employed. The midstroke pressure test pressurizes both sides of the piston with pressure at a ratio equal to the area difference of both sides. (Full piston area versus piston area minus rod area.)

Considering both test methods, let us assume test pressure is 3000 psi. The midstroke load test produces a measured leak at 3000 psi. Now, the same test pressure used in the midstroke pressure test would produce a leak of lesser flow rate with a test pressure of 3000 psi. This is because the opposite side is also pressurized at a pressure equivalent to the area ratio. A ratio of 3:1 would result in 1000 psi on the opposite side of the piston. Therefore, we only are testing at 2000 psi, thus producing a lower leak rate through the same opening.

3.8.4 Rod Seal Leakage

The rod seal leakage test is generally accomplished by cycling a cylinder at maximum pressure and visually inspecting the rod seal for signs of leakage. This applies to both new and rebuilt cylinders.

Prior to this simple examination, the rod seal had to be designed and applied properly. A very thorough test procedure has been developed to define the many factors involved in rod seal performance.

ANSI B93.62M-1982, Reciprocating Dynamic Sealing Devices in Linear Actuators–Method of Testing, Measuring and Reporting Leakage, depicts four test fixture designs used to cycle the seals as they would be in a cylinder design. Each fixture resembles a cylinder piston, rod, and body with actual movement created by an external force.

Once the seal has been assembled into the fixture, the proper test is selected from a numbered test selection guide. The test factors involved include number of cycles, test pressure, test temperature, rod or bore diameter, stroke length, velocity, dynamic surface finish, cold soak temperature, and test fluid. Copies of this exact procedure are available from the NFPA.

3.8.5 Contaminant Ingression

One of the common sources of system contamination is from the cylinder rod seal. As the cylinder cycles, ambient contaminants are drawn into the system during cylinder retraction. Special wiper seals are usually supplied to scrape the rod prior to contaminants reaching the pressure seal. After a period of time, the elastomer begins to fail for various reasons and the contaminates have an easier path into the system.

This is an area that is difficult to predict owing to the wide variety of applications possible. The worst case applications generally include mobile equipment and environments similar to and including steel mill use. Generally, the cylinder is cycled similar to the rod seal leak test, except that the rod is exposed to the contaminant desired. The actual ambient conditions are simulated as close to the application as possible. Testing can be accelerated by adding more contaminates. After a designated number of cycles, the cylinder fluid is sampled and analyzed and the surface finish and seals are visually inspected.

3.8.6 Cycle Life/Endurance Test

The cycle life/endurance test is difficult to perform owing to the various applications available. This test is usually conducted to determine a cylinder's tolerance to a specific set of parameters.

SAE J214, Hydraulic Cylinder Test Procedure, depicts a typical test setup recommended for endurance testing. The circuit consists of two pump circuits: one to power the test cylinder and one to supply a slave cylinder used to load the test cylinder. The test defines a general procedure that can be applied based on the desire test goals. It is suggested that at the conclusion of the test, an internal leak test should be conducted. It is also recommended that external leakage tests be conducted between selected cycles so that accumulated leakage can be analyzed and recorded.

3.8.7 Pressure Rating/Test Methods

Cylinders and rotary actuators are considered pressure vessels, and proper design techniques are applied to ensure that the correct safety factor is applied. In the early days of hydraulics, a safety factor of 3–4 was applied. The component was designed and tested for proof and burst pressure. Proof pressure was designated as the pressure that a component could be subjected to (over its normal operating range) without permanent failure or distortion. Burst pressure meant just that—burst.

In an effort to standardize the rating method, SAE J214, Hydraulic Cylinder Test Procedure, outlines a method to pressurize the test cylinder with the piston midstroke, with visual examination after a 30 sec pressure cycle.

ANSI B93.10M-1969 (R1982) (NFPA/T3.6.5M-1968), Static Pressure Rating Methods of Square Head Fluid Power Cylinders, defines the requirements for static pressure rating, properties of materials, method of marking, and assembly and workmanship. It is stated that the procedure is not intended to determine the maximum operating pressure or safety factor of a cylinder. Also, upcoming procedures will outline the problems encountered with various mounting styles and rod buckling, both of which affect the pressure rating of a cylinder.

Calculations are outlined for determining tubing, head and cap, tie rod, and female tie rod attachment thread failure pressures. Material specifications for tubes, heads and caps, and tie rods and nuts are detailed. Copies of this procedure can be obtained from the NFPA.

3.8.8 Rod Buckling Test

The rod buckling test defines the maximum unsupported load a cylinder rod can safely overcome. Most manufacturers publish simplified tables that define the maximum forces allowed on various combinations of cylinder rod diameter and unsupported rod length. It is a destructive test.

The cylinder is designed with the intent of adding a safety factor above and beyond the maximum allowances calculated from the Euler formula. (See Chapter 1 for additional information.)

3.8.9 Mounting Style Strength

The mounting style strength test is a destructive test where a cylinder is loaded until failure of the cylinder mount or mounting hardware occurs. In many cases, the mounting style is the weakest point in the cylinder design. This also depends on the application and actual cylinder mount on the equipment.

Industry standards are currently being developed to define a standard method.

3.8.10 Rotary Actuator Backlash

The rotary actuator backlash test defines the movement characteristics of the output shaft of an actuator after pressure is vented. During the translation of linear motion to rotary motion, certain manufacturing tolerances come into play, which allow a slight movement of the rotary output shaft. If gears (rack and pinion) are used, the clearance between intermeshing teeth is a factor. Chain-driven designs require chain tension prior to movement. These factors in addition to mechanical elongation change the position of the rotary output shaft whether pressurized or not. The total amount of movement in degrees between maximum operating pressure and zero pressure defines the total actuator backlash.

3.9 Leakage Tests

One of the most misunderstood and misapplied test procedures is the leak test. There are several common methods used based on test pressure, test media, accuracy required, and, most important, the leak rate itself. Leakage testing falls into two categories: internal leakage and external leakage. External leakage is sometimes jokingly referred to as measuring PSOs (Pressure squirting out). Based on proper application and complete knowledge of a specific leak test, results can be quite accurate and repeatable. This section will provide a description of the more common test methods in use for detection of internal and external leaks.

3.9.1 Flow Rate Method

This test method is based on the fact that the leak rate measured falls within the realm of measurable flow rate. Many advances have been made to develop flowmeters that are capable of measuring extremely small flow rates for both fluids and gases. This category, however, does not necessarily limit itself to minute flow rate detection.

At the high-flow end of the test spectrum we have spool leakage. This test is used to define the amount of fluid lost through a valve owing to the annular clearance between the spool and spool bore at various pressures. Any spool valve has leakage inherent to its design. The exception to the rule is the spool with elastomer seals, but these are not common.

There are several options used here because the flow rate generally falls within a range of 0.1–20 in.^{3}/min with oil as a test media, and 0.1–10 cc/min with air as a test media. In both cases, flowmeters are readily available to measure these amounts. Oil leaks can easily be measured with a stopwatch and a beaker, or with a positive displacement flowmeter that will read in actual flow rate. Air leaks of this magnitude can be easily monitored with a mass flowmeter.

In either case it is important to define the leakage characteristic desired. The flow meter and circuitry involved should cause as little

CENTERED   **UNCENTERED**

Fig. 3.63. Centered and uncentered spool within sleeve.

pressure drop as possible. Return line pressure will deduct from test component inlet pressure, thus reducing actual test pressure. Most important, in fluid tests, the fluid viscosity must be defined. Some manufacturers test with higher viscosity fluids. When these components are applied within low viscosity systems, leakage may become a problem.

A calculation used to predict spool leakage when the spool is centered in the sleeve (Fig. 3.63) is

$$Q = \frac{D}{L} \times \frac{h^3 P}{u} \times 0.1075$$

When the spool is uncentered, as encountered during pressure loading, the calculation used is

$$Q = \frac{D}{L} \times \frac{h^3 P}{u} \times 0.215$$

where

Q = leakage, in.^{3}

D = sleeve ID

L = land length

h = clearance between spool and sleeve in mils (0.001 in.)

P = psi

u = viscosity in centipoise

3.9.2 Bubble Test Method

This leak test is employed when gas is used as a test media. Many valves are defined as having "bubble tight" closure with other similar terms such as "leakproof," "drip-tight," and "zero-leak." The intent of these terms is understood, but in critical applications, the general term may not be sufficient, since there is usually some degree of leakage in any device.

First of all, will someone please define a bubble? What is a bubble? How big is a bubble? My point is best understood by telling a true story. A valve manufacturer purchased a bubble test machine, and the downstream side of the valve was connected to a 1/4 in. tube that dropped into a clear water viewing chamber. Pressure applied to the valve inlet would create a reduced pressure on the valve outlet if leakage on the seat was present. This downstream pressure would create a bubble on the tube outlet under the water. As the bubble grew, it eventually broke loose from the vertical tube and floated to the water surface. This was considered one bubble.

Weeks later, the manufacturer claimed that the test stand did not agree with tests conducted on two older test stands. The other two stands both used different methods. One stand used a 1/4 in. tube under water, the tube opening was under the water, but horizontal to the surface. A bubble, smaller than the tube ID was able to roll along the ID and immediately float to the surface once reaching the tube end. This was considered a bubble. The new test stand required that the bubble grow larger than the tube OD prior to escaping to the surface. The other method used was to fill the top of a horizontally mounted valve with water, so that when the bottom was pressurized with air, a bubble could be seen at the seat seal. Once the bubble formed, and was large enough to break loose, it was considered a bubble. Each machine used the bubble method, but each bubble diameter was different.

Disregarding the lack of standardized test procedures, the test is conducted by pressurizing a component with air and watching for air bubbles under water. A shutoff valve can be seat tested using this method, provided that all downstream components and tubing are zero-leak. It is also important that the downstream side be purged of all fluid between tests to prevent inconsistency of test results. In some procedures, the test item is fully submerged under water to detect any external leaks caused by casting porosity, seal failure, or improper assembly.

3.9.3 Pressure Decay Method

The pressure decay method is one of the most common procedures used to find both internal and external leaks. There are definite problems associated with this technique, but applied properly, repeatable test results are attained.

The pressure decay technique can be used with both air and fluid as a test media. This method is based on the principle of compressibility of the media. A test component has a certain internal volume that can be either calculated or determined by fluid displacement.

This internal volume and the volume of the pressurized portion of the test circuit are considered the test volume. This test volume provides us with the volume of the media at atmospheric pressure. As fluid or gas pressure is increased, more of the media is forced into the same volume. This is what causes the increase in pressure, an increase of media volume in the same fixed test volume.

Some components are designed for both fluid or gas service and can be tested with either medium. The selection of medium type is based on the ability of the test component to operate properly using the medium, the size of the leak and the amount of test time available.

Testing with a gas is a very stable method owing to the high compressibility of gases. One frequent problem encountered is the effect on compressed volume due to temperature fluctuations. A gas can be heated while entering a test component owing to internal friction caused by pressure drop. As the component is tested, the gas can cool, thus changing the volume. A change in volume also means a change in pressure, thus the accuracy of the test is affected. Several formulas used to express pressure and temperature effects are:

$$P_1 \times V_1 = P_2 \times V_2 \text{ (Isothermic)}$$

$$P_1 \times T_2 = P_2 \times T_1 \text{ (Isochoric)}$$

$$V_1 \times T_2 = V_2 \times T_1 \text{ (Isoharic)}$$

$$P_1 \times V_n^1 = P_2 \times V_n^2$$

$$\frac{T_2}{T_1} = \left(\frac{V_1}{V_2}\right)^{n-1} = \left(\frac{P_2}{P_1}\right)^{n-1/n}$$

where

P_1 = original pressure (psia)

P_2 = final pressure (psia)

V_1 = original volume

V_2 = final volume

T_1 = original temperature

T_2 = final temperature

n = an exponent dependent on the type of expansion and the gas used: $n = 1$ for isothermic expansion, $n = 1.4$ for adiabatic expansion using diatomic gases such as air and nitrogen, $n = 1.7$ for monatomic gases such as helium

Liquids, on the other hand, are not very compressible. As a rule of thumb, liquid volume increases by 0.5% per 1000 psi, which allows accurate testing of very small leaks, but there are also problems with this method.

The pressure decay test circuit can be used in one of two configurations. In both cases the system must be absolutely leakproof and nonexpanding for the following calculations to be correct. If this is unavoidable, compensation in the pressure decay value must be made.

The first configuration consists of a shutoff valve, a pressure transducer, and the test item. Pressure is applied through the shutoff valve until test pressure is reached. The shutoff valve is closed, and a timer is started. The pressure transducer will depict a drop in pressure as test media leaks from the test item. When testing for external leaks, the outlet of the test component (if one exists) should be plugged if testing with a gas. If testing with a fluid, proper flushing is necessary to remove all air prior to the start of the test. When testing for internal leaks, the downstream side of the potential leak path must be vented to atmosphere.

Fig. 3.64. Pressure decay system. *(Courtesy Heco Division Barker Rockford Co.)*

The second configuration (Fig. 3.64) consists of an additional shutoff valve, a pressure reference chamber, and a differential pressure transducer. This method is used when greater accuracy is needed either due to extremely small leak detection or high test pressures, where the full-scale accuracy of the pressure transducer provides too large an error. The reference chamber is pressurized at the same time as the test item, with the high side of the differential transducer connected to the chamber and the low side connected to the test item. At start of test, the transducer reads zero. The differential pressure will increase as a leak drops pressure in the test item. Accuracy is increased by using a transducer with a lower range. For example, a system pressure transducer with a 0–5000 psi range and accuracy of ± 0.25% full scale equates to an accuracy of ± 12.5 psi. If a leak produced 2 psi decay, the transducer would be useless. By using a differential transducer with a range of 0–100 psi and accuracy of ± 0.25% full scale, we achieve an accuracy of ± 0.25 psi. It is important to make sure the transducer is protected from the maximum test pressure to which it is subjected.

Now, considering all the above, if gas is selected as the test media, the following information will help determine pressure decays and test times required for your test.

GAS LEAK TEST

The following values must first be selected to determine the pressure decay rate:

Air leak rate (cc/min) =
Test volume (in.^{3}) =
Test pressure (psig) =
Test time (sec) =

From these data, an estimate of expected pressure decay can be calculated by first determining compressed volume:

compressed volume (in.^{3})

$$= \frac{\text{test volume (in.}^3) \times [\text{test pressure (psig)} + 14.7]}{14.7}$$

From this answer we must now determine the ratio of compressed volume to the leak rate:

decayed volume ratio

$$= \frac{\text{compressed volume (in.}^3)}{\text{compressed volume (in.}^3) - [\text{air leak rate (cc/min)} \times 0.06102]}$$

Using this number we can now calculate the pressure decay expected in 1 min:

$$\text{pressure decay (1 min)} = \frac{\text{test pressure (psig)}}{\text{decayed volume ratio}}$$

Actual pressure decay in the specified test time can now be solved:

Pressure decay =
$$\frac{\text{Test pressure (psig)} - \text{Pressure decay (1 min)}}{60} \times \text{Test time (sec)}$$

As an example let us assume:

Air leak rate (cc/min) = 5
Test volume (in.^{3}) = 2
Test pressure (psig) = 80
Test time (sec) = 30

Therefore,

$$\frac{2 \times (80 + 14.7)}{14.7} = 12.884 \text{ in.}^3 \text{ compressed volume}$$

$$\frac{12.884}{12.884 - (5 \times 0.06102)} = 1.024 \text{ decay volume ratio}$$

$$\frac{80}{1.024} = 78.125 \text{ psi}$$

$$\frac{80 - 78.125}{60} \times 30 = 0.9375 \text{ psid in 30 sec}$$

If fluid is selected as the test media, the following section is helpful in predicting pressure decay.

FLUID LEAK TEST

The following values must first be selected to determine the amount of pressure decay:

Fluid leak rate (cc) =
Test volume (in.^{3}) =
Test pressure (psig) =

From these data, an ideal situation can be calculated to determine resultant pressure decay. First, compressed volume must be calculated:

compressed volume (in.^{3})

$$= \text{test volume (in.}^3) \times \text{compressibility factor}$$

The compressibility factor is selected from the following table based on fluid type and test pressure:

Test Pressure	Oil	Water
1000	0.0056	0.0033
2000	0.0104	0.0067
3000	0.0147	0.0094
4000	0.0189	0.0125
5000	0.0230	0.0154
6000	0.0267	0.0183
7000	0.0304	0.0213
8000	0.0340	0.0238
9000	0.0377	0.0368
10,000	0.0411	0.0290

Once the compressed volume is calculated, pressure decay can be calculated by the formula:

pressure decay (psid)

$$= \frac{\text{leak rate (cc)} \times 0.06102}{\text{compressed volume (in.}^3)} \times \text{test pressure (psig)}$$

As an example, let us assume

Fluid leak rate (cc) = 0.05
Test volume (in.^{3}) = 10
Test pressure (psig) = 4000
Test media = Water

Therefore

$$10 \times 0.0125 = 0.125 \text{ in.}^3$$

$$\frac{0.05 \times 0.06102}{0.125} \times 4000 = 97.92 \text{ psid}$$

Using this method has both advantages and disadvantages. Owing to the low volume compressibility of fluids, very small leaks can easily be detected by standard gages or pressure transducers. As demonstrated by the preceding formula, a 0.05 cc leak in a 10 in.^{3} test volume produces a pressure decay of 97.92 psi when starting from 4000 psi. Usually, 0.05 cc is equivalent to one drop of fluid. If the same component were tested with air, the results would be quite different. Assuming a 0.05 cc leak, a test volume of 10 in.^{3}, a test pressure of 100 psig, and a 30 sec test with air, the pressure decay would be 0.001955 psid. Quite a difference!

By now your wondering, "But what's the catch"? Well, there are several catches. The first problem is the extreme sensitivity of the system. False readings are common owing to seal movement caused by pressurization. Consider the volume displaced by a seal under pressure. If a 0.05 cc drop of fluid will create a pressure decay of 97.92 psi, the same can be expected if the seal were to compress at an equivalent volume. In many cases, the first test of a component will fail because sufficient stabilization time is not allowed to negate seal expansion. When the test is repeated, the component will then be depicted as a "pass."

Another problem is the design of valves used to isolate the test component and trap compressed fluid in the test component. In some valve designs, the test pressure can increase or decrease just due to the movement of the valve poppet. If the test valve leaks, the test fails, because the test is actually conducted on the test component as well as the test system exposed to the trapped fluid.

If hoses are used to connect the test component, they are often part of the test. Hoses then act as small accumulators, storing fluid, and thus causing calculation to be almost useless.

Trapped air within the component now acts as a storage volume and the compressibility of the fluid and of the air must now be accounted for. This greatly affects the sensitivity of this test and sometimes makes the test useless or time consuming. Proper prefill of the test component is very important to ensure that all the air is removed. In some cases, usually owing to test component mounting position, air pockets are unavoidable.

The only positive method to determine the actual pressure decay expected for a particular leak rate and pressure is to test a known "leaker." Another practice is to substitute an acceptable component with a fine metering needle valve mounted to it. The needle valve is vented to atmosphere and adjusted for the drops/minute leak rate desired, and pressure decay is monitored and set at that point. This procedure may often be time consuming but is usually worth the time and effort.

3.9.4 Helium Leak Detection

Helium leak detection is used to detect very small leaks, which are not possible to detect with previously described techniques. When pressure decay, bubble test, or mass flowmeter lower limits are reached, leak detection becomes costly if any degree of accuracy is required. Helium leak detection works on the principle that a tracer gas is injected in the area of a leak, and the resulting leak mixes this gas with normal ambient gas where the resulting variation is detected. The equipment used to detect this variation is relatively expensive compared to previous methods discussed.

There are three basic methods used in helium leak detection: vacuum test, pressure test, and bombing (or backfilling) method. Helium is the most frequently used tracer gas, because it is inert, nonflammable, nontoxic, and inexpensive, and with high permeability.

The vacuum test requires the test part to be evaluated with a vacuum pump or Roots blower system. With the helium probe inserted in the test component, helium can be sprayed in a certain test area of the part or the part can be placed in a sealed container that is filled with helium. The vacuum inside the component will draw in the tracer gas where it is detected by the probe.

The pressure test is basically the reverse of the vacuum test. Tracer gas is inserted within the component that is pressurized. Internal pressure forces the gas to exit at the leak point. A probe can be passed around the test item and the leak point can be detected. Another method involves a sealed container in which the test item is placed. As helium escapes into the chamber, the accumulated leak can be determined by the amount of helium within the chamber.

The bombing (or backfilling) method involves placing a part within an evacuated chamber that is pressurized with a tracer gas. The part is then removed and tracer gas absorbed by the walls of the part is eliminated. The part is then placed in a vacuum container where residual tracer gas is detected by a tracer gas analyzer while the part is under vacuum.

3.9.5 Ultrasonic Leak Detection

Ultrasonic leak detection involves the use of a sound generator and sonic probe. The generator transmits a specific ultrasonic frequency band, which is placed at one side of the potential leak in the test part. The sound penetrates small holes and cracks and is detected by the probe.

Several probe designs are available to detect emitted sound. Probes can be used to find sound transmitted along rigid structures or housings, or to pinpoint the exact source with various focusing probe envelopes. Other designs use the leaking gas as a sound source and require no sound generator. The probe is designed to detect the frequencies generated by the escaping gas.

3.9.6 Soap Bubble Method

An external gas leak can easily be detected by using a mixture of liquid soap and water. Brushing this soapy solution over potential leak points, such as fitting connections, will produce bubbles that pinpoint the location of the leak. Although leak rate is impossible to measure, very small leaks can be found with this method.

In some industries, the use of soap and water are no longer allowed, but commercially available synthetic bubble solutions are used. These solutions can be supplied for a wide variety of applications with leak detection capability down to 1×10^{-6} (0.000001) standard cc/sec.

3.10 Cyclic Shock Tests

Components that are to be subjected to severe temperature or pressure cycles are tested to determine the resulting effects caused by these conditions. These tests reveal certain mechanical deficiencies or design flaws, as well as moving part interaction problems and resulting leakage.

It is important to maintain operator safety during these tests, since mechanical rupture and potential scalding from hot fluids is possible. In most cases the test item is mounted within an enclosure with a viewing window. The enclosure is designed to withstand explosion of the component and contain fluid spraying from cracked component surfaces.

Cycle test stands are normally designed to operate 24 hr a day, since cycles may be well into millions of cycles. This requires that the test stand have sufficient safety monitoring and self-protection to prevent damage to itself in an unsupervised mode. This includes features such as loss of fluid, overpressure, overtemperature, and component rupture. It should be designed to maintain the count of cycles at time of shutdown in order to preclude a complete rejection of current cycle test data.

The system should be designed with sufficient safety factor, in other words, so that the system should last longer than the test article. This is not always the case, if proper system design techniques are not applied.

3.10.1 Pressure Cycle Testing

Pressure cycle testing subjects a component to repeated pressure pulses to determine fatigue life. The component is subjected to pulses from zero (or close to it) to a predetermined maximum pressure. The actual pressure curve is generally applied as shown in Fig. 3.65. One complete impulse cycle consists of 50% at pressure and 50% at vent pressure. A 15% pressure rise time and 5% decay time are the maximum allowable limits. Test pressure regulation is defined as ± 5% and vent pressure as high as 250 psi is acceptable.

The cycle time is based on the number of cycles required and the time allowed to conduct the test. Millions of cycles are usually applied to a component, so cycle time is generally kept as short as practical. Generally, the governing factors are the amount of pump flow and maximum HP available to raise pressure from zero to test pressure. The 15% maximum pressure rise time governs the amount of flow necessary. This all relates to the compressibility of the test fluid. At first, you would think that because this is a static test (no flow), flow requirement would be minimal; if the component under test is small and pressure is low, this may be the case. But, let us look at a fairly common application.

Assume we are testing a filter housing that has an internal volume

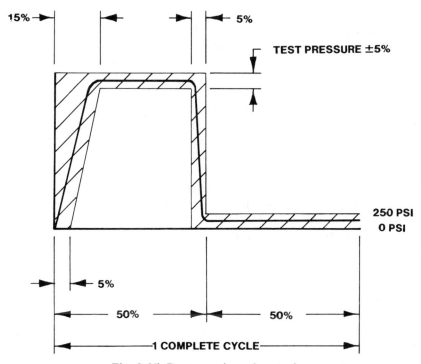

Fig. 3.65. Pressure impulse cycle.

of 100 in.^{3} and test pressure is determined to be 3000 psi. In order to save time, we desire a cycle rate of 4 Hz. The fluid used is a standard fluid with a compressibility rate of 1.47% at 3000 psi. To determine flow rate required, we calculate the following:

    (a) 4 Hz = 1 cycle every 0.25 sec

    (b) 15% (maximum pressure rise) of 0.25 sec = 0.0375 seconds

    (c) 100 in.^{3} × 1.47 % = 1.47 in.^{3} in 0.0375 sec

    (d) $1.47 \text{ in.}^3 \times \dfrac{60}{0.0375} \div 231 = 10.18 \text{ gpm minimum}$

With the test at 10.18 gpm at 3000 psi (85% efficiency), we calculate that 21 HP is required. This does not take into account any system leakage, pressure rise was figured at the maximum limit, and internal volume of the pressurized portion of the test circuit was not included. With all this input, we also must remove a minimum of 50% of total system HP heat input continuously. An effective way to reduce the system size is to fill the test component with steel balls. This reduces test volume by an appreciable amount. If we cut interval volume by 50%, we cut input flow and horsepower by 50%. All this for a no flow situation.

In addition to these high-power demands, we must also consider the pressure spikes caused by continuous interruption of flow. Remember, the maximum pressure variation allowed is ± 5%. This can usually be dampened through the use of an accumulator.

The pressure cycle curve was originally developed by the SAE for testing hydraulic hose assemblies and is outlined in procedure SAE J343a Tests and Procedures for SAE 100R Series Hydraulic Hose and Hose Assemblies. This curve has been adapted by most for testing most hydraulic components including hoses.

NFPA T2.6.1-1974 (R1982), Method for Verifying the Fatigue and Static Pressure Ratings of the Pressure Containing Envelope of a Metal Fluid Power Component, details a standard procedure for pressure cycle testing. In the procedure cyclic test pressure (CTP) is defined as a product of two test parameter factors (K_n and K_v) and the rated fatigue pressure (RFP), calculated as follows: CTP = RFP × (K_n × K_v), where K_n is test duration factor based on the minimum number of test cycles (1 or 10 million) and the type of metal (ferrous or nonferrous) and K_v is a variability factor based on the number of test units (1, 2, or 3), the type of metal, and the assurance level.

This formula is used to define the minimum and maximum pressures used during the test.

Copies of exact procedures can be obtained through the NFPA.

3.10.2 Temperature Cycle Testing

Temperature cycle testing is used to define a component's ability to maintain mechanical integrity under internal temperature shock conditions. Thermal shock can cause many problems such as stress cracking, work hardening of surfaces, uneven expansion/contraction of dissimilar materials, jamming of parts, and component failure.

This test is generally accomplished with two separate fluid loops: one hot and one cold. Test temperature extremes are usually selected due to a particular application. Cycle time is also dependent on application, with accelerated cycles employed to reduce total test time.

The hot loop is accomplished by circulating fluid through a heater, sized to maintain maximum temperature. In many circuits, test temperature must be obtained in a single pass through the circuit. Heat can be generated by an electric circulation or immersion heater, or by using heating fluid circulation systems.

Conversely, the cold loop must cool the fluid to minimum test temperature in a single pass. Depending on the temperature required, the fluid is passed through a heat exchanger with a specific cooling medium. The cooling medium selected depends on the temperature required. For starters, water- or air-cooled heat exchangers are used in applications where the minimum test temperature is 10–20°F above ambient air temperature or available water temperature. This also depends on the water flow available as well as the heat removal capacity of the heat exchanger.

If this is unacceptable, the next step is to pass water flow (or a cooling fluid) through a chilling unit. This unit acts similar to an air conditioner, where Freon is used as the heat-transfer medium. Cooling fluid can be chilled at a temperature lower than ambient conditions, thus increasing the efficiency of the heat exchanger used.

Generally, the final step taken is the use of liquid nitrogen (or a similar coolant). This allows a heat exchanger system to remove large amounts of input heat in a single pass. The main drawback in a system of this type, of course, is cost, as well as viscosity swing of the coolant fluid and test fluid. Excessive pressure drop can deem a complete test circuit useless if pressure drop exceeds the test pressure. In a system of this type, this is not unusual.

Both hot and cold loops are alternately fed through the test item. Temperature must be monitored at the component inlet to ensure that proper test temperature is provided and must have sufficient response to depict true test temperature. Temperature monitoring in each loop is accurate for that loop only. The temperature mix in the common circuitry may cause the temperature to increase or decrease at the component inlet owing to change from uninsulated piping or rapid change in exposed circuitry. It is advisable in most cases to insulate all circuitry to prevent inefficient heating or cooling.

The test component outlet now vents to a common reservoir. This mixes the hot and cold loops creating a common fluid mix at a mean temperature between the high- and low-temperature extremes. This is, of course, including effects of ambient temperature and system inefficiencies. Fluid must now be drawn from a common reservoir by both loops and fluid heated or cooled to the proper temperature prior to entering the test component.

3.11 Environmental Test Methods

Fluid power components are often used within systems that are subjected to ambient conditions outside the normal factory environment. When this occurs, a manufacturer or the end user requires testing to determine if extreme ambient conditions will degrade the performance of a component. Components may be tested alone or as part of an overall package.

One of the most demanding applications is that used for military equipment. Figures 3.66 and 3.67 depict the types of general tests performed to a component or system and the situation the component or system faces during these ambient conditions. The situations are shipping/transportation, storage, and actual use. The figures depicted are taken from MIL-STD-810D, Environmental Test Methods and Engineering Guidelines. This standard is used for both military and industrial testing owing to its thorough coverage of the subject. The following test descriptions are brief summaries of the applicable tests covered in this standard.

3.11.1 Low Pressure (Altitude)—Method 500.2

Method 500.2 details the guidelines used for determining test procedures and test conditions based on the application. Three conditions are defined: storage test, operation test, and rapid decompression test.

Examples of some problems that could occur due to low pressure are:

    a. Leakage of gases or fluids from gasket-sealed enclosures.

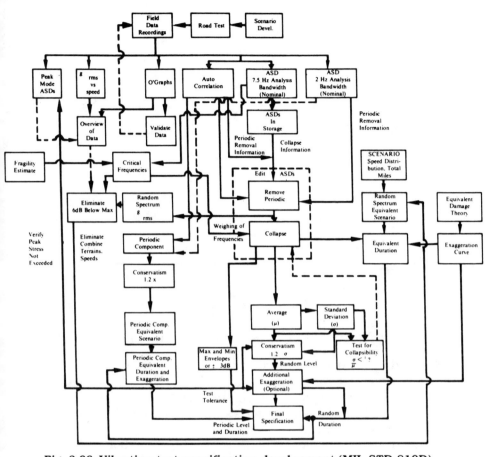

Fig. 3.66. Vibration test specification development (MIL-STD-810D).

   b. Rupture or explosion of sealed containers.

   c. Change in physical and chemical properties of low-density materials.

   d. Erratic operation or malfunction of equipment resulting from arcing or corona.

   e. Overheating of equipment due to reduced heat transfer.

   f. Evaporation of lubricants.

   g. Erratic starting and combustion of engines.

   h. Failure of hermetic seals.

One main example in the fluid power industry would be the suction characteristics of a pump. Reduced ambient pressure results in increased suction requirement.

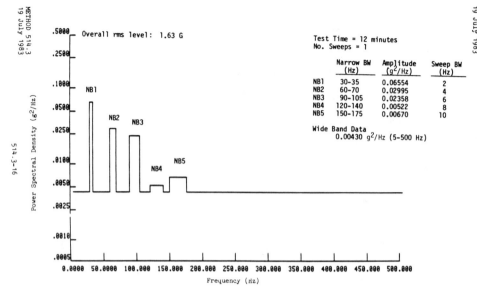

Fig. 3.67. Vertical axis test, phase I (MIL-STD-810D).

3.11.2 High Temperature—Method 501.2

Method 501.2 details procedures used when a component is exposed to excessive temperatures caused by extreme climate conditions, solar radiation, lack of ventilation, etc. This is especially useful in determining seal characteristics, as well as reduced viscosity and heat removal capacity.

Some of the problems that could occur due to high ambient temperatures include:

a. Parts binding due to expansion/contraction.
b. Lubricants and fluids becoming less viscous.
c. Materials changing dimension, either totally or selectively.
d. Elastomers distorting, binding, or failing.
e. Gaskets displaying permanent set.
f. Fixed resistors changing values.
g. Electronic circuitry stability variation.
h. Transformers and electromechanical components overheating.
i. Shortened component lifetime.
j. High internal pressures caused by thermal expansion.
k. Organic materials tending to discolor crack or craze.

3.11.3 Low Temperature—Method 502.2

Method 502.2 details the procedure for testing a component for proper operation after storage in a cold environment, as well as operation and operation safety in a cold environment. Also discussed is the handling required to make the test item operational in a cold environment. This generally applies to systems exposed to long periods of storage between use, such as aircraft construction equipment, etc. Most military test requirements in general procedures use −65°F as test temperature.

Problems encountered in this type environment include:

  a. Loss of lubrication or flow due to increased viscosity.
  b. Changes in electrical components.
  c. Changes in electromechanical components.
  d. Stiffening or binding of moving components.
  e. Condensation and freezing of water.
  f. Change of burning rates.

3.11.4 Temperature Shock—Method 503.2

Method 503.2 is used to define component performance when subjected to rapid ambient temperature changes. This can be caused by aircraft ascent from a desert airfield to high altitude, airdrop from high altitude to a desert environment, and transfer of equipment from hot to cold environments. (See Fig. 3.68.)

Fig. 3.68. Thermal shock test chamber. *(Courtesy Tenney Engineering, Inc.)*

Typical problems caused by this change include:

a. Shattering of brittle components.
b. Binding or slackening of moving parts.
c. Separation of constituents.
d. Changes in electronic components.
e. Electronic or mechanical failure due to rapid water or frost formation.
f. Differential contraction or expansion of dissimilar materials.
g. Deformation or fracture of components.
h. Cracking of surface coatings
i. Leaking from failed seals.

3.11.5 Solar Radiation (Sunshine)—Method 505.2

Method 505.2 defines effects caused by solar radiation, which differ from high temperature. The amount of heat absorbed or reflected depends on the roughness and color of the surface on which the radiation is incident. Intensity of the radiation can cause differential expansion and contraction of components, thus leading to severe stresses and loss of structural integrity.

Some effects caused by excessive radiation include:

a. Jamming or loosening of moving parts.
b. Weakening of solder joints and glued parts.
c. Change in strength and elasticity.
d. Loss of calibration or malfunction of linkage devices.
e. Loss of seal integrity.
f. Changes in electrical and electronic components.
g. Premature actuation of electrical contacts.
h. Fading or discoloring due to ultraviolet exposure.
i. Changes in elastomer characteristics.
j. Blistering and peeling of paint and coatings.
k. Softening of potting compounds.

3.11.6 Rain—Method 506.2

Method 506.2 details the possible effects to components exposed to rain or the aftereffect that could lead to the following problems:

a. Degrades the strength of some materials.
b. Promotes corrosion of metals.

c. Deteriorates surface coatings.

d. Failure of electronic components (inoperative or dangerous conditions).

e. Freezing after penetration, causing cracking and swelling of parts.

f. Causes high humidity, which encourages corrosion and fungal growth.

3.11.7 Humidity—Method 507.2

Method 507.2 is performed to determine the resistance of material to the effects of a warm, humid atmosphere. Moisture can cause physical and chemical deterioration, while the added effect of temperature changes can cause condensation. Figure 3.69 depicts a

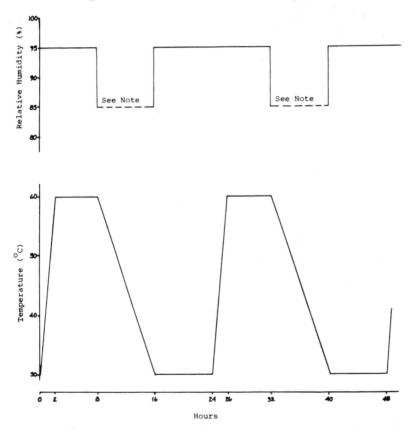

Note: Relative humidity maintained above 85% during temperature drops.

Fig. 3.69. Temperature/humidity cycle (MIL-STD-810D).

Fig. 3.70. Temperature/humidity test chamber. *(Courtesy Tenney Engineering, Inc.)*

typical aggravated temperature/humidity cycle over a period of 48 hours. (The temperature/humidity test chamber is shown in Fig. 3.70.)

Common problems associated in this area include:

a. Swelling of materials.
b. Loss of physical strength.
c. Changes in mechanical properties.
d. Degradation of insulating materials.
e. Electrical shorts due to condensation.
f. Binding of moving parts.
g. Oxidation or galvanic corrosion of metals.
h. Loss of plasticity.
i. Accelerated chemical reactions.
j. Organic surface coating breakdown.
k. Deterioration of hydroscopic materials.

3.11.8 Fungus—Method 508.3

Method 508.3 helps evaluate the extent to which the test item will support fungal growth or how fungal growth will affect the test item's performance. This test is conducted for at least 28 days with a recommended 84 day test to achieve test results with a greater degree of certainty.

Five species of test fungi are selected owing to their ability to degrade materials, their worldwide distribution, and their stability. The test item is placed in a chamber where temperature, humidity, and air circulation can be monitored and controlled.

3.11.9 Salt Fog—Method 509.2

Method 509.2 determines a component's resistance to a salt atmosphere such as experienced at sea. This method states that the procedures outlined do not necessarily duplicate the corrosive conditions of a marine atmosphere. However, this test is frequently used to derive a general idea of a component's tolerance to corrosion and buildup of salt deposits.

The test procedure is based on test duration, cycle of exposure and drying periods, salt concentration, and test item configuration. A minimum exposure of 48 hours is recommended followed by a 48 hr drying period. Also, a more severe test outlined is alternating 24 hr periods of salt fog exposure and ambient drying conditions for a minimum of four 24 hr periods.

The salt solution used is (depicted in Fig. 3.71) adjusted and

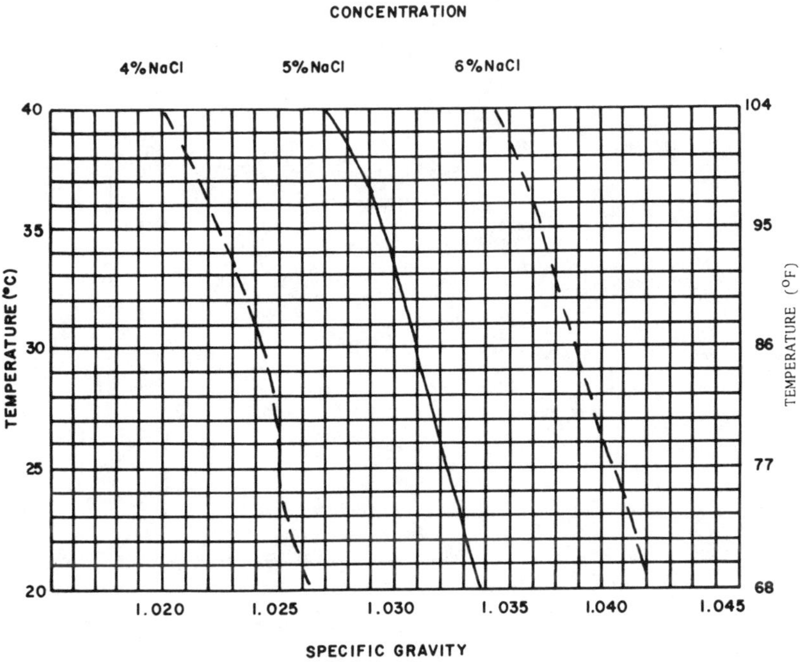

Fig. 3.71. Salt characteristics versus temperature (MIL-STD-810D).

maintained at a specific gravity by using the measured temperature and density of the solution. Water used is from steam or distilled, deionized, or demineralized. The solution is sprayed into the chamber with atomizing nozzles.

3.11.10 Sand and Dust—Method 510.2

Method 510.2 is divided into two categories owing to the difference each type of airborne contaminates have and the resulting problems they cause. Two types of blowing sand facilities are used and are depicted in Figs. 3.72 and 3.73. The variables associated with this type of test include air velocity, temperature, test item configuration and orientation, sand and dust concentration and composition, test duration, and additional guidelines as appropriate.

Basically this test defines the problems created such as:

- a. System contaminant ingression.
- b. Abrasion of surfaces.
- c. Penetration of seals.

MIL-STD-810D
19 July 1983

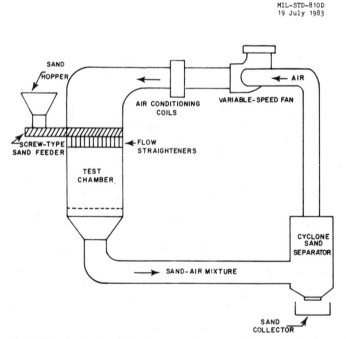

Fig. 3.72. Vertical flow blowing sand test facility (MIL-STD-810D).

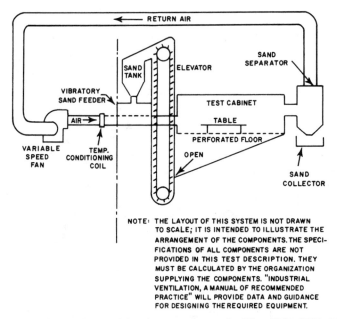

NOTE: THE LAYOUT OF THIS SYSTEM IS NOT DRAWN
TO SCALE; IT IS INTENDED TO ILLUSTRATE THE
ARRANGEMENT OF THE COMPONENTS. THE SPECI-
FICATIONS OF ALL COMPONENTS ARE NOT
PROVIDED IN THIS TEST DESCRIPTION. THEY
MUST BE CALCULATED BY THE ORGANIZATION
SUPPLYING THE COMPONENTS. "INDUSTRIAL
VENTILATION, A MANUAL OF RECOMMENDED
PRACTICE" WILL PROVIDE DATA AND GUIDANCE
FOR DESIGNING THE REQUIRED EQUIPMENT.

Fig. 3.73. Horizontal flow blowing sand test facility (MIL-STD-810D).

d. Erosion of surfaces.
e. Degradation of electrical circuits.
f. Clogging of openings and filters.
g. Physical interference with mating parts.
h. Fouling of moving parts.

3.11.11 Explosive Atmosphere–Method 511.2

Method 511.2 determines a component's ability to function after
an explosion, or the ability for a component to operate in a flammable environment without creating an explosion. Any electrical device
or mechanical friction can cause a spark. In certain areas, such as
aircraft fueling, ordnance manufacturing, and petrochemical facilities the environment may contain flammable gases. Fluid power and
other equipment operating in these areas must not create a catastrophic condition.

The component is mounted in a test chamber as depicted in Fig.
3.74. A mixture of air and n-hexane is injected into the chamber
and the component is operated in simulated operating conditions,
usually worst case. In addition, the test item may be subjected to an
explosion by igniters located within the test chamber.

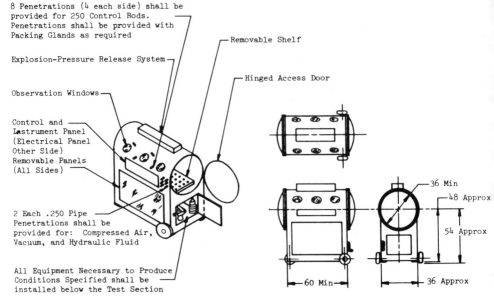

8 Penetrations (4 each side) shall be
provided for 250 Control Rods.
Penetrations shall be provided with
Packing Glands as required

Removable Shelf

Explosion-Pressure Release System

Hinged Access Door

Observation Windows

Control and
Instrument Panel
(Electrical Panel
Other Side)
Removable Panels
(All Sides)

36 Min

48 Approx

54 Approx

2 Each .250 Pipe
Penetrations shall be
provided for: Compressed Air,
Vacuum, and Hydraulic Fluid

All Equipment Necessary to Produce
Conditions Specified shall be
installed below the Test Section

60 Min

36 Approx

Facility shall be designed to Withstand Transportation by Sling or Forklift

Dimensions in Inches

Fig. 3.74. Flammable atmosphere test chamber (MIL-STD-810D).

It is recommended that the test item be subjected to vibration and/or temperature testing prior to this test to potentially reduce the effectiveness of seals.

3.11.12 Leakage (Immersion)—Method 512.2

Method 512.2 basically determines the seal effectiveness of a component when immersed in water at a depth of 1 m over the uppermost point of the test item.

The water is maintained at a specific temperature (64 ± 18°F), and the test item is maintained at 49 ± 4°F above the temperature of the water.

The test item is visually inspected and test results are recorded.

3.11.13 Acceleration—Method 513.3

Method 513.3 is performed to ensure that equipment can structurally withstand the g forces that are encountered onboard aircraft, helicopters, manned aerospace vehicles, air-carried stores, and

Vehicle Category		Forward Acceleration A in g's 1/	Test Level					
			Direction of Vehicle Acceleration (See figure 513.3-1)					
			Fore	Aft	Up	Down	Lateral	
							Left	Right
Aircraft 2/, 3/		2.0	1.5A	4.5A	6.75A	2.25A	3.0A	3.0A
Helicopters		4/	4.0	4.0	10.5	4.5	6.0	6.0
Manned Aerospace Vehicles		6.0 to 12.0 5/	1.5A	0.5A	2.25A	0.75A	1.0A	1.0A
Aircraft Stores	Wing/ Sponson Mounted	2.0	7.5A	7.5A	9.0A	4.9A	5.6A	5.6A
	Fuselage Mounted	2.0	5.25A	6.0A	6.75A	4.1A	2.25A	2.25A
Ground-Launched Missiles		6/, 8/	1.2A	0.5A	1.2A' 7/	1.2A' 7/	1.2A' 7/	1.2A' 7/

1/ Levels in this column should be used when forward acceleration is unknown. When the forward acceleration of the vehicle is known, that value shall be used for A.
2/ For carrier-based aircraft, the minimum value to be used for A is 4, representing a basic condition associated with catapult launches.
3/ For attack and fighter aircraft, add pitch, yaw, and roll accelerations as applicable.
4/ For helicopters, forward acceleration is unrelated to acceleration in other directions. Test levels are based on current and near future helicopter design requirements.
5/ When forward acceleration is not known, the high value of the acceleration range should be used.
6/ A is derived from the thrust curve data for maximum firing temperature.
7/ Where A' is the maximum maneuver acceleration.
8/ In some cases, the maximum maneuver acceleration and the maximum longitudinal acceleration will occur at the same time. When this occurs, the test item should be tested with the appropriate factors using the orientation and levels for the maximum (vectorial) acceleration.

Fig. 3.75. Suggested G levels for procedure I—structural test (MIL-STD-810D).

ground-launched vehicles. There are two test procedures. Procedure I (Structural Test) is used to determine that equipment will structually withstand loads induced by in-service accelerations. Procedure II (Operational Test) is used to determine that equipment will operate successfully both during and after being subjected to in-service accelerations. Suggested g levels for both Procedures I and II are depicted in Figs. 3.75 and 3.76. Figure 3.77 depicts the acceleration direction relative to an aircraft.

Vehicle Category		Forward Acceleration A in g's 1/	Test Level					
			Direction of Vehicle Acceleration (See figure 513.3-1)					
			Fore	Aft	Up	Down	Lateral	
							Left	Right
Aircraft 2/, 3/		2.0	1.0A	3.0A	4.5A	1.5A	2.0A	2.0A
Helicopters		4/	2.0	2.0	7.0	3.0	4.0	4.0
Manned Aerospace Vehicles		6.0 to 12.0 5/	1.0A	0.33A	1.5A	0.5A	0.66A	0.66A
Aircraft Stores	Wing/ Sponson Mounted	2.0	5.0A	5.0A	6.0A	3.25A	3.75A	3.75A
	Fuselage Mounted	2.0	3.5A	4.0A	4.5A	2.7A	1.5A	1.5A
Ground-Launched Missiles		6/, 8/	1.1A	0.33A	1.1A' 7/	1.1A' 7/	1.1A' 7/	1.1A' 7/

1/ Levels in this column should be used when forward acceleration is unknown. When the forward acceleration of the vehicle is known, that value shall be used for A.
2/ For carrier-based aircraft, the minimum value to be used for A is 4, representing a basic condition associated with catapult launches.
3/ For attack and fighter aircraft, add pitch, yaw, and roll accelerations as applicable.
4/ For helicopters, forward acceleration is unrelated to acceleration in other directions. Test levels are based on current and near future helicopter design requirements.
5/ When forward acceleration is not known, the high value of the acceleration range should be used.
6/ A is derived from the thrust curve data for maximum firing temperature.
7/ Where A' is the maximum maneuver acceleration.
8/ In some cases, the maximum maneuver acceleration and the maximum longitudinal acceleration will occur at the same time. When this occurs, the test item should be tested with the appropriate factors using the orientation and levels for the maximum (vectorial) acceleration.

Fig. 3.76. Suggested G levels for procedure II—operational test (MIL-STD-810D).

3.11.14 Vibration—Method 514.3

Method 514.3 is used to determine a component's ability to withstand vibration incurred during shipment or from the application itself. Figure 3.78 depicts the various categories covered under this section.

Typically, the complete system is tested to reveal any mechanical component that may fail due to induced stresses. Each category is covered with detailed information regarding test procedure, accept-

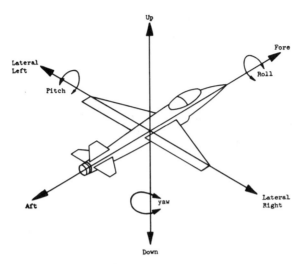

Fig. 3.77. Directions of vehicle acceleration (MIL-STD-810D).

Vibration environment categories.[1]

DIVISION	CATEGORY	DESCRIPTION	TEST PROCEDURE	TEST CONDITIONS[2]
Transportation/ Cargo-Induced Vibration	1. Basic Transportation	Equipment carried as secured cargo.	I	I-3.2.1
	2. Large Assembly Transport	Very large shelters, van, & trailer systems as an alternative to shaker testing.	III	I-3.2.2
	3. Loose Cargo Transport	Equipment carried on ground vehicles as unrestrained cargo.	II	I-3.2.3
Application-Induced Vibration	4. Propeller Aircraft	Equipment installed in propeller aircraft manned and unmanned.	I	I-3.2.4
	5. Jet Aircraft/ Tactical Missiles	Equipment installed in jet aircraft, manned and unmanned, and installed in tactical missiles - free flight phase.	I	I-3.2.5
	6. Helicopter	Equipment installed in helicopters.	I	I-3.2.6
	7A. External Stores	Assembled stores externally carried on jet aircraft (including captive missile flight).	IV	I-3.2.7
	7B. External Stores	Equipment installed in stores externally carried on jet aircraft.	I	I-3.2.8
	7C. External Stores	Assembled stores externally carried on helicopters.	I	I-3.2.9
	8. Ground Mobile	Equipment installed in wheeled vehicles, trailers, and tracked vehicles.	I	I-3.2.10
	9. Marine	Equipment installed in ships or other naval watercraft.	I	I-3.2.11
	10. Minimum Integrity Test	a. All other. b. Vibration-isolated equipment.	I	I-3.2.12

[1] Also referred to as "equipment categories".

[2] The provisions of section I-4 apply to all vibration tests.

Fig. 3.78. Vibration environment categories (MIL-STD-810D).

able test results, and information concerning typical vibration environments.

3.11.15 Acoustic Noise—Method 515.3

Method 515.3 is used to verify a component's (typically electrical) ability to withstand vibration caused by airborne noise. Unlike mechanical vibration, acoustic vibration can affect even components that use mechanical vibration mounting, because pressure fluctuations impinge directly on the equipment.

Figure 3.79 depicts categories used to define the various types of acoustic noise tested for. These tests are usually performed using a reverberant test chamber with a volume of at least 10 times the test item volume. In some cases the noise field behind a jet engine can be used. This method is difficult in achieving a uniform acoustical environment.

3.11.16 Shock—Method 516.3

Method 516.3 outlines the procedures used to determine the ability of a component to withstand the relatively infrequent nonrepetitive shocks or transient vibrations encountered during handling, transportation, and service environments. Mechanical shock can cause failures due to increased or decreased friction, interference between parts, changes in electrical characteristics, permanent deformation, or increased fatiguing.

Basically, this procedure is divided into nine procedures, which include:

a. Functional shock—which causes damage to a component during normal operation.
b. Equipment to be packaged—used when a test item requires a shipping container.
c. Fragility—used to determine the type of packaging required for a sensitive component.
d. Transit drop—defines the capability of a component to withstand the shocks encountered during loading/unloading.
e. Crash hazard—intended for testing a component mounted in a ground or air vehicle that could break loose during a crash, presenting a hazard to vehicle occupants.
f. Bench handling—defines a component's ability to withstand the shock encountered during bench-type maintenance.
g. Pyrotechnic shock—intended for components subjected to shock from explosive devices. This test is also used for simi-

MIL-STD-810D
19 July 1983

Category[1]	Suggested Test Overall Sound Pressure Level (dB[2] [3])	Equipment Application	Vehicle Source	Equipment Location	Suggested Exposure Time [4] (minutes)
A	165	Ground Based	Rocket	On launch site	8
B	150	Ground Based	Aircraft	Near runway/in jet engine run-up pads	30
C	150	Airborne	Aircraft	Near noise source and separated by thin partition	30
D	160	Airborne	Aircraft	Near noise source or in noise cone of aircraft	30
E	160	Airborne	Rocket	Majority of locations, exclusive of booster or engine compartments	8
F	165	Airborne	Rocket	Booster or engine compartment	8
G	140	Airborne	Aircraft	Majority of locations	30
H	See Table 515.3-III	Airborne	Aircraft	Near or in open cavities exposed to the airstream	See Table 515.3-III
I	See Table 515.3-II	Airborne	Aircraft	Externally-carried stores	See Table 515.3-II

1/ In the qualification test, the pressure levels and exposure times for categories A thru G are for the functional test. No separate endurance test is required.
2/ Reference 20 μPa (2×10^{-4} dynes/cm^2).
3/ Already adjusted for a reverberant test environment, see I-3.2.5.1.
4/ Use only 10 minutes of exposure for environmental worthiness test. If suggested duration is less than 10 minutes, use shorter duration.

Fig. 3.79. Acoustic noise test categories.

lar types of shock caused by atmospheric reentry, water entry of missles, and high velocity aerodynamic buffeting of high-performance weapons systems.

h. Rail impact—used to verify a component's tolerance to normal railroad car impacts that occur during rail shipment.

i. Catapult launch/arrested landing—used to test a component mounted in or on fixed winged aircraft that are subjected to catapult launches and arrested landings.

Another test often applied to fluid power components and systems is MIL-S-901.

3.11.17 Aircraft Gunfire Vibration—Method 519.3

Method 519.3 defines a component's ability to withstand vibration induced by onboard gun firing. The vibration resulting from repetitive blasts is roughly two orders of magnitude above normal flight vibration levels. This vibration can cause intermittent electrical contact, catastrophic electrical failures, hydraulic malfunctions, and structural fatigue failures. Considering the fact that many of the aircraft guns in use are hydraulically driven, this test is particularly important.

Vibration occurs from overpressure pulses emanating from the muzzle, recoil kick of the gun on its mounts, and the motion of ammunition handling mechanisms as the gun is firing. Testing is generally done on the aircraft under worst case conditions. A calculation procedure is also defined to approximate the potential vibration possible.

3.11.18 Temperature, Humidity, Vibration, Altitude— Method 520.0

Method 520.0 is mainly intended for testing electronic equipment installed in aircraft. The combination or individual effects or both of these factors account for all but 12% of the environmentally induced failures in the field.

The test cycle is developed around the mission profile of the aircraft. The various phases of flight such as takeoff, cruise, combat, or low-level penetration are accounted for. During the test cycle, temperature, vibration, humidity, altitude, and cooling airflow are varied based on the requirement.

3.11.19 Icing/Freezing Rain—Method 521.0

Method 521.0 is used to define problems created by freezing rain. This causes binding, weight addition, moving part clearance changes, structural failures, etc. Since many moving parts on air-craft are hydraulically actuated, this procedure is somewhat related.

3.11.20 Vibroacoustic, Temperature—Method 523.0

Method 523.0 is based on the combined effects of temperature, vibration and other operating stresses experienced by an externally carried aircraft store. These combined effects may interact to give effects that are not predictable from the results of single environment tests, but which occur in actual service use.

This test is based on mission profile, temperature profile, and vibration profile over a mission time base.

3.12 Fluid Power Connectors

Interconnection of hydraulic components is accomplished with fittings, tubing, and hoses. These connectors must be rated to maintain structural integrity with some degree of safety margin. Many of the fittings and hoses used in hydraulic systems are manufactured to SAE standards. These standards also govern certain tests that must be applied to each connector.

3.12.1 Fitting Tests

The following tests are required for fittings to meet SAE standards:

Wrenching Test—Steel nuts when assembled without tubing into mating brass fittings that are held securely by a suitable means, such as in a vise, shall be capable of being tightened by means of a standard open end wrench, having an opening as listed to the minimum torque values without failure (rounding) of the hexagon corner. (Torques listed are for testing only, not installation torque.)

Nominal Tube OD (in.)	Minimum Torque (in.-lbf)	Maximum Wrench Opening (in.)
1/8	60	0.322
3/16	120	0.384
1/4	150	0.446
5/16	180	0.510
3/8	210	0.636
7/16	300	0.699
1/2	400	0.763
5/8	500	0.888
3/4	650	1.077

(These ratings apply to Automotive Inverted-Flare design fittings.)

Dimensions, Tolerances, Surface Finish, Workmanship and other specifications outline the minimum requirements for fittings to be classified as per SAE standards.

3.12.2 Hose Tests

The following tests are outlined for hydraulic hose per SAE J517b.

Qualification Tests—For qualification to this specification hose or hose assemblies or both shall conform to the following tests and requirements:

1. Dimensional Check Test (all samples)—shall conform to dimensions listed.

2. Proof Test (all samples)—shall not leak at the proof pressure.

3a. Change in Length Test (one sample)—shall not exceed +2% to −4% change when pressurized to operating pressure.

3b. Change in Length Test (one sample)—shall not exceed +2% to −4% change on hose sizes ¼ in. nominal ID and smaller nor exceed +2% to −4% change on hose sizes 5/16 in. nominal ID and larger, when pressurized to operating pressure.

3c. Change in Length Test (one sample)—shall not exceed ±3% change when pressurized to operating pressure.

4. Burst Test (one 18 in. assembly)—shall not leak or fail below the minimum burst pressure.

5. Leakage Test (two 12 in. assemblies)—shall not leak or fail.

6a. Cold Flexibility Test (one assembly)—shall exhibit no cover cracks or leakage. Exposure shall be at −40°F.

6b. Cold Flexibility Test (one assembly)—Shall exhibit no cover cracks or leakage. Exposure shall be at −40°F. On sizes larger than ¾ in., tube and cover samples may be substituted for the hose flexing test.

7a. Oil Resistance Test—after 70 hr immersion at 212°F in ASTM No. 3 oil, the volume change of hose inner tube and cover specimens shall be between 0% and +100%.

7b. Oil Resistance Test—after 70 hr immersion at 212°F in ASTM No. 3 oil, the volume change of the hose inner tube and cover specimens shall be between −5% and +35%.

7c. Oil Resistance Test—after 70 hr immersion at 212°F in ASTM No. 3 oil, the volume change of the hose inner tube and cover specimens shall be between −15% and +35%.

8. Ozone Resistance Test (two samples)—after 70 hr exposure in an atmosphere comprised of 50 parts ozone per 100 million parts of air at an ambient temperature of 100°F, specimens shall not show evidence of cracking or deterioration when viewed with seven power magnification while still in a stressed condition.

9. Resistance to Vacuum Test (one sample)—after exposure for 5 min at 25 in. Hg (absolute pressure of 17 kPa), there shall be no evidence of hose blistering or collapse.

10. Corrosion Test (two samples)—shall not leak or fail below the minimum burst pressure.

11. Electrical Conductivity Test—the maximum leakage shall not exceed 50 μA when subjected to 75 kV/ft (246 kV/m) for 5 min. (This test shall not be applicable to hose with pinpricked outer cover.)

12a. Impulse Test (four unaged assemblies)—hose assemblies, when tested at 125% of operating pressure for hose sizes 1 in. nominal ID and smaller and 100% of operating pressure for hose sizes 1¼ in. nominal ID and larger, with 200°F circulating petroleum base test fluid, shall withstand a minimum of 150,000 cycles without leakage or other malfunction.

12b. Impulse Test (four unaged assemblies)—hose assemblies, when tested at 133% of operating pressure (except no test pressure in excess of 5000 psi to be used), with 200°F circulating petroleum base test fluid, shall withstand a minimum of 200,000 cycles without leakage or other malfunction.

12c. Impulse Test (four unaged assemblies)—hose assemblies, when tested at 133% of operating pressure, with 200°F circulating petroleum base test fluid, shall withstand a minimum of 200,000 cycles without leakage or other malfunction.

12d. Impulse Test (four unaged assemblies)—hose assemblies, when tested at 125% of operating pressure for hose size 7/8 in. nominal ID and smaller and 100% of operating pressure for hose sizes 1⅛ in. nominal ID and larger, with 200°F circulating petroleum base test fluid, shall withstand a minimum of 150,000 cycles for hose sizes ⅞ in. and smaller and minimum of 100,000 cycles for hose sizes 1⅛ in. nominal ID and larger, without leakage or other malfunction. Hose sizes 1⅛ nominal ID and larger shall be tested straight.

12e. Impulse Test (four unaged assemblies)—hose assemblies, when tested at 125% of operating pressure, with 200°F circulating petroleum base test fluid, shall withstand a minimum of 150,000 cycles without leakage or other malfunction.

12f. Impulse Test (four unaged assemblies)—hose assemblies, when tested at 133% of operating pressure with 200°F circulating petroleum base test fluid shall withstand a minimum of 200,000 cycles for sizes ⅜ in. and ½ in., and 300,000 cycles for all other sizes without leakage or other malfunction.

12g. Impulse Test (four unaged assemblies)—hose assemblies, when tested at 133% of operating pressure with 200°F circulating petroleum base fluid shall withstand a minimum of 400,000 cycles without leakage or other malfunction. Hose sizes ³⁄₁₆, ¼, and ⅜ in. are not usually impulsed as these sizes are not recommended for systems with conventional hydraulic surges.

12h. Impulse Test (four unaged assemblies)—hose assemblies, when tested at 133% of operating pressure with 200°F circulating petroleum base test fluid shall withstand a minimum of 400,000 cycles without leakage or other malfunction. Hose sizes ³⁄₁₆, ¼, ⅜, and ½ in. are not usually impulsed as these sizes are not recommended for systems with conventional hydraulic surges.

13. Visual Examination (all samples) Inspection Tests—inspection tests listed below shall be performed on two samples representing each lot of 500–10,000 ft of bulk hose or 100–10,000 assemblies. Lots of less than 500 ft of hose or 100 assemblies need not be subjected to these tests if a lot has been tested and met the requirements within the previous 12 month period. Requirements for all hose classifications shall be the same as for corresponding Qualification Tests:

1. Dimensional Check Test
2. Proof Test
3. Change in Length Test
4. Burst Test

In addition all hose or hose assemblies or both made therefrom shall be subject to visual examination.

3.12.3 Quick Disconnect Tests

Quick disconnects follow the same general test requirements described for fittings and hose as previously discussed. However, a few specialized tests are conducted which apply to this special connector.

Force to Connect and Disconnect Some disconnects can be connected under pressure (those with internal check valves) and some cannot. In either instance the force required at no pressure or rated pressure is tested for to determine operator ease or difficulty of operation.

Connect and Disconnect Cycling This endurance test determines the amount of cycles prior to any degradation or failure of either the male or female half.

Tensile Strength The mechanical design limit encountered under pressure. This is a destructive test, where pressure is continuously increased until failure occurs.

Swiveling The amount of rotation encountered or possible while pressurized or at no pressure.

Air Inclusion During connection and disconnection, it is possible to pull in air or create a void that can be introduced into the hydraulic system. Air in a system can cause excessive wear, cavitation and erratic system performance. This test determines the potential for this condition.

Fluid Loss Another condition that quick disconnects may encounter due to design variations is fluid loss during connection or disconnection. This is usually encountered when volume from the check valve of either half is displaced momentarily either allowing fluid under pressure to escape, or by displacement of the check poppet itself.

Vibration Because this connector is frequently used, it is more subject to potential conditions that affect its overall life. In systems with high degrees of vibration present, overall life can be affected. This test determines the connectors mechanical integrity usually applied at vibration levels which simulate actual field conditions.

Hydraulic Shock or Surge Flow this condition is simulated to test resistance to short duration, pressure, or flow peaks such as seen when a cylinder bottoms out.

Standard Operating Tests these tests are consolidated under this heading because they apply to the overall operating limits of a quick disconnect and have been discussed previously. These tests include maximum and minimum pressures, maximum and minimum temperatures, maximum flow, pressure drop, internal and external leakage, burst pressure, proof pressure, maximum vacuum, and structural integrity.

Component and System Accuracy

We have discussed component tests, instrumentation, and test procedures. Now, it is important to review the overall picture once the test procedures and a schematic have been designed. Instrumentation accuracy is only the initial step in constructing an accurate test system. The minute characteristics of a component under test or a component within the test circuit may create test data that are inconsistent or totally inaccurate.

I stress this point in order to maintain the awareness necessary to find potential errors within a test system design prior to its construction. It is much less expensive to erase an idea on paper than to absorb the cost of an expensive piece of equipment.

Each test system is different. It is not often that two tests are similar. This is due to the changes in flow, pressure, temperature, fluid, component operation, etc. Therefore, each system should be completely reevaluated even if there is a similarity to another working system. This chapter will define the evaluation of a system that is necessary to ensure overall accuracy. In addition, calibration of instrumentation is also included to further define potential error.

4.1 General Accuracy Terms

Most instruments have ratings to define the various effects that cause deviation of accuracy. These ratings usually apply to instruments of different types, so this section will be devoted to explaining what these general terms mean. Any terms that apply specifically to one type of instrument will be discussed in the section pertaining to that instrument.

4.1.1 Accuracy of Reading

The curve illustrated in Fig. 4.1 depicts the combined effects of several factors. An instrument with a rating of percentage of reading will maintain a constant deviation tolerance throughout the full range output. For example, a flowmeter with a range of 10–100 gpm, which has an overall accuracy rating of ± 1% of reading, will have a deviation of ± 1 gpm at 100 gpm and ± 0.1 gpm at 10 gpm.

4.1.2 Ambient Effects

Ambient effects can apply to any external condition surrounding a transducer. The accuracy may be affected by temperature, pressure, humidity, etc. Each transducer rating should be studied carefully if it is to be applied in a situation with variable ambient conditions.

4.1.3 Asymmetry

An asymmetry rating is used to describe the difference in sensitivity on either side of null output. This generally relates to the mechanical differences between both sides of a transducer.

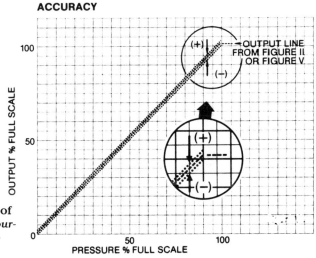

Fig. 4.1. Accuracy of reading curve. *(Courtesy Barksdale Controls.)*

4.1.4 Best Fit through Zero

Best fit through zero defines the maximum deviations expected in a nonlinear output device. This includes the effects of nonlinearity, hysteresis, and nonrepeatibility.

4.1.5 Combined Error

Combined error is the maximum deviation expected from more than one error-causing effect. Some transducers are rated in this manner to provide an easy method of predicting maximum error.

4.1.6 Compensation

Compensation defines a method used to correct for an error-causing effect. Generally, a transducer can be mechanically designed to negate certain effects, or the electronics can allow correction for potential error. A transducer rating will normally describe a range that will be corrected automatically. Outside of this range, the error-causing effects will reduce accuracy.

4.1.7 Creep

Creep is the change of transducer output over a period of time with all external and internal conditions remaining constant.

4.1.8 Creep Recovery

Creep recovery is the time it takes for a transducer to return to its initial no load output after all loads have been removed. This term is usually expressed as a percentage of rated output.

4.1.9 Drift

A change in transducer output with no change in load is called drift. Drift can apply to any characteristic. A common example is pressure transducer zero drift. When power is applied to a transducer at no load, you will read one output value. After 5 or 10 min, still at no load, you will read a different output. (Not all pressure transducers drift; however, this is one area where the problem is common.)

4.1.10 Frequency Response

Frequency response is the range of frequencies that a transducer will follow the sinusoidally varying mechanical input within the specified limits.

4.1.11 Full-Scale Output

An instrument with a rating that reads percentage of full scale output will provide the same accuracy as the flowmeter described in Section 4.1.1, but only at the 100 gpm reading. At 100 gpm, the expected deviation would be ± 1 gpm, but at 10 gpm, the deviation remains ± 1 gpm.

To calculate the actual accuracy of a full-scale-accuracy instrument, you must first determine the maximum output of the device. The percentage accuracy rating is now applied to this maximum value. The resultant value applies to any point within the full range of the instrument. This is a common error found when a transducer is selected for a high-accuracy application. It is often assumed that ± 1% means ± 1% of reading, when in fact the error is much greater at lower output.

4.1.12 Hysteresis

Hysteresis is the difference in transducer output between reading the same value when coming up to the value or going down to the measured value (i.e., increasing load reading versus decreasing load reading).

4.1.13 Nonlinearity

The curves previously shown depict a straight line curve from zero to full output. As an example, say a pressure transducer with a range of 0–100 psi had an output of 0–10 Vdc. At 10 psi, the output would be 1 Vdc, at 50 psi, output would be 5 Vdc, and so on.

Some outputs are not as linear as the curves depicted. These outputs are noted as nonlinear (not a straight line) outputs. In particular, thermocouples and flowmeters have nonlinear outputs. Usually electronic signal conditioners correct this output to increase accuracy.

4.1.14 Output (Sensitivity) Tolerance

A transducer also has a tolerance on full rated output sensitivity. This is defined as a tolerance of plus or minus of a measured point.

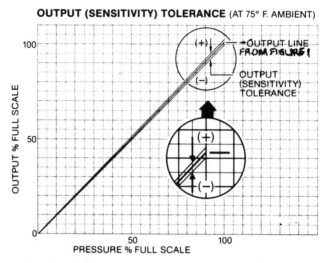

OUTPUT (SENSITIVITY) TOLERANCE (AT 75° F. AMBIENT)

Fig. 4.2. Output (sensitivity) tolerance curve. *(Courtesy Barksdale Controls.)*

Figure 4.2 depicts this curve where the deviation increases as output increases. At full scale output, the greatest degree of deviation occurs, but the actual percentage of reading deviation remains the same.

4.1.15 Repeatability

Repeatability describes a transducer's ability to repeat the same output at the same measured value. Many transducers have mechanical or electrical values which have tolerances that slightly vary the output in a nonpredictable or controllable fashion. For example, if a pressure transducer is brought up to 3000 psi, then the pressure is dropped and brought back up again to 3000 psi, the transducer might read 2997 or 3005 psi. This effect does not encompass many of the other effects described. In fact, most other effects are additive to repeatability.

4.1.16 Resolution

Resolution defines how small a value a transducer will measure. Resolution generally applies to applications where transducers with a wide range are used to measure extremely small incremental values. For example, imagine an LVDT with a stroke of 1 in. If you needed to determine if a component moved 0.001 in., you would need better than 1000:1 resolution. At 1000:1 resolution, your reading of 0.001 in. would be ± 100% accuracy. In order to achieve an accuracy of ± 10% while reading 0.0015 in. you would need resolution at 10,000:1 or ± 0.0001 in.

Resolution applies to the transducer, the signal conditioner, and the display. Each device will have a threshold of reading. As an example of display resolution, consider a tachometer application. A high-speed turbine had a magnetic pickup and a six-tooth gear attached to the output shaft. The turbine was designed to reach speeds of up to 150,000 rpm. At 150,000 rpm the input frequency to the display was 15,000 Hz. The pulses were exactly measured so no resolution factor was applied at this point. The digital display, however, was a 4½ digit display with an additional "dummy" zero used to depict rpm with no multiplier required. The dummy zero did not change, so if the turbine was rotating at 148,995 rpm, the display would read 148,990 rpm. In addition to this, the frequency conversion circuitry would only allow a change of the last digit by 4. So, with the circuitry resolution in combination with the dummy zero, resolution of the display was 40 rpm. This may seem like a lot, but considering that normal speeds were in the 100,000–150,000 rpm speed range, ± 40 rpm was negligible.

Another area to be concerned with is programmable controllers and other digital devices. Rather than sensing an analog output, output consists of steps of voltage or current following a straight line output. (See Section 4.6.1.)

4.1.17 Stabilization Period

The stabilization period is the time required before a reading may be taken without an error caused by slow response characteristics. The response problems can be electrical, mechanical, or component response. In many test systems, readings taken before the system can stabilize are usually incorrect and highly inaccurate. In these systems, readings taken at the same time interval are usually inconsistent and unpredictable.

4.1.18 Standard Test Conditions

Often, transducer manufacturers select a particular reference point at which the calibration of the instrument is conducted. This generally applies to temperature, ambient conditions, and, sometimes, pressure, which are the standard test conditions. If your transducer is going to be applied in a system far from standard calibration, correction factors may be required. You can usually request that your transducer be initially calibrated to your specific conditions.

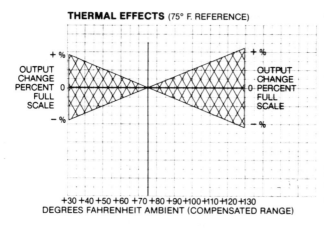

THERMAL EFFECTS (75° F. REFERENCE)

+30 +40 +50 +60 +70 +80 +90 +100 +110 +120 +130
DEGREES FAHRENHEIT AMBIENT (COMPENSATED RANGE)

Fig. 4.3. Thermal effect curve. *(Courtesy Barksdale Controls.)*

4.1.19 Thermal Effects

A transducer's output can be affected by a change in ambient temperature. This change affects both zero balance and sensitivity. Thermal effects are expressed as percentage change of full scale output per degree Fahrenheit ambient change. Figure 4.3 depicts how these percentages relate to the total compensated range. In this case the transducer was calibrated at 75°F. If the ambient temperature was at 75°F, the effect on accuracy due to ambient temperature would be zero.

4.1.20 Thermal Sensitivity Effect

A change in output voltage at full output of the transducer also occurs due to temperature change. This effect is illustrated in Figure 4.4. Again, this effect is expressed as percentage of actual reading. Therefore, the amount of deviation increases at the higher output range, while actual percentage of output remains constant. In other words, at the lower end of the output scale, the deviation is less significant.

4.1.21 Thermal Zero Effect

Figure 4.5 illustrates the effect caused on zero output with a change in ambient temperature. This effect is rated at a plus or minus percentage of output. The curve represents the anticipated deviation of zero output voltage and output line at 75°F.

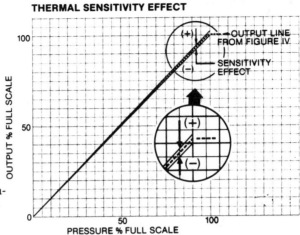

Fig. 4.4. Thermal sensitivity curve. *(Courtesy Barksdale Controls.)*

4.1.22 Zero Balance

Zero balance is a term used to define a transducer output at zero output. A perfect situation is represented as a dotted line in Fig. 4.6 The solid lines represent a plus or minus tolerance. In this case the tolerance is represented as ± 2.5% of full scale output. The area between the lines is the actual potential output deviation through the full range of the transducer that would be acceptable under the specified zero balance tolerance. Therefore, any outside effect that shifts zero balance from the "perfect situation" would cause the deviation as shown.

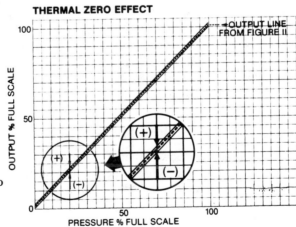

Fig. 4.5. Thermal zero effect curve. *(Courtesy Barksdale Controls.)*

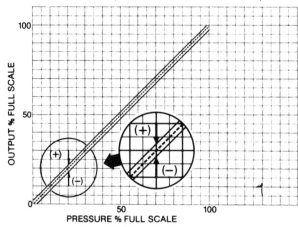

ZERO BALANCE (AT 75° F. AMBIENT TEMPERATURE)

OUTPUT % FULL SCALE

PRESSURE % FULL SCALE

Fig. 4.6. Zero balance curve. *(Courtesy Barksdale Controls.)*

4.2 Flow Measurement

Flowmeters are probably the most misunderstood and misapplied instrument used in fluid power testing. Many techniques have been developed over the years to convert the flow of liquid or gas into known engineering units. The method used to measure flow depends on the flow range desired, the fluid (or gas) type, the accuracy required, and the operating pressure. Other factors such as viscosity swing due to temperature and pressure variations or fluid stability over excessive operating conditions play an important role in flowmeter accuracy. If a flowmeter is calibrated for a specific viscosity and viscosity changes, the calibration is no longer valid. Unfortunately, each flowmeter type and size is affected differently by viscosity change. This makes it difficult to understand exactly what the resultant error will be with a known viscosity change. The effect could be minimal, or it could produce an error too great to be used. Therefore, if fluid changes are suspected, it is advisable to contact the flowmeter manufacturer for information on accuracy change.

4.2.1 Flow Measurement Accuracy

Since most fluidpower testing is conducted with turbine flowmeters, this section will be devoted mainly to this flowmeter type. However, a great deal of discussion will apply to other designs that are affected by viscosity. This includes variable area rotameters and pressure drop meter designs.

As discussed in Chapter 2, the turbine flowmeter incorporates ro-

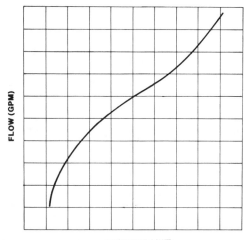

Fig. 4.7. Pulse versus flow
rate curve.

tors that are spun by fluid or gas passing through the meter. Exactly
how the rotor spins depends on the fluid, the flowrate, and the
range of the meter. The rotor blade passes a magnetic pickup, which
produces a pulse. The pulse is proportional to the speed of the rotor
spin (or flow). If the pulses follow a straight line on a pulse versus
flow rate graph (Fig. 4.7), then the unit would be said to have a linear output.

Linear output is not common (if ever) in this type of meter design.
This is an area where flowmeters are commonly misapplied. Signal-
conditioning electronics are used to take this pulse output and correct the readout based on stored information about the curve created by the fluid used in the system. This correction is highly dependent on accurate calibration for the fluid used. Often a flowmeter
calibration laboratory only has one or two fluids used for calibration
and actual data must be extrapolated to define the variation between the calibration fluid and the fluid used.

Pulse data, usually referred to as frequency output (Hz), can be
manipulated to correct the curve by several methods dependent on
how far off the linear curve the application wanders. In some cases,
linearization techniques are not needed, where viscosity remains
fairly constant, and a wide flow range is not needed. Where one flow
point is only necessary, highly nonlinear meters can be used if accurately calibrated at one point. It is in this situation that the repeatability of the meter is important for accurate measurement. Frequency output can be corrected by an equation, but most often, the
signal conditioner has point-by-point calibration. This technique

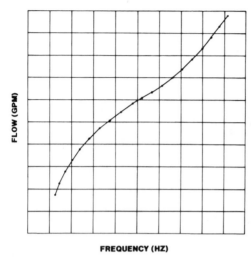

FLOW (GPM)

FREQUENCY (HZ)

Fig. 4.8. Point by point calibration curve.

incorporates multiple points along the curve with a straight line connecting each point (Fig. 4.8). The electronics sense actual frequency, then convert this frequency either digitally or using digital and analog conversion to produce a corrected value.

Flowmeter systems of this type are fairly expensive, and, in many cases, the best accuracy possible is not always financially practical. One method used to overcome the linearization cost is to use a "best fit straight line curve." In this instance the nonlinear curve is intersected by a straight line (Fig. 4.9) that best follows the curve. Accuracy variation is then the distance between the straight line and the part of the curve that is farthest from the straight line. This holds true, of course, with meter calibration at a specific viscosity and with measurement at the exact same viscosity.

Well, now that we have calibrated our meter at one specific viscosity, what happens when temeprature changes? Temperature, of course, affects viscosity. You can easily determine the fluid viscosity change by looking at data provided by oil manufacturers or in a number of hydraulic reference books. Once calibration is accomplished at one viscosity point, the meter can again be calibrated at several other viscosities that relate to the fluid characteristic. These data can then be fed into the signal-conditioning electronics, which also sense actual fluid temperature with an RTD or thermocouple. The curve is then corrected not only for flow rate versus frequency variation, but also for temperature variation. The combination of these two techniques produce high accuracy ($\pm$ ¼% of reading accuracy) flowmeter systems that are unaffected by temperature varia-

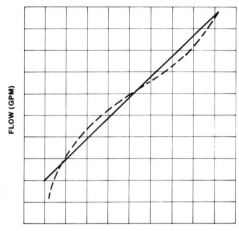

Fig. 4.9. Best fit straight line curve.

tions. This accuracy is typically sufficient for even the most demanding fluid-power test conditions.

Now for the bad news. Viscosity is also affected by pressure variations. This effect, not known by many a few years ago, can be more severe than temperature. This problem can make the high-accuracy systems just discussed highly inaccurate. I presented a paper outlining this problem at the 1984 National Conference on Fluid Power. It was published in the conference proceedings and I again offer a reprint for your review of this problem.

PRESSURE EFFECT ON OIL VISCOSITY
AND THE EXTREME INACCURACIES
OF FLOWMETER READINGS
CAUSED BY THIS EFFECT

Abstract

A problem exists in the application of any viscosity-sensitive flowmeter when testing components or systems under varying pressure conditions. Many flowmeter manufacturers who boast ¼% of reading accuracies of their equipment have admitted to 4–25% error in a system that varies from 0–5800 psi. A letter of mine published in the November 1982 issue of *Hydraulics & Pneumatics* produced information from top technical authorities in our field that was not consistent nor conclusive in the answer to my questions. Since that time, I have conducted a "search for the answer" only to find that our industry does not have the instrumentation necessary to correct this problem, nor do we have a solution to this problem. It is my goal in submitting this paper to make the industry aware of this problem, challenge the existing theories and equations currently being employed, and offer my solutions with the current limited instrumentation now in existence.

Introduction

This paper may read more like a story than a technical presentation, simply because my purpose is only to make this problem known. Relating these events in chronological order may stress the degree of dismay experienced by myself over the past two years (as well as anyone else aware of this problem).

The pressure–viscosity effect was first learned by myself while visiting a manufacturer of hydraulic pumps. They had recently been told by a flowmeter manufacturer that there was a degree of error in flowmeter readings due to changes in pressure. Of course, they were quite concerned that any test results documented with so-called "high-accuracy" flowmeters may not be quite that accurate.

Imagine a piston pump with a 6000 psi maximum pressure capacity tested for flow versus pressure. If the flowmeter were installed in the pressure line, flow readings would be taken anywhere in the 0–6000 psi range. What would this do to overall accuracy? No one was really quite certain.

Initial Investigation

In order to clarify the existence of a problem, several oil companies and a flowmeter manufacturer were contacted. The following information was received from technical support personnel at each company.

Oil Company #1: These people stated that the only problem with varying pressures was the compressibility factor of oils based on the paper "Computation of some Physical Properties of Lubricating Oils at High Pressure," Dow and Fink, *Journal of Applied Physics*, May 1940.

Percentage Reduction of Oils under Pressure

Pressure (psi)	Temperature (°F)			
	60	100	140	180
500	0.208	0.217	0.225	0.231
1000	0.413	0.432	0.448	0.458
5,000	1.93	2.04	2.14	2.20
10,000	3.55	3.81	4.02	4.16
15,000	4.83	5.29	5.65	5.89

Oil Company #2: These people referenced a 1953 ASME paper entitled "Pressure vs. Viscosity Report" that states 10,000 psi will increase density by 8%. One report they had concerning one of their oils stated:

Pressure (psi)	Viscosity (SSU)
0	1,100
5,000	3.500
10,000	8,500

Oil Company #3: A characteristic of one of their oils was:

Pressure (psi)	110°F	120°F	130°F
	Viscosity (Centipoise)		
0	25.26	20.16	16.87
2,500	35.38	27.89	23.09
5,000	48.79	38.00	31.13

Oil Companies # 4, 5, and 6: "We never heard about that problem."

Turbine Meter Manufacturer: These people stated that organic lubricating oils and hydraulic fluids change in viscosity as pressure increases. The following factors apply to these fluids only and do not apply to synthetic fluids.

Centistokes @ psig	Multiply by
0	1.0
1,000	1.15
2,000	1.32
3,000	1.50
4,000	1.75
5,000	2.0
6,000	2.3

"The above information is valuable in the prediction of turbine meter performance and sizing."

After reviewing the preceding information confusion began to set in, so I sent a letter to the editors of *Hydraulics & Pneumatics* magazine. They agreed to publish my dilema in the editorial column. The following is how my letter appeared in the November 1982 issue.

Hydraulic Flow Measurement

Our company has manufactured hydraulic test stands and power systems for the industrial and military aerospace marketplace since 1964. We produce equipment that is extremely accurate in the measurement and control of certain values . . . (which) can be pressure, flow, torque, speed, displacement, velocity, or any other imaginable engineering unit. . . . Therefore, we rely on the manufacturers of these sensing devices for accurate information regarding their equipment and their function within a system.

It has been recently brought to my attention that a problem exists in the relationship of oil viscosity vs. oil pressure. Turbine flowmeter systems have allowed the hydraulic industry to measure flows to accuracies of $\pm \frac{1}{4}\%$ of reading. This is accomplished through special calibration based on the type of oil used (viscosity characteristics) at a specific temperature giving the unit a "best fit" straight line curve over the intended flow range. As we all . . . should know, any variations of oil temperature change the oil viscosity, thus resulting in erroneous or misleading flow readings . . . we have another problem. Pressure also changes oil viscosity. To what degree I have not been able to determine, but it may be substantial.

We contacted several oil companies and (makers of) turbine flowmeters, posed our question, and received information that was either not consistant or pertinent to the question.

What happens when pressure variations act upon temperature variations and the oil starts to break down?

I call on your respected readers to pool their knowledge of this subject, put aside their competitive secrecy and help this industry in its hour of need.

Responses

Several readers either phoned or sent letters detailing the information that they had obtained. Some information was duplicated, but for the most part, the information provided did not solve the problem, and there seemed to be some disagreement on the method of pressure–viscosity correction.

It seems as though a commonly known equation can be used to calculate the change in viscosity due to pressure variations. This equation is

$$u = u_0 e^{\beta p} \quad \text{or} \quad u_p = u_0 e^{\beta p}$$

where
  u = viscosity at pressure p

  u_0 = viscosity at atmospheric pressure

  e = mathematical constant of 2.718

  β = pressure coefficient of viscosity

  p = pressure

  u_p = absolute viscosity at pressure p

This would be a very useful equation if the pressure coefficient of viscosity (β) was known, but it varies based on the type of oil, the temperature of the oil, and so on. This coefficient is not supplied by oil manufacturers nor is it considered an error-causing factor by most in the application of flowmeters. Only through individual experimentation can this factor be determined, for the particular oil to be used.

Another equation used to predict the viscosity increase with pressure is

$$\log_{10}\frac{u_P}{u_0} = c^{P}$$

A fluid that had an increase in viscosity of about 2.25 at 5000 psi results in a value of $c = 7 \times 10^{-4}$/psi. Again this formula is used for approximation purposes only.

Figure 4.10 depicts various pressure versus viscosity characteristics of different typical hydraulic oils (temperature = 100°F, viscosities in centipoise, pressures in psi).

For further evaluation of the pressure effect the viscosity characteristics of Gulf Harmon 78EP 2190 TEP oil were used. Figure 4.11 depicts this oil's particular characteristics.

Some other interesting points to consider:

1. Viscosity increases with increasing pressure, the rate of increase being greater the lower the temperature.
2. Variation of viscosity with pressure is greatly influenced by the chemical composition of the fluid. Petroleum and castor oil base fluids generally show an appreciable increase in viscosity with pressure while water base fluids show small changes.
3. The differences between a California oil and a Pennsylvania oil with the same viscosity at atmospheric pressure can be significant. For example, several sources indicate

$p=14.7$	$p=1,000$		$p=5,000$		$p=10,000$	
u_o	u	$B \times 10^4$	u	$B \times 10^4$	u	$B \times 10^4$
28.3	33.4	1.65	60.0	1.49	121	1.45
46.4	56.6	1.99	199	1.89	293	1.84
83.1	101	1.99	215	1.90	522	1.84
122	151	2.15	345	2.08	933	2.03
288	351	1.99	714	1.80	1,560	1.69
422	515	1.99	1,050	1.80	2,290	1.69
579	730	2.30	1,630	2.16	4,070	1.96

Fig. 4.10. Pressure versus viscosity characteristics.

that California crude may be twice as viscous at 10,000 psi.

4. Hydraulic fluids employing viscosity-index improvers, such as polymeric thickeners, have complex non-Newtonian viscosity behavior. Either temporary or permanent viscosity losses can take place when the fluid is subjected to high shear rates. Temporary loss occurs in laminar flow when the long chain polymeric molecules become aligned and stretched, and thus lose their thickening power. Removal of the shear stress permits the molecules to revert to their original random configuration, and the fluid regains its initial viscosity. This process is fully reversible.

PRESSURE		VISCOSITY (CENTIPOISE)	
(Bar)	PSI	$22^{\circ}C$ $71.6^{\circ}F$	$40^{\circ}C$ $104^{\circ}F$
0	0	197.9	69.1
100	1450.4	251.0	86.1
200	2900.8	318.6	107.4
300	4351.2	404.4	134.0
400	5801.6	513.3	167.2
600	8702.4	827.2	260.3

Fig. 4.11. Viscosity characteristics.

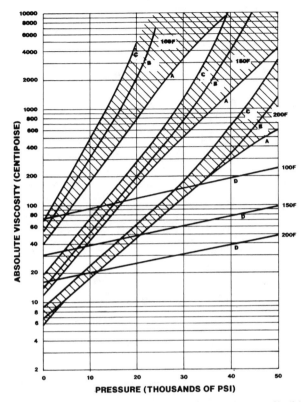

Fig. 4.12. Pressure–viscosity isotherms: (a) petroleum oil, (b) phosphate ester, (c) phosphate ester base, (d) water–glycol base.

The temporary loss decreases as temperatures increase. Permanent viscosity loss results from severe turbulent flow, cavitation, large pressure drops across sharp edged orifices, and scission between mating parts. These conditions can stress molecular chains to the breaking point, creating shorter less viscous fractions. The rate of permanent viscosity loss is dependent upon the severity of operating conditions and the stability of the polymeric molecule. Sheared fluid has greater viscosity–temperature dependence than new, unsheared fluid.

5. Entrained or dissolved air will cause certain fluids to exhibit different viscosity/temperature values. This poses a serious problem as there is some disagreement among fluid experts concerning how to determine that all intermolecular air has been removed.

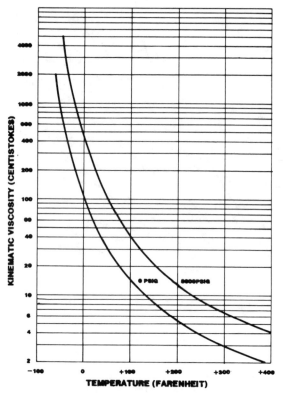

Fig. 4.13. Viscosity characteristics (temperature change).

The Effect on Flowmeters

Before continuing, it is important to understand how viscosity affects a turbine flowmeter's performance. This performance varies based on the individual flowmeter itself and on the flow ranges passed through any particular size meter. At lower viscosities, laminar flow is usually present and because of this, the rotor and other critical surfaces have a low drag coefficient. As viscosity is increased, however, the oil may become turbulent, greatly increasing drag. This increased drag will reduce the number of pulses from the flowmeter at the same flow rate. These pulses are converted to read actual flow based on the number of pulses per gallon of flow.

It was estimated by a flowmeter manufacturer that a meter calibrated at 10.0 gpm at 69.1 centipoise would read approximately 7–7.5 gpm at 167.2 centipoise. If that same flowmeter were cali-

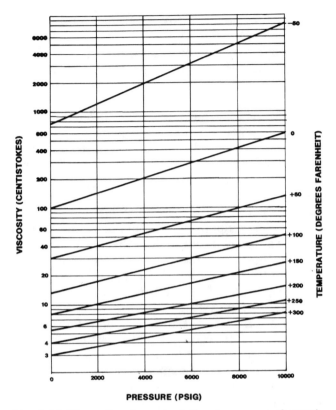

Fig. 4.14. Viscosity characteristics (pressure change).

brated at 10.0 gpm at 167.2 centipoise, it would read approximately 13 gpm at 69.1 centipoise. This is an incredible change in the flow reading of an oil which is at a constant temperature. Now consider what readings would be seen with the additive effect of temperature upon the oil during these pressure variations of 0 to 5801.6 psi.

Two other sources estimated that the error in reading would be approximately 4–8% depending on the size of the turbine block and where the reading was taken along the actual range of that particular meter.

It is also important to note at this point that this problem not only applies to turbine flowmeters, but any flowmeter which works on the principle of pressure drop such as a variable-area glass tube rotameter or a differential pressure measurement device across a fixed orifice.

Meters that do not have this inherent problem are fixed displacement meters. They don't care about what viscosity the fluid is, just how much volume passes through over a period of time. Unfortunately they don't solve all problems. They generally have higher pressure drops, are noisy at higher flows, and are more expensive.

The main problem is that most applications incorporating viscosity-sensitive flowmeters, don't compensate for fluctuating viscosities. Most technical publications will have sections devoted to the change of viscosity due to temperature influences, but very few even mention the pressure–viscosity effect. Those publications that do provide limited information are so general that the figures are virtually useless.

The problem is compounded by the fact that no instrumentation is available to continually adjust the flowmeter reading based on viscosity samples. Currently, viscosity measuring devices are rated up to 1000 psi maximum, while high-pressure viscosity measuring devices, such as the Ruska viscometer, are designed to take "one-time" readings.

From this information it seems as though the effect of pressure versus viscosity should be considered a greater factor in the selection of a flowmeter than temperature versus viscosity. Most production tests on components such as pumps are conducted with oil temperatures somewhat stabilized, while flow is checked at various pressures.

Solutions to the Problem

Several methods could be applied to correct this problem:

1. When ordering a flowmeter, first test the viscosity of your oil at the exact temperature AND pressure at which you would use it. Have the meter calibrated at that viscosity and test only at that temperature and pressure. (This certainly limits the versatility of your test capabilities.)

2. Install your viscosity-sensitive flowmeter in your return line. This will almost completely eliminate pressure/viscosity problems while temperature variations are still present. Turbine flowmeters are generally too small for return lines (velocity sizing of plumbing ID) causing one problem, while component leakage on the pressure side of the circuit may also cause inaccurate readings.

3. Request correction factors from the flowmeter manufacturer. These correction factors will only be applicable to the flowmeter in your possession as each meter will have

a different correction factor. Employing this method will require you to test your oil at various temperatures and pressures to determine viscosity so that once a reading has been taken, you can refer to your test results to apply the proper correction factor.

4. Purchase a temperature-correction module and test at only one pressure. This will greatly reduce erroneous readings by limiting and correcting for viscosity fluctuations. (Note: Not all flowmeter manufacturers offer temperature modules.)

5. Wait until pressure-correction modules are available on the market. These modules, when used in conjunction with temperature modules, should offer automatic compensation for all changes in viscosity. The only problem with this concept is that the pressure coefficient for viscosity (β) would have to be programmed into the module for your particular oil's characteristics. (See Fig. 4.15.)

6. Wait for a module that will actually measure viscosity and correct the reading for any given situation. This one module would replace both temperature and pressure correc-

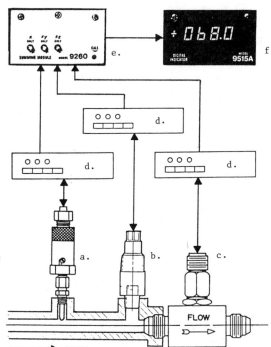

Fig. 4.15. Flowmeter with temperature and pressure compensation:
(A) thermocouple,
(B) pressure transducer,
(C) flow turbine,
(D) signal conditioner,
(E) summation module,
(F) digital indicator.

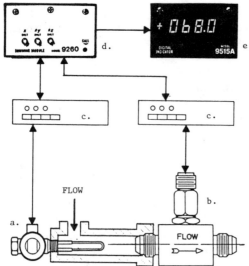

Fig. 4.16. Flowmeter with viscosity compensation:
(a) viscosity sensor,
(b) flow turbine,
(c) signal conditioner,
(d) summation module,
(e) digital indicator.

tion modules and not require any pressure coefficient programming. (See Fig. 4.16). In addition, this device would not require periodic viscosity testing of your oil. This testing would be required to maintain calibration of your flowmeter in the event of permanent viscosity changes caused by oil breakdown, etc.

It is my hope that in compiling these facts, that someone in the instrumentation or hydraulic fluid industry will initiate a program to provide our industry with either the proper instrumentation of fluid characteristics when the pressure–viscosity effect is present. As has been pointed out, there is not total agreement on the methods of solving this problem or even the awareness of this problem's existence.

Acknowledgments

I wish to thank the following people for sharing their knowledge of this subject:

1. James I. Morgan, President, National Fluid Power Association.
2. Paul K. Schacht, Manager, Fluid Power Laboratories, Rexnord Corporate Research and Innovation Group.
3. Bob Haning, Rockwell International.
4. Walter Ernst, Ernst & Associates.
5. Edward A. Uchno, Product Development Engineer, Fenner Stone, Inc.

6. F. Cracknell, Sr. Hydraulic Engineer, DAF Indal, Ltd.
7. W. K. Wilcox, Head Submarine Hydraulic Control Systems Branch, Ship Control Systems and Equipment Division, Naval Sea Systems Command, Department of the Navy.

Bibliography

The following publications reference information concerning the pressure–viscosity relationship.

1. L. B. Sargent, *Lubrication Engineering*, Chapter 11, page 249 (1955).
2. *Lubrication*, Volume 47, The Texaco Co. (1961).
3. James O'Conner, *Standard Handbook of Lubrication Engineering*, Chapter 10, page 8 and Chapter 12, page 10 (1968).
4. Roger Hatton, *Hydraulic Fluids*, Chapter 4, page 99.
5. Backburn, Reethof and Shearer, *Fluid Power Technology*, M.I.T. Press (1960).
6. Fluid Power Research Center (OSU) Report No. 40778 (7 April 1978) Viscosity Versus Pressure and Temperature.
7. Walter Ernst, *Oil Hydraulic Power and Its Industrial Applications*, McGraw Hill (1960).
8. H. H. Zuidema, *Performance of Lubricating Oils*, Reinhold (1959).
9. R. B. Dow, The Effects of Pressure and Temperature on the Viscosity of Lubricating Oils, *Journal of Applied Physics* **8**, pp. 367–372 (1937).
10. SAE Aerospace Information Report No. AIR 1362 Physical Properties of Hydraulic Fluid.

The problem just described is applicable to any viscosity-sensitive meter. Rotameters and other common fluid power meters are all subject to inaccuracy caused by viscosity swing.

One problem many meters also have is slow response. This is another area where flowmeters are frequently misapplied. Turbine flowmeters, in particular, require up to several seconds for updated information. Test systems that do not allow sufficient time for response are acquiring incorrect test results. A system using a turbine flowmeter must be allowed to stabilize prior to data collection.

Turbine meters have their own individual characteristics based on size, flow rate, and manufacturing tolerances. Often catalog data define turbine block response around 50 msec dependent on the meter design and the system itself. The system varies due to loaded or unloaded conditions, where the compressibility of fluid affects response. In addition, response depends on how long it takes a column of fluid to begin moving and then to stabilize. Once the system finally stabilizes, the flowmeter pulses must be measured and the time compared to produce a stable reading. Flowmeter electronics can usually be adjusted to vary the sample time (of the pulse), but too short of an adjustment will produce a reading that bounces, making reading difficult or impossible. In other words, the pulses must be averaged out over a period sufficient to make the reading stable. This timed period can be as low as 100 msec to 3 sec or higher depending on the method of pulse to readout conversion.

Another important factor in maintaining turbine flowmeter accuracy is the use of straight tubing or flow straighteners in and out of the turbine block. Turbulent flow created by elbows or sharp turns in tubing will increase the drag on the rotor, thus affecting pulse output. The actual length of straight tubing required varies from one manufacturer to another, but a general rule of thumb is 10 times the pipe ID in length both in and out. Certain meter designs are unaffected by this phenomenon, but *all* turbine meters require flow straightening.

4.2.2 Flow Calibration Methods

Flow calibration is not an easy task. It is an area that requires years of experience before any accuracy of method or design can be attained. Flowmeters can be tested for on-site calibration verification or must be sent out for calibration to the highest accuracy possible. In both cases equipment used for calibration is matched to standards at National Institute of Standards and Technology (formerly the NBS).

The on-site calibrators vary in style, method, and price (see Fig. 4.17). Some units produce a pulse output to simulate turbine pulses. This will verify that the signal-conditioning electronics is producing a reading appropriate to the frequency input. Some units bypass the signal conditioner, taking a direct reading from the turbine block and reading frequency. Conversion of frequency to flow rate is simple, provided turbine K factor is known. (The K factor is another term for pulses versus flow rate.) In addition, calibrated secondary flowmeter systems are often installed in a bypass loop around on-board flowmeters so that the reading between the two can be compared. But, comparators, as well as individual flowmeters, must be calibrated with highly sophisticated equipment to ensure the best possible accuracy at a specific fluid condition.

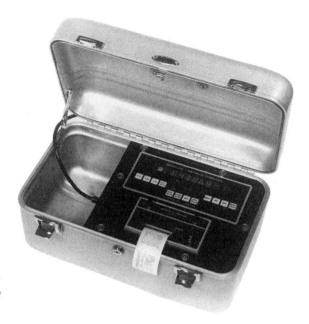

Fig. 4.17. Flowmeter
calibrator. *(Courtesy
Datarate, Inc.)*

There are two basic methods used to calibrate flowmeter systems: static and dynamic. Both systems work with weight (or mass) of the fluid over a timed period. Each portion of the circuit must be extremely accurate, nearly friction free, and absolute. A good example of how this type of system can be misapplied is described in Section 4.9.2.

The Static Weighing Calibration System incorporates a diverter valve, timer, and a weigh tank on a scale. The system is designed to maintain a specific temperature of fluid at a constant flow rate. The diverter valve is designed to channel flow back to system storage or to the weigh tank. Most importantly, the valve is designed not to disturb flow rate during its shift. Any changes in back pressure would create an unstable period, thus affecting total accuracy. The meter under test is installed just prior to the diverter valve so that flow passes through it at any point during calibration.

Flow initially circulates back to system storage until flow, pressure, and temperature are stable. At this point the diverter valve is shifted, and a timer starts during the shift. Tare weight in the scale is stored, and flow now enters the weigh tank until a sufficient volume is collected. At this point the diverter valve shifts, the timer stops, and the resultant end weight is stored and subtracted from initial tare weight. Flowmeter output is then compared to total fluid weight over the timed period. This value is converted to desired flow rate or totalized flow to determine meter accuracy.

There are several obvious areas in a system of this type that require extreme accuracy in order to arrive at the accuracy required for calibration. First, controlled flow rate is mandatory. A system designed to produce constant, steady flow rate is mandatory. Any deviation is directly proportional to overall accuracy. Second, the scale system must be responsive enough to not cause error due to slow response. Next, the timer actuation must be precise so that it does not start too early or too late. Error at this point will create an additive error at the start or finish of the test. Any system leaks will contribute to error as will system turbulence and entrained gas in the fluid. Finally, the system mechanics and collection methods play a vital role in accuracy.

The Dynamic Weighing Calibration System (Fig. 4.18) uses a method where a predetermined weight of fluid is selected and the time needed to accumulate this total weight is recorded. Again, temperature is an important factor, since fluid weight changes with temperature. Figure 4.19 depicts the hydraulic schematic used in the Cox calibration systems. A centrifugal pump provides a constant flow rate through a filter, heat exchanger, flow control, the test meter, and a back pressure control. The back pressure control permits

Fig. 4.18. Flow calibration laboratory. *(Courtesy AMETEK, Inc., Cox Instruments.)*

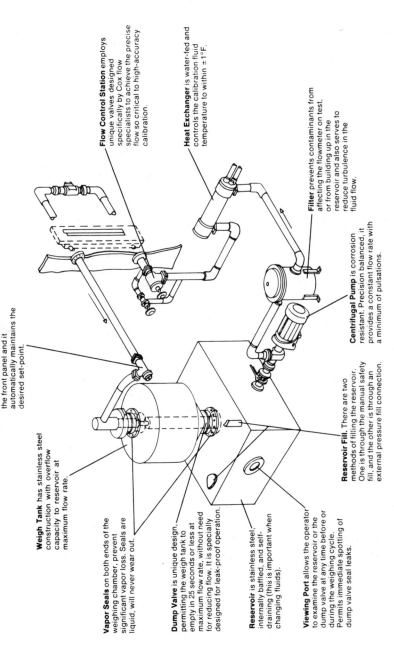

Back Pressure Control prevents cavitation that can occur with "light" fluids. It is adjusted from the front panel and it automatically maintains the desired set-point.

Flow Control Station employs unique valves designed specifically by Cox flow specialists to achieve the precise flow so critical to high-accuracy calibration.

Heat Exchanger is water-fed and controls the calibration fluid temperature to within ±1°F.

Filter prevents contaminants from affecting the flowmeter on test, or from building up in the reservoir and also serves to reduce turbulence in the fluid flow.

Centrifugal Pump is corrosion resistant. Precision balanced, it provides a constant flow rate with a minimum of pulsations.

Reservoir Fill. There are two methods of filling the reservoir. One is through the manual safety fill, and the other is through an external pressure fill connection.

Weigh Tank has stainless steel construction with overflow capacity to reservoir at maximum flow rate.

Vapor Seals on both ends of the weighing chamber, prevent significant vapor loss. Seals are liquid, will never wear out.

Dump Valve is unique design, permitting the weigh tank to empty in 25 seconds or less at maximum flow rate, without need for reducing flow. It is specially designed for leak-proof operation.

Reservoir is stainless steel, internally baffled, and self-draining (this is important when changing fluids).

Viewing Port allows the operator to examine the reservoir or the dump valve at any time before or during the weighing cycle. Permits immediate spotting of dump valve seal leaks.

Fig. 4.19. Hydraulic schematic of calibration system. *(Courtesy AMETEK, Inc., Cox Instruments.)*

fluid to enter the weigh tank where fluid is either collected or dumped into the main reservoir.

Fluid is initially circulated until fluid temperature has been maintained within ± 1°F. Flow control valves are adjusted for desired flow and a tare weight is placed on the scale pan. The weigh tank is suspended on a scale beam with the scale pan on the opposite end along with a timer/actuator. At the start of the test, a prefill cycle runs until the weight of the added fluid matches that of the tare weight. At this point, calibration begins, and the timer starts. During the cycle, a precision weight is added to the scale pan. This weight is equal to the total calibrated weight desired.

Once the weigh tank matches the weight of the added calibration weight, the timer stops and the time to produce this fluid weight is displayed with resolution to 1 msec. The dump valve is opened, the weigh tank is drained, and the calibrator is ready for a new test.

This type of system is again dependent on several areas to achieve the accuracy required. The scale pivot mechanism must exhibit exceptionally low friction to negate mechanical error effects. Also, fluid forces generated during the fill vary based on how full the weigh tank is. The more filled the tank the lower the impact force. This force is counterbalanced by the additional weight of the fluid column falling into the tank, and the two generally cancel each other out. Finally, all the effects covered in the static system also apply.

There are other methods used to accomplish the dynamic weighing calibration, but the one just described is the most common. The other methods basically consist of nonmoving containers of a known volume with sensing electrodes or photocells at various locations. Once the container is prefilled, the bottom electrode starts a timer and another stops the timer. Fluid weight can be sensed by scale measurement or resultant head pressure created by a vertical column of fluid.

An important factor to remember in flowmeter calibration is that the procedure should be repeated several times to ensure consistency. This negates any one time problem occasionally encountered in sophisticated systems of this type. Reputable calibration laboratories generally run 10 repeat cycles for each point measured and calibrated.

4.3 Pressure Measurement

Pressure measurement is the most common test data used in fluid power testing. When relating a component's function, pressure defines its output, efficiency, maximum and minimum limits, etc.

Therefore, with pressure relating to many secondary functions of a component, accurate readings are mandatory.

Instrument accuracy does not always relate to electronic error, or mechanical friction, but can be related to error by test personnel. Pressure gages can be misread if attention is not paid to the actual pointer position. Graduations may not be marked and, in reading the gage, one graduation may be mistaken for another. Parallax view is also a factor. If an operator reads a gage at an angle (not standing directly in front of the gage), the pointer shifts to the right or left of the actual reading owing to the gap between the pointer and face of the gage. Some high-accuracy gages are available with a narrow mirror near the scale divisions to correct for this problem. This is one reason why many engineers prefer digital gages (transducers). It allows one less potential for error. Or does it? Many people in the test area still feel uncomfortable with transducers. You would be surprised at the number of systems built that have a transducer display and pressure gage side-by-side for comparison purposes.

4.3.1 Pressure Accuracy

Pressure-measuring devices are fairly easy to maintain as far as accuracy is concerned. This is fortunate, since these devices are the most frequently calibrated instrument. Span and zero drift caused by mechanical fatigue, temperature swing, and other effects eventually require a device to be recalibrated. Then, there is the all-too-often pressure spike that completely destroys all calibration or the device itself.

Special care should be taken not to subject a pressure gage or transducer to any pressure spikes or mechanical vibration. These factors are generally the main cause for premature failure or accuracy deviation. Also, wide temperature variations can cause significant deviation. Most transducers have built in temperature compensation over a range of operation. However, this range is not always as wide as your system range, so careful attention should be paid when selecting an instrument for an application. Section 4.1, "General Accuracy Terms," should be reviewed for the discussion of specific instrument ratings.

Another important consideration is the effect of pressure on viscosity. There is little information supplied concerning an oil's reaction to pressure changes. This is partially due to inconsistent variations of the same oil (oil manufacturing tolerances), as well as a lack of understanding by most people in the petroleum and fluid power industries.

The important thing to remember is that pressure does affect viscosity. When any component or instrument is viscosity sensitive, a relationship of pressure and viscosity of the oil used should be known. In particular, this effect can cause extreme inaccuracy in flowmeter readings. A detailed description of this effect is in Section 4.2.1, "Flow Measurement Accuracy."

4.3.2 Pressure Calibration

Frequent calibration of pressure instrumentation is recommended. There are many factors that cause drift, permanent distortion, and fatigue of internal mechanical components of the instrument. The following details some common methods used for pressure calibration.

DEADWEIGHT TESTERS

There are several methods used to calibrate pressure instrumentation, the most common method being the deadweight system. A typical deadweight tester is shown in Fig. 4.20. A series of weights are placed on a piston mechanism to generate a pressure, which is proportional to weight. The ratio of weight to pressure is based on the area of the deadweight piston and is expressed as

$$P = \frac{F}{A}$$

where

P = pressure (psia)

F = force (pounds)

A = piston area (in.^{2})

Weights of various calibrated divisions are placed on the pressure-producing piston. The pressure instrument is attached to the test media side where pressure is produced by the weight. The test medium is generally air, oil, or water. The pressure instrument is then subjected to several pressures throughout its range to verify that the device is accurate at all points. If a transducer or gage were only tested at one point, the calibration is often worthless because any nonlinearity of the device would not be detected.

The accuracy of the deadweight tester is usually sufficient for any type of fluid power testing. Published accuracies exceed ±0.0025% of reading at lower ranges and ±0.005% at higher pressure ranges.

Fig. 4.20. Deadweight tester.
(Courtesy DH Instruments, Inc.)

There are three main areas that cause a deadweight tester to degrade in accuracy:

1. *Piston Binding*—A deadweight tester is designed so that the pressure-producing piston does not contribute inaccuracy due to friction. The piston is lubricated to eliminate friction and the deadweight tester is generally leveled to maintain the piston in a true vertical condition. Often, the piston is rotated constantly by a motor or by hand to further eliminate error-causing effects.

2. *Change in Weight*—Since pressure is directly proportional to weight, the weights should be subjected to calibration. If a weight is scratched or nicked, this affects overall accuracy.

3. *Change in Piston Area*—If a deadweight tester is used frequently, wear on the piston could cause a change in piston area. This change is again proportional to pressure change.

Deadweight testers are calibrated to standards traceable to the NIST. They usually have valving for ease of operation and piston-location indicators to signal the tester when the piston must be restroked.

Taking the basic deadweight tester another step, the fully automatic primary pressure standard (Fig. 4.21) uses the same basic principle with computer control. This system allows easy weight changeover and communication with remote devices for automated pressure calibration.

Fig. 4.21. Automatic primary pressure standard. *(Courtesy DH Instruments, Inc.)*

PORTABLE PRESSURE CALIBRATORS

Recently, an influx of portable pressure calibrators have been introduced. These devices are generally rated at 0.05% to 0.1% full scale accuracy and are intended to be hooked in parallel with the pressure instrument being monitored. In some designs, they also act as a pressure source, producing a precise pressure to which the pressure transducer is calibrated. Figure 4.22 depicts a typical design used for calibration of pressure instruments.

These devices are usually calibrated to a deadweight tester. Zero and span information during calibration is stored and automatic compensation of zero and span during operation is accomplished with internal electronics. This compensation is required to maintain the accuracy of the higher-rated accuracy units.

These devices are available for calibration of absolute, gage differential, and vacuum types of pressure instruments.

4.4 Temperature Measurement

Temperature measurement is necessary for two reasons. First, accurate measurement is needed when subjecting a component to var-

Fig. 4.22. Portable pressure calibrator. *(Courtesy Consolidated Controls Corp.)*

ious temperature extremes, both internal (media temperature) and external (ambient temperature). Media temperature defines the limits to which a component can still operate properly. This includes slow rise/decay cycles as well as temperature shock. Media temperature changes can affect seal capability, cause component parts to expand and contract, and cause eventual failure of certain internal parts. Ambient temperatures can cause premature degradation of component parts. Excessive temperatures can fatigue elastomer components, damage external finish, and cause early component failure.

The second main temperature measurement required is fluid temperature for viscosity identification. Without an accurate measurement, viscosity may not be identifiable. Most components are viscosity sensitive. That is, some portion of their operating characteristic is affected by viscosity of the fluid used. This affect can be on pressure drop, pilot flow through orifices, response time, and many other characteristics. Of course, without an accurate chart on fluid viscosity versus temperature, an accurate temperature measurement will not be helpful either.

In both cases control of the temperature (both internal and external) is important. Test stand operators often allow a temperature swing of ±20 or 30°F, while conducting viscosity sensitive tests. These tests may be acceptable, if extreme accuracy is unimportant. But many more times accuracy is important. There are several methods used to maintain the temperature of system fluid, the most basic of which is the use of an on–off temperature switch controlling either a water solenoid valve on a water/oil heat exchanger, or a motor on an air/oil heat exchanger. If these cooling systems are sized properly, you can usually expect a control of ± 10°F.

Another common method is the use of a modulating water valve to control water flow through a water/oil heat exchanger. This valve senses temperature with a remote bulb and meters water at a rate proportional to the difference between sensed temperature and set temperature. The finer control offered with this type of system offers temperature control of ± 5°F or better if the system is sized properly.

Electronic temperature controllers are available for exact temperature control. These devices can be programmed for a specific correction of temperature cycles. If a recurring temperature change is known, the controller can follow and proportionately control a water solenoid valve or air/oil heat exchanger motor, maintaining excellent accuracy. The designation for a controller with this capability is generally defined as a PID controller. (See Chapter 2 under temperature controllers for more details on this controller.)

System heating is generally accomplished with immersion heaters, circulation heaters, or steam/oil heat exchangers. Again, these devices are controlled as described previously. In systems where both heating and cooling is required, it is necessary to use a controller that has dual set-point control. The use of two separate controllers often produces a system with heating and cooling crossover. This sometimes creates a situation where the heating and cooling devices are on at the same time. Also, too large a gap may be experienced between the two set points. Both cases decrease the accuracy of temperature control.

Fluid temperature should be maintained within set tolerances to maintain accuracy. When a test system is designed, sufficient capacity should be allowed for worst case conditions. Cooling water (water/oil systems) or ambient air (air/oil systems) should be provided at least 10–20°F cooler than the fluid temperature desired. Lesser differential temperatures could involve extremely large coolers or require the use of refrigeration systems. In extremely cold tests where a high degree of heat is input into the system, liquid nitrogen or a similar medium is used for cooling.

4.4.1 Temperature Accuracy

Since most equipment is now being monitored with thermocouple systems for accuracy tests, this section will deal only with the potential problem areas associated with thermocouples.

Thermocouples are manufactured to the standards of the NIST and each type has tolerances for their specific temperature ranges.

(See Chapter 2, Thermocouples.) These standards are based on the fact that the materials used in the manufacture are pure and no contamination is present. This problem does surface at times. In addition to impurities, a thermocouple that has oxidized will also deviate from the standards.

A thermocouple that has been subjected to extreme temperature, even for a short time, can deviate significantly from prescribed tolerances. If a thermocouple has been exposed to these extremes, calibration would be conducted before starting other tests.

Another common problem occurs with additive resistance from thermocouple wire. Additional resistance can reduce the millivolt output of the thermocouple, providing a lower temperature reading. The resistance of this wire should be reviewed prior to use so that a correction factor can be applied. In all cases use the wire recommended by the manufacturer.

Thermocouple sensitivity can affect a reading. This problem varies based on the thermocouple type. For example, an S type thermocouple has an average sensitivity of 7 $\mu V/°F$, while a J type is 30–32 $\mu V/°F$. If very small changes in temperature are being monitored, the proper thermocouple type should be selected.

Thermocouple output is not linear. The millivolt signal increases with increased temperature, but not in a straight line. Therefore, a direct conversion of millivolts to temperature is not possible. Linearization is required to compensate for the variation. Along with linearization comes conformity error. Multiple points along the output curve are memorized digitally and the curve is duplicated somewhat. However, only a certain amount of stored points are practical. So between points, actual output and translated output will deviate. This deviation is called conformity error.

4.4.2 Temperature Calibration

Temperature calibration techniques depend on the instrument type and accuracy desired. One method involves comparing the instrument to a parallel instrument, while another requires the test instrument to be adjusted to a traced standard.

CALIBRATION BATHS

These devices are used to provide a source of controlled temperature (Fig. 4.23). The bulb of a thermometer or a thermocouple can be inserted into the bath and allowed to stabilize, and then can be monitored for temperature reading variation between the instrument and bath readout. One process is called fluidization where low-

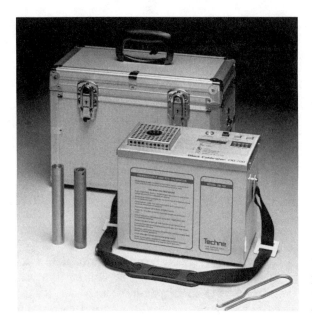

Fig. 4.23. Tempera-
ture calibration bath.
(Courtesy Techne Corp.)

pressure air or nitrogen is allowed to flow upward through particles
of aluminum oxide within the bath chamber. Flow is set so that the
gas separates the particles, allows them to move, and then suspends
them. These particles exhibit excellent heat-transfer characteristics, which allows the precise control required for a temperature
standard.

Other bath designs incorporate oil or cooling salt where the process is relatively similar, whereby the bath medium is precisely controlled by heating or cooling elements.

Calibration baths allow a temperature instrument to be compared
to a known temperature. The nice feature about this device is that
a wide range of temperatures can be selected, so that nearly any temperature instrument can be calibrated.

I C E   P O I N T   R E F E R E N C E S   A N D   B A T H S   (0° C)

These devices provide ice point reference and cold junction compensation for thermocouples. Thermocouples are calibrated with
this reference and accuracy ratings are based on this reference
point. (See Chapter 2 Thermocouples.) The device (Figure 4.24)
allows a thermocouple to be connected and a millivolt signal is produced equivalent to the thermocouple output at a selected reference
temperature.

Fig. 4.24. Ice point reference. *(Courtesy Transmation, Inc.)*

THERMOCOUPLE CALIBRATOR

These devices are portable and allow on-site testing of thermocouple output. The device (Fig. 4.25) allows connection to a thermocouple, and the output is displayed. The calibrator display is then compared to the instrument to which the thermocouple is normally connected.

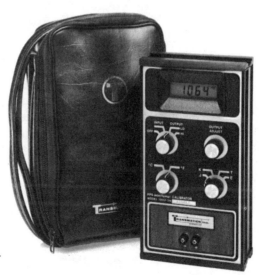

Fig. 4.25. Thermocouple calibrator. *(Courtesy Transmation, Inc.)*

4.5 Torque/Force Measurement

Force measurement defines cylinder output, torque output of a motor, horsepower requirement of a pump, as well as efficiency of any force-generating component. Fortunately, calibration of such instruments is relatively easy if proper instrumentation is used. However, application of the transducer and mounting considerations can cause unexpected error if all factors of accuracy are not reviewed. The following outlines the major topics of concern in their application.

4.5.1 Force Measurement Accuracy

Linear and rotary force instrumentation each have similar and different operating characteristics that must be considered when high-accuracy testing is required. In addition, variations between dynamic and static test conditions should also be analyzed.

Dynamic response is a factor when measuring fast-response torque or force variations in relationship to other variables. For example, if an engine were tested for output torque, a slow-response system would provide an accurate average output torque. If the actual torque curve were required to depict the output force of each piston acting on the crankshaft, we have a different situation. If the true relationship between piston position and output torque were required, then the torque system would have to provide instantaneous output to accomplish this. Torque transducers, however, have two factors that can affect this situation. The first factor is the response of the strain gage network, and the second factor is the transducer structure itself. The actual effect created by the type of input should be reviewed along with data supplied by the transducer manufacturer. Changes in speed, acceleration, and any time-related function should also be reviewed.

Natural frequency of the transducer also plays an important factor in some applications. When an external force is applied to a structure, the resultant counterforce creates a natural oscillation within the structure that is called the natural frequency. This is normal, and all structures including force transducers, react in this fashion when a load is applied. When the force is stopped, the structure will decrease in oscillation until it comes to rest. During some test conditions, the force measured may generate a frequency close to the natural frequency of the instrument. In this case, resonance occurs, providing erratic test data. Equations exist to approximate frequencies for certain applications, and the transducer manufacturer should be contacted if this effect is present.

Windage is an effect caused by a rotating transducer opposing the frictional forces of the ambient air. As speed increases, so does friction.

Bearing friction of the transducer and associated fixturing can also add frictional torque that does not relate to the force being measured, which can sometimes be negated during calibration of the instrument at no load.

Brush friction can be an error factor when using brush-type transducers. Proper installation and calibration can also negate this effect.

Locked-in torque is a frictional effect caused by improper alignment between the transducer and the shaft to which it is coupled.

Bearing temperature is another factor to be considered when conducting tests for long periods of time. Although this may not affect test data initially, eventual catastrophic failure may result causing unsafe conditions.

Other stress-related functions such as fatigue loading, dynamic or impact loading, and stress concentration should be reviewed in applications that apply greater shock load values than the transducer is designed to accommodate. These factors can greatly reduce transducer life and affect accuracy. Again, the transducer manufacturer should be consulted if these problems are suspected.

The critical speed of a rotary torque instrument should be analyzed; this speed is a result of a combination of several factors including actual rpm, load conditions, and transducer support. The basic factor is the relationship of rpm to the natural circular frequency of transverse vibration. With the transducer shafts supported at each end, the center section is pulled down by gravity. It is this force that causes this problem. The mechanical design of the instrument can cause the supported section to oscillate owing to these factors. Most rotary transducers, when operated within their normal specified limits, will not approach critical speed. However, when they are modified for high-speed service, this effect is most prevalent.

One of the most important factors to be considered when applying torque transducers is the mechanical mounting in relationship to the load and the method of attachment. Proper fixturing design should provide concentric and parallel connections to the rotating load. In addition, the types of flexible coupling and mounting of components is essential if long life and accurate data are desired. Figures 4.26, best installations, 4.27, acceptable installations, and 4.28, incorrect installations, depict the "do's and don't's" of transducer mounting.

BEST INSTALLATIONS

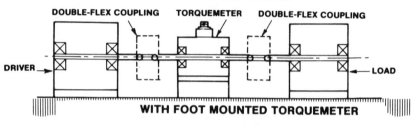

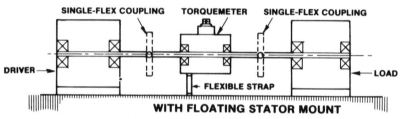

Fig. 4.26. Best installation for torque transducers. *(Courtesy S. Himmelstein and Co.)*

ACCEPTABLE INSTALLATIONS

(MORE CARE REQUIRED IN SET-UP TO CONTROL POSSIBLE RUNOUTS)

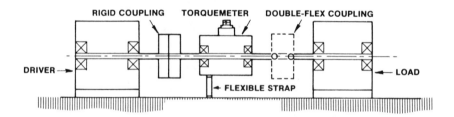

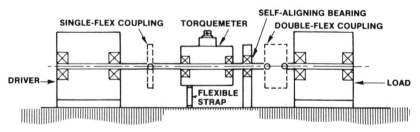

Fig. 4.27. Acceptable installation for torque transducers. *(Courtesy S. Himmelstein and Co.)*

INCORRECT INSTALLATIONS

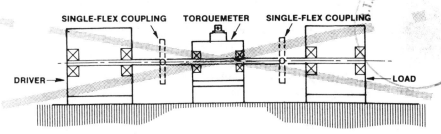

SINGLE-FLEX COUPLINGS IN THE ABOVE INSTALLATION WILL NOT ACCEPT THE PARALLEL SHAFT MISALIGNMENT THAT WILL EXIST. FORCING THE INSTALLATION WILL RESULT IN BENDING MOMENTS ON THE SHAFT AND PREMATURE SYSTEM FAILURE.

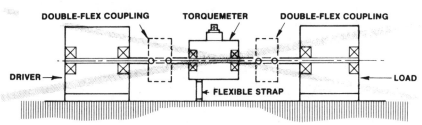

THE TORQUEMETER IN THE ABOVE INSTALLATION IS NOT CONSTRAINED IN THE RADIAL DIRECTION. THIS LACK OF CONSTRAINT WILL ALLOW THE SHAFT SYSTEM TO ROTATE ECCENTRICALLY, WITH RESULTING HIGH VIBRATIONS AND MOST LIKELY CATASTROPHIC FAILURE.

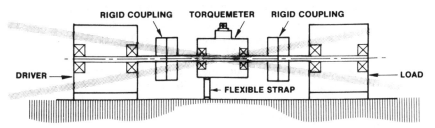

RIGID COUPLINGS AS USED ABOVE FORCE THE SHAFTS INTO ALIGNMENT BY BENDING THEM. THE DEGREE OF BENDING IS UNKNOWN, BUT COULD BE CONSIDERABLE IF MISALIGNMENT IS HIGH OR IF ALL COMPONENTS ARE EXTREMELY RIGID AND WILL CHANGE WITH TEMPERATURE GROWTH. THIS BENDING LOAD WILL CAUSE EXCESSIVE WEAR AND PREMATURE FAILURE.

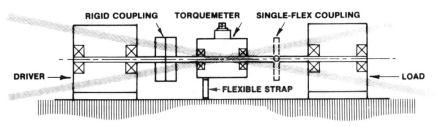

THE SUBSTITUTION OF ONE SINGLE-FLEX COUPLING AS SHOWN ABOVE IN PLACE OF A RIGID COUPLING IN THE PREVIOUS EXAMPLE DOES NOT IMPROVE THE INSTALLATION. THE SHAFTS ARE STILL FORCED INTO ALIGNMENT BY BENDING THEM.

Fig. 4.28. Incorrect installation for torque transducers. *(Courtesy S. Himmelstein and Co.)*

4.5.2 Force Instrument Calibration

Both rotary and linear force transducers are calibrated by calibrated weights traceable to the NIST. This is a relatively simple task as long as a few factors are considered.

Transducer output is defined as millivolt per unit of scale. For example, a load cell with a full scale output of 5000 lbf and an output of 0–5 Vdc would theoretically produce 1 Vdc at 1000 lbf, 2 Vdc at 2000 lbf, and so on. Owing to manufacturing tolerances, each individual transducer has its own specific output, where 4.997 Vdc may represent 5000 lbf and 2.998 Vdc may represent 3000 lbf.

During calibration exact weights are applied and the resultant output of the transducer at that point is recorded. This output must then be programmed into the signal-conditioning equipment to correct for this deviation.

Tension and compression load cells have forces applied by either calibrated weights or by special force-generation machines. The machines incorporate specially calibrated load cells used as the reference standard. These reference load cells must be frequently calibrated to NIST standards for verification.

Rotary force transducers are calibrated in a similar fashion. Either deadweight test equipment or rotary dynamometers are used to produce a controllable loading force. Deadweight testers generally include a balanced force beam to which calibrated weights are applied. One end of the torque transducer is mechanically locked, while the other is attached to the center of the load beam. Special attention is paid so that the load forces do not transmit side loading to the transducer shaft. The load beam is leveled to a horizontal plane so that no angular error from cosine is produced. This ensures that the calibrated load remains at the proper distance from transducer centerline under the influence of gravity. This test provides calibrated information with the transducer in a static condition only.

Dynamic calibration is accomplished with the dynamometer setup (or similar load-producing system). Again, the dynamometer includes a specially calibrated torque transducer used as a reference standard.

In all cases of load transducer calibration, several points throughout the transducer range should be verified to uncover any deviation of linearity or change in output from previous calibration.

4.6 Electrical/Electronic Measurement

This section could easily be covered in a few volumes detailing every aspect of electronic accuracy and calibration methods. It is not the intent of this book to delve into the theories of electronics. However, there are a few areas of importance normally associated with fluid power testing. These areas are the related inaccuracies in data conversion.

4.6.1 Electronic Accuracy

Most instrumentation provides an output of ± 5 Vdc, ± 10 Vdc, 4–20 mA, or BCD, which corresponds to measured signals from the transducer. These signals are typically used to drive data-collecting instruments such as plotters or recorders for hardcopy of the test results. In other sections we have discussed the combined accuracy factors of the instrument and its related signal conditioner. An increasing use of programmable controllers and computer systems interfaced to these instruments requires a better understanding of electronics by fluid power test engineers.

Analog information from the transducers must be converted into digital form within the analog input models of programmable controllers, while analog output modules must convert digital back to analog for powering or driving analog-controlled devices such as servovalves, flow controllers, or pressure converters. This process is called A to D or D to A conversion.

The most important factor to consider when designing a test system using a programmable controller is to determine the resolution needed for the test versus resolution capability of the programmable controller. In order to explain this, Fig. 4.29 depicts an analog signal versus a digital signal. Rather than a smooth analog output, the

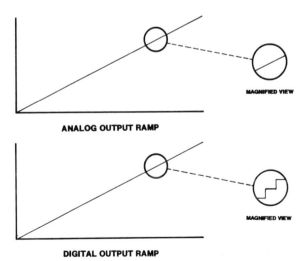

Fig. 4.29. Analog versus digital output.

digital system employs a technique that starts and arrives at the same point through a series of steps.

Programmable controllers and other digital devices define resolution accuracy by applying a term called "bits." The total number of bits resolution defines how many steps are required from zero output to full scale output of the device. In order to clarify this, the following list translates bit information to real numbers:

1 bit = 2	11 bits = 2,048
2 bits = 4	12 bits = 4,096
3 bits = 8	13 bits = 8,192
4 bits = 16	14 bits = 16,384
5 bits = 32	15 bits = 32,768
6 bits = 64	16 bits = 65,536
7 bits = 128	17 bits = 131,072
8 bits = 256	18 bits = 262,144
9 bits = 512	19 bits = 524,288
10 bits = 1,024	20 bits = 1,048,576

Now, let us take this information and apply it to a specific application.

A pressure test requires that a pressure regulator with an output of 0–1000 psi be measured with an accuracy of ± 0.25 psi. A pressure transducer with a range of 0–1000 psi is used, and for the moment we will overlook its accuracy and other system inaccuracies. An analog input module of a programmable controller is used and programmed to monitor output pressure at 750 psi with a deadband of ± 5 psi. Any pressure not within the deadband range of 745 to 755 psi is cause for test failure.

In selecting a programmable controller, we have the option of an 8-bit, 12-bit, or 16-bit resolution unit with more money required for higher resolution. An 8-bit resolution unit will provide resolution of 1000 ÷ 256 or ± 3.91 psi. This is not close enough. A 12-bit resolution unit will provide 1000 ÷ 4096 or ± 0.244 psi resolution. Is this good enough? Considering that an accuracy of ± 0.25 psi is required, the resolution of the A to D conversion alone only gives us ± 100 % accuracy of a single point. This does not even take into account all the other errors. Therefore, the 16-bit resolution unit is chosen as its accuracy is 1000 ÷ 65,536 or ± 0.015 psi.

Multiplexing techniques are also applied to test systems with many transducers. This technique allows data collection from many sources by one data logger, and also introduces new aspects into overall accuracy, as response time and transient signals can provide false information. During multiplexing, several discrete lines of information are converted to a common signal. Each discrete line is sampled, converted, sent to the data logger, extracted, and stored. Synchronization between the sample and extraction intervals is mandatory, otherwise an overlap of two signals may be erroneously stored. In addition, while one transducer signal is being stored, the other signals are not received. Any high-speed, short duration pulse or bit of information may be missed because of this. Multiplexing does have an advantage—reduced system cost.

It is strongly recommended that a book on multiplexing errors be studied prior to installation of a system of this type. Also, describe your application *in detail* to an instrumentation manufacturer so that all small details affecting overall accuracy can be evaluated prior to purchasing a system, which may produce incorrect information.

Data averaging is another technique used to negate transient peaks typically encountered with signals caused by fast-response data. Many instruments are available with dampening circuitry that either slows the circuitry or spreads a sample period over a wider range. This allows short-term sampling data such as that encountered during the multiplexing to be more representative of what is actually happening in the system. This also prevents a digital display from bouncing so fast that it interferes with an accurate reading. Caution must also be applied to prevent any fast-response data necessary for the test from being overlooked by a slow system.

4.6.2 Electronic Calibration

There exists many choices when it comes to the selection of electronic calibration devices. The type of signal, accuracy, voltage range, current range, frequency, phase, amplitude, etc., all are factors in the proper selection of a desired measurement. This section depicts a small sample of the various portable calibration devices and techniques available for the calibration of instrumentation signals.

Shunt resistors provide a simple means to verify that a signal conditioner is adjusted to the transducer output. After calibration data are collected during transducer calibration, an "equivalent number" is calculated for that specific transducer. This number is then used to adjust the zero and span of the signal conditioner. The signal conditioner can usually be supplied with a built-in calibration resistor that bypasses the transducer. By depressing a calibration button, the equivalent number can be reviewed to determine if the transducer signal conditioner is still adjusted properly. This method

Fig. 4.30. Frequency generator and counter. *(Courtesy Transmation, Inc.)*

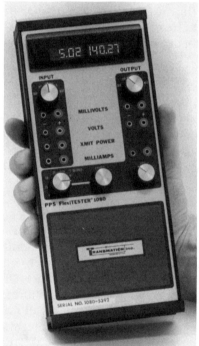

Fig. 4.31. Voltage calibrator. *(Courtesy Transmation, Inc.)*

Fig. 4.32. Milliamp flow calibrator. *(Courtesy Transmation, Inc.)*

is simple and accurate until actual transducer output characteristics change after initial use.

There are many hand-held devices available to simulate transducer output. Figure 4.30 depicts a frequency generator and counter that can be used to measure or calibrate flowmeters, tachometers, and any other device which works on a frequency principle. Figure 4.31 depicts a voltage calibrator with readout to measure and generate voltage signals typically generated by instrumentation. This particular device can generate 0–9.999 Vdc output. Figure 4.32 depicts a milliamp flow calibrator used to generate 0–22 mA output and measure signals in a range of ± 99.99 mA.

4.7 Particle Counter Accuracy

As you probably have gathered during your term in the fluid power industry, or reviewing this book, the most attention focused on any single component is on the filter. Its filtration capacity, the rating methods, and the devices and systems used to rate filters have been under scrutiny for years. Filtration is most important, and everyone wants to be assured that the filter they are using is performing according to their expectations.

Because the particle counter is the main device used today in the performance evaluation of a filter, it too is under considerable scutiny. Frequent articles are found concerning the calibration methods used for counter calibration, and the common mistakes made during calibration and use of particle counters.

The following information is extracted from Alvin Lieberman's article "Fluid Contamination Analysis—Using Particle Counters," originally published in the *BFPR Journal*. Used with permission of the Fluid Power Research Center, Oklahoma State University, Stillwater, OK.

4.7.1. Particle Counter Calibration Requirements

Proprietary calibration procedures are generally accepted for most particle counting devices. At this time, however, the International Standards Organization (ISO) has defined only one calibration procedure for optical particle counters. This is ISO 4402, a procedure for optical particle counters based on calibration with an AC Fine Test Dust for sizing particles to 100 micrometers in diameter. At this time, the particle size distribution that is used for ISO 4402 is being questioned below 10 micrometers in size. The particle size distribution function in use is a log-log squared distribution. A log probability distribution is presently believed to be more accurate. The effect of changing from a log-log squared to a log probability distribution is that the number of particles below approximately 10 micrometers in diameter increases with the increase at 1 or 2 micrometers almost two orders of magnitude. The numbered particles in the size ranges up to 80 micrometers would decrease slightly. The net effect is that optical particle counters reporting 5 micrometers would then report 7 micrometers; those reporting 80 micrometers would report 65 micrometers; and those reporting 100 micrometers would then report 80 micrometers.

The procedure for ISO 4402 calibration requires preparing a suspension of AC Fine Test Dust in MIL-H-5606 hydraulic fluid and determining the threshold level detection settings for the optical particle counter needed to produce cumulative concentration data in accordance with the tabulated concentration data in ISO 4402. Note that any accepted particle size distribution tabulation can be used.

The assumed similarity between AC Fine Test Dust and the contaminants that can be expected in hydraulic fluids may be justifiably questioned. AC Fine Test Dust is basically granite powder with an aspect ratio that varies from near one for particles below 3 to 5 micrometers (present distribution function) to an aspect ratio of approximately three for particles of 40 to 50 micrometers and larger. Contaminants normally found in hydraulic fluid range from ingested atmospheric particles from industrial processes, reentrained soil debris (normally quite similar to AC Fine Test Dust) to internally generated contaminants of metal fragments, O-ring and seal fragments, and polymers and reaction products produced from the hydraulic fluid itself.

In addition to the ISO 4402 calibration procedure, the American Society for Testing and Materials has recently formulated a recommended practice for calibration of optical particle counters using monosized spherical particulate materials. The particles are presized microscopically, and the threshold levels for the particle counter are set to correspond to an equivalent spherical diameter. In addition, the concentration in the sample being analyzed is determined by using a referee method and by comparing the particle counter concentration data to that obtained by the referee method.

This recommended practice has the advantage of referring particle counter data to a more absolute size criterion. However, that size criterion refers to a diameter for a sphere of equivalent diameter and optical properties. Actual contaminant particles differ from a spherical calibration base material in shape factor and optical properties.

4.7.2. Counting Accuracy

Accuracy can be defined as the agreement between the experimentally determined value and the accepted reference value. This agreement must be applied to both sizing accuracy and number or concentration counting accuracy. For good instrument operation, accuracy is necessary. Precision is also required.

Precision is defined as the degree of agreement of repeated determinations of a measurement expressed in terms of the dispersion of test results around an arithmetic mean result obtained by repetitive testing of a homogeneous sample. The precision can be expressed as a standard deviation computed from the results of a series of controlled determinations. Frequently, the term precision is expressed as "repeatability." Another convenient term for precision is coefficient of variance which can be defined as the ratio of standard deviation to the mean value from a series of measurements.

Good particle counting accuracy requires that the instrument or technique also be capable of high resolution because the particle count data include measurements of particle numbers in a range of particle sizes. Particle counting instruments describe the particle size by converting a pulse or voltage level to a particle size. If single size particles can produce a range of pulse levels, the instrument resolution is degraded as pulse height variation increases. Since most particle size measurements must include both size and concentration within a size, cross channel sensitivity or pulse height dispersion for a single size degrades the sizing and counting accuracy at any size. Specifications for sizing resolution are defined as the relative standard deviation of pulse height no more than plus or minus 20 percent at the smallest size range, and no more than plus or minus 5 percent at any of the size ranges.

In addition to the instrument limitations, a number of system operating parameters exist that control accuracy. These parameters are related to the nature of the sample that is being measured, to the operation of the instrument, and to personnel technique. When the concentration of material in the sample decreases, the counting accuracy also decreases. In general, if particles are distributed in a sample with a Gaussian distribution, the variance is normally equal to the square root of the mean. Thus, if a sample taken from a larger population has a particle count of 100 particles, the variance is plus or minus 10 percent. If the sample has a particle count of 1,000 particles, the variance is plus or minus 3 percent.

For most particle counting instruments, the sample flow rate affects the response of the instrument because the bandwidth of the system can affect the pulse height produced. If the sample flow rate is not uniform, then the pulse height varies inversely with flow rate. In addition, the coincidence errors for any instrument can affect the counting accuracy. If two or more particles are present within the sensing volume at any time, they are counted as one particle of somewhat larger size than either of the two that are present. If the concentration is increased to the point where the pulse generation rate exceeds the capability of the electronic system, then the particle counter is unable to count at the required counting rate.

If the sample container or sample handling systems produce excessive debris or result in removal of particulate materials from the sample, errors are produced. Sizing error and instrument resolution affect the counting accuracy as well as sizing accuracy.

Another important parameter that controls counting and sizing accuracy is the state of the sample being analyzed. Samples should be well mixed with a reasonably uniform dispersion of particulate material throughout the sample. This is especially important in the situation where aliquots are measured. If the particulate material is not well mixed throughout the sample, variability from one measurement to another will be excessive, and the standard deviation of the measurements will be so high that the reliability of any of the measurements will be in doubt. If the concentration of particles is quite low, a relatively large sample must be measured to achieve a sufficiently large number of data points for good statistical reliability. On the basis that the variance in a Gaussian distribution of measurements is equal to the square root of the mean of the measurements, any measurement that contains less than 100 particles has a variance of plus or minus 10 percent. At levels where 1,000 particles are measured, the variance is reduced to plus or minus 3 percent. Thus, larger samples improve repeatability both within and between sample measurements.

To maximize counting accuracy, several parameters mentioned in the preceding section must be controlled. A liquid volumetric sample feeding and measuring system with flow rate control is highly desirable. In addition, it is necessary that any particle count data be obtained after sample flow is started and before the sample flow is stopped. Otherwise, startup and shutdown variations in the flow rate will affect the sizing accuracy of the particle counter.

For the most part, manufacturers indicate coincidence error limits at a maximum recommended concentration. The coincidence error is defined as a percent of the concentration at that level. For minimum sizing and counting error, it is recommended that the particle counter be operated at a concentration where the coincidence error is no more than 2 to 3 percent.

4.7.3. Conclusions

Every effort should be made to prevent the introduction of artifacts in the fluid specimen. In addition, counting instruments should be properly calibrated and maintained, both optically and electronically. With these precautions and the use of standardized procedures, good data can be obtained when measurements of fluid contamination are made.

4.7.4. Problem Areas

In-line particle counters are subject to problem areas that can be avoided with batch counting. The first is interferences from the presence of immiscible mixtures in the hydraulic fluid. Either water or air can be included in the hydraulic fluid to be measured by the in-line particle counter. In either case, problems arise because the particle counter cannot differentiate between solid contaminant particles and liquid droplets or air bubbles suspended in the hydraulic fluid. In many cases, the only way that immiscible material can be detected is due to the atypical particle size distribution that is produced. Because the droplet and/or air bubble size distribution function frequently differs from that of the solid contaminant, the particle counter may report excessive contaminant particles in either an overly large or overly small size range. If the concentration of immiscible material is very high, the coincidence limits of the particle counter may be exceeded and the readout may be immobilized.

Where water may be present in the in-line hydraulic oil sample, the best remedial action is, of course, to ensure that water does not enter the hydraulic system. If this is not possible, sample point location and line configuration should be selected to permit some opportunity for water to settle out of the oil and permit the operator to drain water from the lines at intervals. If air or vapor bubbles are captured in the oil sample, sample point and sample line configuration should be chosen to permit the air bubbles to rise from the oil as much as possible before measurements are made. In some cases, making measurements in-line at pressures up to 2,000 to 3,000 psi result in most gas bubbles going into solution in the oil before measurements are made. Problems are more severe for on-board systems than for test stand systems.

If dilution is required for an in-line particle counter application, some precautions are necessary. These needs arise from the changing particle concentration. The particle size distribution function that is reported can also be changed. Most particle size distributions are power law functions; that is, the concentration can be described as an inverse power function of size. Normally, the power function is a third or a fourth power. Thus, if one wishes to reduce the concentration of small particles to a point where the particle counter in use will not be overwhelmed by excessive particle concentrations, it is possible that the large particle concentration may be reduced to the point where data are no longer statistically valid. In other words, it is necessary to ensure at all times that data in all size ranges are adequate for good reliability.

In-line dilution systems have been developed where a recirculating fluid is passed through a filter that removes essentially all particles in the size range of concern. The clean fluid is mixed with sample fluid with a simple tee mixing chamber, and the diluted material is then passed through the particle counter. Effluent can then be pumped through the filter for continued use a diluent. Hardware and materials for dilution systems include an accurate flow measuring and control device to permit accurate dilution ratio definition. If dilution ratios greater than 100 : 1 are required, it may be necessary to use a two-stage diluter because flow measurements of reasonable quantities of diluent may require accurate flow measurements at very low flow rates for samples.

A third problem that must be handled for an in-line particle counter is controlling sampling flow rate. With an in-line particle counter, continuous flow may be used with time as the basis for defining sample volume. If flow rate control is inadequate, sample volume errors may be a reason for producing inaccurate concentration data. A more subtle problem associated with flow rate control is that changes in flow rate may produce changes in pulse height reported by the particle counter. In this case, particle sizing error is added to particle concentration error. Most particle counter manufacturers indicate the tolerance to flow rate variability before significant particle size error is produced.

Means for controlling flow rate are available. For example, a pressure-compensated flow control valve may be used downstream of the particle counter and immediately after a filter to protect the flow control valve element from plugging. A second means of controlling flow rate is to use a metering pump with motor speed control as a controlling device for flow. Another possibility is to use a flow transducer with a motor driven control valve. The flow rate transducer may be a turbine device or a differential pressure indicator. In any case, the flow rate control and measuring device along with any protective filter system should be located after the particle counter has determined the particle content of the fluid.

4.8 Viscosity's Effect on Accuracy

Throughout this chapter (and others) it has been stressed that fluid viscosity is greatly affected by temperature and pressure. There are so many tests and instruments affected by viscosity, and yet many component manufacturers or repair houses neglect this factor.

I do not intend to list all problem areas in this section, since this would be redundant to many sections of the book. This section has been included as a reminder that viscosity's effect should be considered at all times. In particular, flowmeters are especially sensitive to viscosity changes, as are pressure drop and response time tests.

4.9 Component and Instrument (System) Interaction

There are several terms used to describe the compounding inaccuracies found in fluidpower test systems. Some terms are not that pleasant to hear. This area is probably one of the most important aspects of system design. The overall picture is sometimes overlooked, while a great deal of analysis is conducted on individual components or instruments.

If you have been involved in fluidpower testing, you would probably agree that very few tests are similar. Even though standards have been established in many areas, each product has its own set of peculiarities that separates it from the "norm." It is because of this that some components are not properly tested.

Probably the easiest way to describe this problem is to offer specific examples. The examples described are based on actual systems designed for a specific test or set of tests. Keep in mind that these three descriptions are only three out of thousands of potential test scenarios. Each test should be reviewed with the same degree of analysis.

4.9.1 Accumulator Precharge Test Analysis

A company made small accumulators rated at 5000 psi. A production test stand was required for a 10 in.^{3} volume unit that had a fixed (nontamper) gas precharge. The test stand was to verify precharge pressure and check for external and internal (bladder) leaks.

An older test system being used verified precharge pressure by increasing air pressure to the accumulator while monitoring pressure rise versus time. The precharge required was 1100 ± 35 psig. Their system was not consistent because of several reasons. First, imagine an accumulator with an 1100 psi precharge. As the unit is slowly filled, pressure increases, at a fairly proportional time rate. However, once system pressure reaches precharge pressure, the pressure versus time rate changes sharply. It is this point that they used to detect precharge pressure.

Their problems included a microprocessor with 20 msec data

sampling rate, unsynchronized data sampling, and total system interface accumulation. These factors when evaluated, could produce 20–60 msec actual response time for detecting that single point on the curve. The variation of 20–60 msec was also based on when the data gathering process began in relationship to the scan rate of the system. Therefore, the unpredictable sampling would be the main culprit in system repeatibility.

The system could have been improved by changing to a synchronized data sampling system with faster data collection. In fact, a 3 msec system was considered, but was not used, since another problem may have surfaced—too high of a sensitivity. The system could have been buffed to slow down response if needed, but then you approach the current problem and solve no problems.

A completely different approach was considered, which offered excellent repeatibility. A known volume of fluid was injected into the accumulator and the resultant pressure was measured. During the evaluation of this system, the chart depicted in Fig. 4.33 was developed to clarify the proposed method. Again, the desired precharge pressure was 1100 ± 34 psig. Therefore, resultant pressure values for 1065, 1100, and 1135 psig were calculated. In addition, each of the three pressure values were evaluated at an injected volume of 1–9 in.^{3} with a potential injected volume variation of ± 0.05 in.^{3} as a safety factor. Of course, it was important that the injected volume displace sufficient volume to overcome the precharge pressure. Insufficient volume would produce a worthless test.

An injected volume of 4 in.^{3} seemed to be a good number to work with. Given the 1100 ± 35 psig precharge, resultant pressure would be 1784.8–1901.5 psig with 1843.1 psig resultant pressure occurring if precharge were exactly 1100 psig. Even with variations in injected volume, the resultant pressures were not that big of a problem. Using a leakproof, positive displacement injection piston, injection volume was exact and repeatable.

The moral of this example is that many test procedures can be changed to improve accuracy. Procedures developed a "long time ago" are often considered "the way things are done, no questions asked." If by resourceful thinking a new method is suggested and it seems feasible, evaluate it. It may save time, money, and improve the product. This situation occurred at an established firm. Many objected to the possibility of a new test procedure even before hearing the full description of the procedure. Once they had a chance to review the test and look at the reduced cost for this simpler method, it was not difficult to make a change in procedure.

Fluid Volume (in.^{3})	1065 psig Precharge		
	-0.05 in.^{3}	Exact Volume	$+0.05$ in.^{3}
1	1,178.3	1,184.9	1,191.7
2	1,326.5	1,334.9	1,343.4
3	1,516.8	1,527.7	1,538.8
4	1,769.9	1,784.8	1,799.9
5	2,123.3	2,144.7	2,166.5
6	2,651.2	2,684.5	2,718.7
7	3,525.3	3,584.3	3,645.3
8	5,252.1	5,383.8	5,522.2
9	10,268.1	10,782.3	11,350.6

Fluid Volume (in.^{3})	1100 psig Precharge		
	-0.05 in.^{3}	Exact Volume	$+0.05$ in.^{3}
1	1,217.0	1,223.8	1,230.8
2	1,370.0	1,378.7	1,387.4
3	1,566.4	1,577.7	1,589.2
4	1,827.8	1,843.1	1,858.7
5	2,192.6	2,214.7	2,237.2
6	2,737.6	2,772.0	2,807.3
7	3,640.0	3,700.9	3,763.9
8	5,422.9	5,558.8	5,701.7
9	10,601.5	11,132.3	11,718.9

Fluid Volume (in.^{3})	1135 psig Precharge		
	-0.05 in.^{3}	Exact Volume	$+0.05$ in.^{3}
1	1,255.7	1,262.7	1,269.9
2	1,413.5	1,422.4	1,431.5
3	1,616.1	1,627.7	1,639.5
4	1,885.6	1,901.5	1,917.6
5	2,261.9	2,284.7	2,307.9
6	2,824.1	2,859.5	2,895.9
7	3,754.8	3,817.6	3,882.6
8	5,593.6	5,733.8	5,881.2
9	10,934.8	11,482.3	12,087.4

This chart is based on an approximate accumulator volume of 10 in.^{3} charged at 1100 ± 35 psig (nitrogen/helium) ($2\frac{1}{2}$ in. i d, $v = \frac{4}{3} \pi r^3$). The figures depicted are based on the resultant fluid pressure when a certain amount of fluid displaces the gas ($P_1 \times V_1 = P_2 \times V_2$). This does not take into account thermal expansion from initial compression—stabilization period will be determined.

Fig. 4.33. Accumulator precharge/volume chart

4.9.2 Solenoid Valve Flow Test Analysis

Here is a situation where a company measures flow rate through their solenoid valve to calibrate their flowmeter. The valve is used in water systems and, therefore, accurate flowmeter systems are hard to come by at inexpensive prices. Their method incorporates a timer and a high precision scale. The solenoid valve is energized and a timer starts. Water flows through the valve into a container on the scale. The scale reads in grams and measures the weight of the water over the timed period. The timer stops, the solenoid valve deenergizes, and the scale reading is stored. The difference between sensed scale weight at the start and finish of the timed period determines the exact flow rate the valve passed. The weight of water converted to volume over the time period equals exact flow rate. Right? Wrong.

It is important to remember that this "timed shot" of fluid is used for calibration of a flowmeter. The flowmeter is used in the production testing of a high-volume solenoid valve. The valve is passed or failed based on its comparison to the flowmeter. Desired accuracy of the calibration is ± 0.1 g from 0 to 2000 g and ± 1 g to 22,000 g.

Let us look at the problem areas here. First, the solenoid valve response is 20–40 msec. The flowmeter system was a dual system to provide high accuracy over a wide range. The low flow meter range is 0.2–2 gpm and the high flow meter range is 2–20 gpm, both providing a total range of 0.2–20 gpm. Using the upper range of each meter and converting to grams we find that 2 gpm = 7577.53 g/min and 20 gpm = 75,775.32 g/min. This translates to 126.29 and 1262.9 g/sec, respectively. Now a solenoid valve response time of 20–40 msec translates to an actual error of 2.53–5.05 g error at 2 gpm and 25.3–50.5 g error at 20 gpm. So far, we have exceeded desired accuracy error by 50 times or more.

The problem continues by evaluation of the scale response time. In discussing the actual scale response with the resident scale company scientist, I was informed that the scale uses an operating principle that requires a 2–3 sec stabilization period before an accurate reading can be taken. In fact, the required stabilization was a nonlinear event, so that running data were virtually useless. Giving the benefit of best possible accuracy, a 2 sec stabilization time converts to 252.58 g error at 2 gpm and 2525.8 g error at 20 gpm. We have now successfully increased accuracy error 2500 times above desired accuracy. Sampling scale weight while filling and additive solenoid response time proved to make this method ridiculous.

Now for the moral. But first, it is important to understand the circumstances. I was asked to quote a system to accomplish this "timed shot" calibration system. After reviewing this portion of the system and informing the industrial engineering group of the absence of any accuracy whatsoever, they bought a machine from someone else. Someone who wouldn't attack their long-standing calibration methods.

Now for the moral. Do not neglect accuracy to save face. Most important is a quality product. A quality product is only as good as the quality of manufacture and test. Admit error if a better product can be produced.

4.9.3 Centrifugal Pump Test Analysis

This analysis will outline the stack up of tolerances that occurred in a pump test stand. To further complicate the situation, the pump is a centrifugal design. The system is to perform a variety of tasks, but the main area we will look at is the relationship of output pressure versus flow rate.

Figure 4.34 depicts a typical pressure versus flow curve of a centrifugal pump. As you can see, flow drops sharply with an increase in pressure. In order to define the output characteristics, a point on the curve must be evaluated. The question is, Do you control flow to read pressure or control pressure to read flow? This pump is designed for water service and water is used as the test medium. Is the selection getting easier?

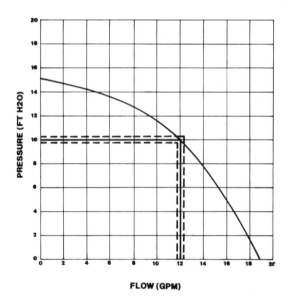

FLOW (GPM)

Fig. 4.34. Centrifugal pump output curve.

It was determined that pressure was the obvious choice for back-pressure control. An air-operated throttling valve was used to control pressure with a worst case regulation tolerance of $\pm$ 0.1 ft H_2O. Although this is rather close regulation, it is important to realize what happens if desired pressure is not exact.

The system is designed as a go–no-go operation, and upper and lower limits are set for flowmeter deadband tolerance. In order to maintain accuracy, pressure regulation was important, otherwise the fixed flowmeter trip points would shift. For example, looking at the curve, imagine pressure regulated at 10 ft H_2O and flow at 12 gpm. If pressure drifts to 10.1, flow would decrease to about 11.9 gpm. This shift of 0.1 ft H_2O pressure changes flow output. If a flow deadband of 11.9 to 12.1 gpm were desired as a "perfect situation," then the shift in pressure would reject the test. Pressure is not monitored, while flow is the monitored value in determining if the pump is acceptable. (There are better ways to accomplish this, but owing to the fact that 16 test stations were required, the cost effectiveness of the system was important.)

So starts the tolerance stack up. Now the flowmeter accuracy must be evaluated. The system used provided an accuracy of $\pm$ ½% full scale with a flow range of 5–20 gpm. This accuracy translates to 0.1 gpm, which now increases the inaccuracy of the system nearly twofold.

Now the problem of sensing the flow output with external voltage sensing circuits occurs. The signal conditioner used with the turbine block provided a 0–5 Vdc unisolated analog output. A 0 V output represented 5 gpm, while 5 Vdc was equivalent to 20 gpm. The zero and span of the unit allowed adjustment of the output, so that maximum resolution could be attained. However, because the output was unisolated, a tracking error of $\pm$ ¼% of reading could be possible in another "worst case" condition. This signal was fed into a voltage-sensing circuit, which was calibrated to detect the voltage equivalent of 11.8 gpm on the low end and 12.3 gpm on the high end. The voltage-sensing circuit exhibited negligible error, but calibration could be off if the proper equipment was not used.

For ease, we will neglect possible error factors due to improper calibration, zero drift, ambient compensation, etc., and review the main error contributors:

1. The pressure regulator = $\pm$ 0.1 ft H_2O at 10 ft H_2O, which equals $\pm$ 1%.
2. The flowmeter = $\pm$ ½% full scale at 12 gpm, which equals $\pm$ 0.83%.
3. Analog output tracking error = $\pm$ ¼% of reading.

The combined inaccuracy can be determined by using a method called the root-sum of the squares (or RSS). Therefore,

$$\sqrt{1^2 + 0.83^2 + 0.25^2} = 1.32\% \text{ total error at 12 gpm}$$

Other test engineers evaluate the total potential error as the addition of all tolerances. Therefore, $1 + 0.83 + 0.25 = 2.08\%$ total error at 12 gpm

It is usually recommended to evaluate a system in worst case conditions for every aspect and component of the system. When in doubt, allow sufficient safety factor in order to avoid a system not sensitive enough for the application.

System Automation

What criteria are used in the selection of a test stand design? The main consideration, of course, is the procedure used on the item to be tested. Some test components will not allow automated testing, and other components require too much operator intervention for adjustments, visual checks or whatever. A review of exact procedures required must be conducted before the question of automation is approached.

Budget is usually the limiting factor in any machine purchase. Often, convenience, accuracy, and automation are sacrificed because of expenditure limits. This is unfortunate, but also a fact of life. Time studies are conducted to determine the overall cost of the test stand and support equipment needed. This overall cost is compared to the savings gained from the shorter operator test time required. This time gained can then be applied to other productive tasks for the operator. This evaluation produces the number of years in which the test stand will pay itself off and is termed "payback period."

Many companies have their own procedure set up to determine payback. This includes all aspects of costs involved, such as operator salary, overhead factors, time study reviews on operator efficiency, etc. There are several reasons why a test machine is purchased. In all cases, the system is designed to minimize load/unloading time, test time, and increase accuracy.

Figure 5.1 depicts a typical evaluation of a necessity purchase. By necessity we mean the system was purchased either to allow testing

Itemized Operation Cost	Year 1	Year 2	Year 3	Year 4	Year 5
Machine Cost	50,000	0	0	0	0
Support Equipment	6,500	1,000	750	750	1,000
Installation Cost	1,750	0	0	0	0
Electric Power Consumption	1,200	1,300	1,400	1,500	1,600
Cooling Water Consumption	175	180	195	195	200
Maintenance Parts	100	500	1,000	1,000	3,000
Average Maintenance Hours (4.0 hr/yr @ $15.00 hr avg) (+ $0.75 hr increase per year)	600	630	660	690	720
Maintenance Overhead (182%)	1,092	1,147	1,201	1,255	1,310
Machine Operator Cost (2080 hr/yr @ $7.50 hr avg) (+ $0.50 hr increase per year)	15,600	16,640	17,680	18,720	19,760
Operator Overhead (167%)	26,052	27,789	29,526	31,262	32,999
Yearly Cost of Operation	103,069	49,186	52,412	55,372	60,589

First year 5% warranty return for a $10,000,000/year company = $500,000 return

Warranty repair costs	5% Return	3% Return	1% Return	1% Return	1% Return
Repair Parts Cost	50,000	30,000	10,000	10,000	10,000
Repair Hours (5 yr avg) (4160 hr/yr @ $8.50 hr)	35,360	21,216	7,072	7,072	7,072
Repair Overhead (167%)	59,051	35,431	11,810	11,810	11,810
Total Cost for Repairs	144,411	86,647	28,882	28,882	28,882
Yearly Cost Savings	0	57,764	115,529	115,529	115,529
Cost Differential Operation vs Reduced Warranty	−103,069	+8,578	+63,117	+60,157	+54,940

Fig. 5.1. Necessity purchase.

not currently conducted or to increase testing capacity. The initial purchase requires a test stand cost of $50,000 plus $6,500 for support equipment such as fixturing adapters, hoses, and fittings, and an installation cost of $1,750. These fixed amounts are added to the regular cost of operation items such as yearly electric and water consumption, maintenance costs, and the cost of a full-time operator. The first year will require a total investment of $103,069, with the next four years figured at cost of operation.

This may seem like a great deal of money, but let us assume that the company has a total sales volume of $10,000,000, with a 5% rejection rate (warranty returns). This equals $500,000 of goods returned for rework. As a justification it was assumed that the first year would still be at 5% rejection, year 2 would be 3% rejection, and finally, each year thereafter, a 1% rejection would be average. Reviewing the cost of repair parts needed and the cost of a repair technician provides us with the total cost for repairs each year. Yearly cost savings depicts the money saved each year from decreased warranty return.

Now, the yearly cost savings subtracted from yearly cost of operation provided us with the amount of money lost or gained from the purchase. In year 1, we have lost $103,069. But, as year 2 starts, we are starting payback. The addition of years 2,3,4,and 5 equals $186,792. So payback resulted somewhere in year 4, with all savings thereafter applied to increased profits.

Figure 5.2 depicts the purchase of the same test stand as a replacement for an existing system. In the justification process, it is determined that no increases for maintenance, operator cost, and utilities will be required. Therefore, the word "same" signifies this tradeoff. The purchase of the test stand, support equipment, and installation are the only additional investments required.

It was calculated that owing to increased set up time from fixture design, a 25% increase in operator productivity would result. This translates to the cost savings depicted. The yearly cost of operation, minus the yearly cost savings, provides the cost differential that results from the purchase. Year 1 results in a loss of $47,837, while the total of years 2 through 5 equals a gain of $48,594. So, payback results in about 5 years.

Figure 5.3 depicts an evaluation of a system identical to Fig. 5.2 replacement purchase, except for automated features. This system will initially cost more, and there is a slight yearly operational cost increase. But, look at the cost differential. This differential was based on the fact that an operator would only have to be present for 25% of the time. The remaining 75% could be devoted to other tasks

Itemized Operation Cost	Year 1	Year 2	Year 3	Year 4	Year 5
Machine Cost	50,000	0	0	0	0
Support Equipment	6,500	Same	Same	Same	Same
Installation Cost	1,750	0	0	0	0
Electric Power Consumption	Same	Same	Same	Same	Same
Cooling Water Consumption	Same	Same	Same	Same	Same
Maintenance Parts	Same	Same	Same	Same	Same
Maintenance Hours (0 hr/yr @ $15.00 hr)	Same	Same	Same	Same	Same
Maintenance Overhead	Same	Same	Same	Same	Same
Machine Operator Cost (0 hr/yr @ $7.50 hr)	Same	Same	Same	Same	Same
Operator Overhead	Same	Same	Same	Same	Same
Yearly Cost of Operation	58,250	0	0	0	0

This results in a 25% increase in operator productivity, so:

	Year 1	Year 2	Year 3	Year 4	Year 5
25% Machine Operator Cost (520 hr/yr @ $7.50 hr) (with $0.50 hr increase per yr)	3,900	4,160	4,420	4,680	4,940
Operator Overhead (167%)	6,513	6,947	7,381	7,816	8,250
Yearly Cost Savings	10,413	11,107	11,801	12,496	13,190
Cost Differential Operation with Increased Productivity	−47,837	+11,107	+11,801	+12,496	+13,190

Fig. 5.2. Replacement purchase.

Itemized Operation Cost	Year 1	Year 2	Year 3	Year 4	Year 5
Machine Cost	80,000	0	0	0	0
Support Equipment	10,500	Same	Same	Same	Same
Installation Cost	1,750	0	0	0	0
Electric Power Consumption	+500	+550	+600	+650	+700
Cooling Water Consumption	Same	Same	Same	Same	Same
Maintenance Parts	+500	+500	+1,000	+500	+1,000
Maintenance Hours Adder (20 hr/yr @ $15.00 hr) (+ $0.75 hr increase per yr)	+300	+315	+330	+345	+360
Maintenance Overhead (182%)	+546	+573	+601	+628	+655
Machine Operator Cost (0 hr/yr @ $7.50 hr)	Same	Same	Same	Same	Same
Operator Overhead	Same	Same	Same	Same	Same
Yearly Cost of Operation	94,096	1,938	2,531	2,123	2,715

This results in a 75% increase in operator productivity, so:

	Year 1	Year 2	Year 3	Year 4	Year 5
75% Machine Operator Cost (1560 hr/yr @ $7.50 hr)	11,700	12,480	13,260	14,040	14,820
(+ $0.50 hr increase per yr) Operator Overhead (167%)	19,539	20,842	22,144	23,447	24,749
Yearly Cost Savings	31,239	33,322	35,404	37,487	39,569
Cost Differential	−62,857	+31,384	+32,873	+35,364	+36,854

Fig. 5.3. Automated replacement purchase.

or increased test turnaround time. The net result is increased productivity with payback within 3 years.

Note that the examples provided take into account time and materials only. Added benefits and increased payback schedules can also result from tax benefits and depreciation calculations for capital equipment purchases.

Based on budget allowance, a decision is made as to what type of test equipment can be purchased. If a system is fully automated, a review must be conducted to determine the practical limit of the extent of automation. In many instances a machine is purchased that is far too sophisticated. This can be due to lack of working knowledge by the company's engineering or maintenance staff, or simply because a company is expecting too much from one machine. Complicated adjustment procedures, automated test component loading/unloading and high-level programming languages can be reasons that are intended to be a benefit, which end up being a headache for everyone involved.

5.1 Manual Testing

Manual testing can provide highly accurate test results, provided that test procedures are written properly and that the operator understands the reasons for the procedure. It is very common to see shortcuts taken by an operator conducting production tests. Often, these operators are paid a fixed rate for quota, then a bonus for all tasks performed over quota. This applies to mainly production test systems because of the production scramble to ship parts. Manual systems used for engineering evaluation typically provide accurate data, because the system is used to analyze a component's operating characteristics, rather than acceptance of these characteristics.

A manual system can be the most versatile test machine if used properly. However, increased set-up time and test time is often the main drawback. Attempting to adjust one parameter often changes others. So, getting the right combination of values to the point of taking a reading can, sometimes, be frustrating. Designing a production test stand with too much versatility can provide a machine that is too complicated for an operator to understand.

Three examples of manual test stands follow. Review of the system design of each will demonstrate the range of complexity that can be encountered in systems of this type.

5.1.1 Oil Pump Test Stand

A good example of a manual test stand is the Oil Pump Test Stand depicted in Figs. 5.4 (photograph) and 5.5 (schematic). This system

Fig. 5.4. Oil pump test stand. *(Courtesy Heco Division/Barker Rockford Co.)*

was designed to test aircraft pumps that circulate lube oil within a jet turbine engine. Several pump types were to be tested, so the circuit was designed to accommodate all the tests required.

PUMP DRIVE SYSTEM

Because some pumps were driven directly from the turbine engine, speeds required to drive the pump reached 13,000 rpm, with torques up to 200 in.-lbf. The first step was to select a variable speed drive that would provide good speed regulation at infinite set points with manual input. As described in the pump test section, any point along the flow/pressure/speed relationship may be required to determine pump performance. Using an electric drive meant using a speed-increasing gearbox because of the maximum speed output rating of the drive. Speed regulation accuracy then decreases proportionally to the ratio of the gearbox, and low end stability is sacrificed for the same reason.

A fixed-displacement, bent-axis hydraulic motor was powered by a variable volume pump. Pump displacement was controlled by an electric motor drive on the maximum volume stop. This method allowed the operator to hold an increase/decrease button, while

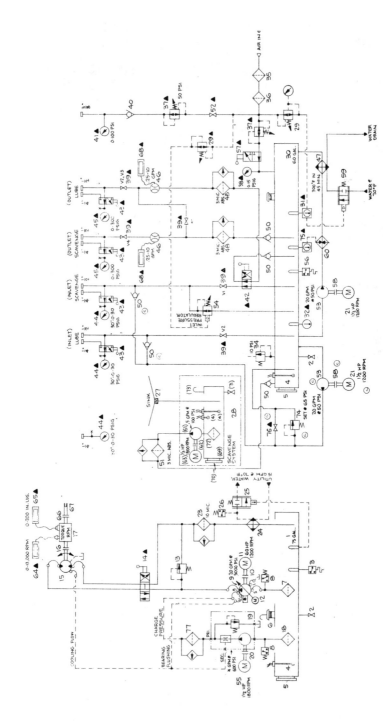

Fig. 5.5. Oil pump test stand schematic. (Courtesy Heco Division/Barker Rockford Co.)

watching the drive tachometer. When the proper speed was attained, release of the button locked the pump at that speed. Once locked, the only speed variance experienced was drive motor efficiency due to pressure changes from varying drive loads, and this was minimal. Temperature effect on motor efficiency was negligible due to a temperature control system within the drive system that maintained oil temperature within 5°F of set point. Also, because of the high speeds required, a steady flow of cool oil was circulated through the case of the motor. This prevented excessive mechanical heat from also affecting motor efficiency.

The drive motor displacement was 0.3 in.^{3}/rev, so in order to attain 13,000 rpm we calculated 13,000 rpm × 0.3 in.^{3}/rev ÷ 231 = 16.88 gpm theoretical required. As a safety margin, a 20 gpm pump was selected. Maximum torque required was 200 in.-lbf. The motor was rated at 47 in.-lbf torque/1000 psi theoretical, so (200 in.-lbf ÷ 47) × 1000 = 4255 psi theoretical pressure required to attain 200 in.-lbf output torque. Again, as a safety factor, 5000 psi pump output pressure was selected to overcome system inefficiency. A 60 hp, 1200 rpm electric motor was selected to drive the pump. Pump efficiency was 95% at the worst case, so 20 gpm × 5000 psi ÷ (1714 × 0.95) = 61.4 hp. Although we required 61.4 hp, a 60 hp motor was sufficient owing to its 1.15 service factor. This allowed total continuous output of 69 hp without overload degradation of the motor.

The drive circuit also included a relief valve set at 5000 psi, a 10 micron absolute return filter, a water-cooled heat exchanger, and a modulating water control valve. The relief valve never needed adjustment because a pump only produces the amount of pressure required to perform the work. It was only used as maximum pressure protection. A zero-leak, four-way, three-position valve allowed clockwise, counterclockwise, and neutral selection of motor control. Again, a zero-leak design was chosen to prevent variable leakage within the valve at different operating pressures. This prevented an additive effect on motor speed regulation. The heat exchanger was sized to remove 10 hp heat input (one-sixth of the duty cycle) because the motor circuit only needed heat removal to compensate for heat generated from inefficiency. Even at 10 hp, the system was extremely oversized.

The drive system was supplied with a separate 75 gal reservoir, because the test fluid was to be used at temperatures up to 300°F.

PUMP TEST CIRCUIT

Two independent test loops were provided, because some of the pumps tested were double pumps. In each pump outlet line, a needle valve and pressure gage, turbine flowmeter, and 3 micron absolute filter were provided. The needle valve was sized to allow minimum pressure drop in the full open position, and also to allow precise throttling control for backloading the pump. A 0–300 psi gage with $\frac{1}{4}$% full-scale accuracy was supplied to depict pump outlet pressure.

The turbine flowmeter and digital display provided a range of 0.25 to 10 gpm with $\frac{1}{4}$% of reading accuracy. Owing to wide operating temperature ranges, a temperature-compensation module was included to vary the meter output based on the viscosity swing of the fluid at different temperatures.

One pump output circuit returned directly to the reservoir, while the second circuit could return either to the reservoir or back to the inlet of the pump, through a selector valve.

The pump inlets had a variety of controls to provide restricted inlet or pressurized inlet, depending on the test requirements. A needle valve provided suction control in excess of 15 in. Hg for tests that determined pump suction characteristics. An adjustable air regulator provided supercharge pressure to the sealed 60 gal test circuit reservoir for pressures up to 10 psi. This also acted as a make-up supply for closed-loop pump tests. A 10 psi relief valve on the reservoir provided tank overpressure protection. Some pumps under test needed full flow supercharge up to 50 psi, so an auxiliary 20 gpm, 50 psi pump was added to both inlet lines to provide this function. A compound gage with a range of 30 in. Hg to 100 psi and $\frac{1}{4}$% full-scale accuracy was provided to monitor pump inlet pressure.

A very important consideration in a system such as this is pressure drop, because of the low-pressure testing required on some of the pumps. Careful attention must be paid when selecting components and line sizes on the pump outlet test circuit. Additive pressure drop from fittings, the needle valve, flowmeter, 3 micron absolute filter, etc., could create pressure drops higher than required test pressures at higher flow rates. Therefore, this circuit was plumbed with $\frac{1}{2}$ in. tubing and some components with 1 in. ports, even though most test pump outlets were $\frac{1}{4}$ or $\frac{3}{8}$ in. Because of this problem, the fluid temperature control circuit had to be a separate loop.

FLUID TEMPERATURE CONDITIONING

This loop allowed precise control of the test fluid temperature. It was required that a range of 85 to 300°F adjustment capability be provided. Cooling was accomplished by a water/oil heat exchanger with 10 gpm 70°F water available. The pump test loop created a great deal of heat, because all the power that was produced by the drive was converted into heat by the pump outlet flow across the needle valve. In order to size the cooler properly, it was assumed that all 60 hp could be continuously generating heat in an endurance test situation. Therefore, the cooler was sized to remove 60 hp of heat input and cool incoming oil to 85°F with a water/oil temperature differential of 15°F.

Heating was accomplished by an inline circulation heater rated at 18 kW. The specification required that oil could be heated from 80 to 300°F in 30 min, and 18 kW proved to be quite adequate. Once temperature was reached, the heater was turned off and on within an acceptable control range.

Both the cooler and heater were controlled by a common temperature controller/indicator that had high/low set points. This made for easy operator temperature control.

OTHER SYSTEM FEATURES

The console design included a sealed test chamber with Lexan shields to allow the operator to observe the test pump without the possibility of being scalded by hot oil. The pump drive fixturing allowed for easy pump changeover and included a bidirectional rotary transformer torque sensor with a 60 tooth gear and magnetic pickup for speed sensing. The torque sensor was calibrated for both directions up to 200 in.-lbf and speed was sensed up to 13,000 rpm. The entire rotary group within the drive fixture was dynamically balanced to prevent vibration at high rpm.

Several auxiliary circuits were also provided, which are not shown on the simplified schematic; these included a controlled air circuit, additional pressure gage taps, an automatic scavange system, and drive motor cooling circuit.

System safety monitoring included loss of cooling water, drive system overtemperature, test circuit overtemperature, excessive pump suction, low fluid levels, and filter element pressure drop status.

5.1.2. 8000 psi Aircraft Component Test Stand

Aircraft hydraulic systems, until recently, have operated around 3000 psi. Higher pressures in the range of 6000–12,000 psi have been considered in order to reduce overall system weight. Aircraft manufacturers, component manufacturers, and various branches of the Defense Department have conducted high-pressure test programs at least as far back as 1966. These tests proved that certain fluids and components could be used at 8000 psi or that they required minimal redesign to accommodate this pressure increase.

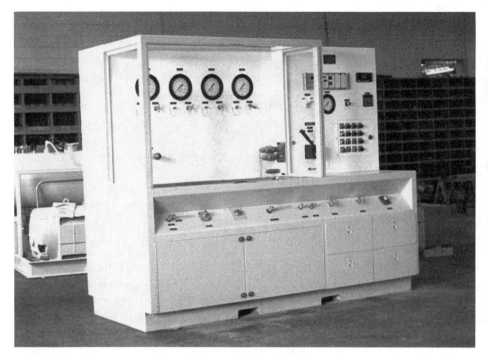

Fig. 5.6. 8000 psi aircraft component test stand. *(Courtesy Heco Division/ Barker Rockford Co.)*

Therefore, at present, 8000 psi has been selected as the mean operating pressure of future aircraft systems.

The test stand depicted in Fig. 5.6 is designed to functionally test aircraft valves and actuators. It supplies up to 20 gpm at 8000 psi with infinite control of flow and pressure. The console and power unit are remote, providing isolation of vibration and objectionable noise levels at the operator station. Connection between both assemblies is accomplished with hose assemblies that further reduce transmitted noise and vibration associated with hard-plumbed systems.

The fluid used was either MIL-H-83282B or CTFE (chlorotrifluoroethylene). At the time this book was written MIL-H-83282B was deemed the replacement for MIL-H-5606 and CTFE was under development stage. CTFE was targeted as a new fluid with increased fire-resistance rating and lower viscosity at low operating temperatures (−65°F). During this stage CTFE was available only from the Air Force at a cost of $200–300 per gallon. Tests performed with this fluid revealed some satisfactory characteristics, while other problems with rust inhibitors and water retention surfaced. Components exposed to air after being wetted with this fluid immediately began to oxidize. Moisture in the air would be absorbed into the fluid creating gelatinous "blobs," which caused more problems. Other problems surfaced, such as the creation of hydrochloric acid vapor at higher temperatures, as testing continued.

CONSOLE ASSEMBLY

The console houses all controls and instrumentation necessary for an operator to conduct tests required. The circuit (Fig. 5.7) shows the inlet from the power unit connected to the inlet of a four-way valve, an accumulator, and a relief valve. The four-way valve allowed selection of pressure or vent to two outlet ports located in the test chamber. Each port contained a bleed valve, quick disconnect, and a 0–10,000 psi pressure gage with needle valve, snubber, and

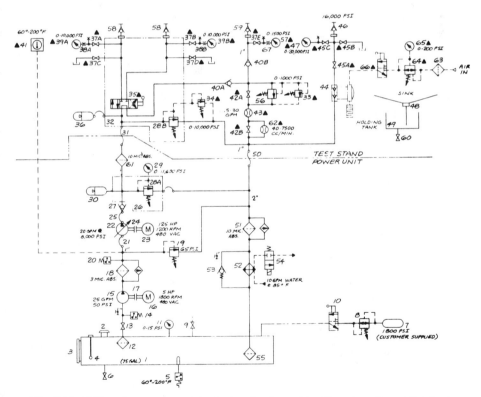

Fig. 5.7. 8000 psi component test stand schematic. (*Courtesy Heco Division/ Barker Rockford Co.*)

test port. The accumulator was sized at 1 gal capacity to reduce potential system shock created by components under test. The relief valve was remotely controlled by a vent relief and provided 0–10,000 psi pressure adjustment.

The return line contained a shut-off valve and relief valve controlled by a vent relief. This allowed back-pressure control up to 1000 psi, while allowing unrestricted return line pressure for tests requiring this feature. Pressure was monitored by a 0–1500 psi pressure gage with needle valve, snubber, and test port. Return flow was monitored by two flowmeters.

A turbine flowmeter with a range of 0.5–30 gpm was supplied to monitor most test results. It included a linearizer, a temperature-compensation module, and 20-point calibration at four viscosity points. Because the meter was installed in the return line, pressure variations affecting accuracy were negligible, but variable temperatures would cause error. The temperature compensator monitored oil temperature and adjusted the display output based on data obtained during calibration at the factory. Also, a separate EPROM was supplied with calibration data for the other fluid used. This allowed easy changeover from MIL-H-83282B to CTFE.

A positive-displacement flowmeter was also provided to monitor flow in the range of 40–7500 cm^3/min. This allowed leakage testing of components as well as providing highly accurate low-range data. Both systems were rated at $\frac{1}{4}$% of reading throughout the entire flow range. A shutoff valve was installed in the return line to divert all return flow through the positive-displacement meter if flow dropped below 0.5 gpm on the turbine meter.

The console also included a 0–16,999 psi air-operated intensifier for static pressure testing. It was rated at 150:1 ratio, so 16,000 psi hydraulic pressure could be attained with an air input pressure of 106.7 psi. The hydraulic circuit contained a shutoff valve, bleed valve, and test port. The air circuit included a shutoff/vent valve, filter, and regulator with gage for operator pressure adjustment.

The console was provided with an enclosed Lexan-shielded test chamber and stainless steel sink. Sink fluid drained to a holding tank, which was drained periodically. All system control and power unit control were easily accomplished from the operator station. Emergency shutdown from overtemperature, low oil level, and low pump suction was provided, as well as indication of filter element status. A digital temperature controller monitored reservoir temperature and was used to operate the cooling water solenoid valve at the heat exchanger.

POWER   UNIT   ASSEMBLY

The main component of the power unit was variable-volume, pressure-compensated pump that produced 20 gpm at 8000 psi and 9400 psi at reduced flow. It was driven by a 125-hp 1800-rpm electric motor. A stroke controller was used to operate remotely the displacement control arm, providing infinite control of flow from the console. The pump outlet included a check valve, 1 gal accumulator, 10,000 psi relief valve, and 10 micron absolute pressure filter.

The main pump required a full flow supercharge of 50 psi, so a fixed displacement vane pump rated at 25 gpm was used. It was driven by a 5-hp 1800-rpm electric motor. Its inlet contained a 100 mesh suction strainer, shutoff valve, and vacuum switch to monitor excessive pump suction. Its outlet contained a 3 micron absolute filter, 50 psi relief valve, and a pressure switch to monitor boost pressure. The main pump motor would not start unless sufficient pressure was present. Also, a timed shutdown of the boost pump motor occurred after the main pump motor was turned off, either by hand or during an emergency shutdown situation. This eliminated the potential for main pump cavitation, which could lead to extremely high repair costs.

The return line of the power unit contained a 10 micron absolute filter and a water-to-oil heat exchanger. The heat exchanger was sized to cool oil 20°F with continuous heat input of 50 hp with water supply at 10 gpm and 85°F maximum temperature. Cooling water was controlled by a solenoid valve, which received its signal from the temperature controller in the console.

The system contained a 75 gal stainless steel reservoir with baffles, cleanout covers, drain valve, sight glass, low oil level switch, and relief valve. The reservoir was charged with 10 psi nitrogen from an 1800 psi supply tank. A shutoff valve and regulator to reduce pressure from 1800 to 10 psi were supplied.

Most components and all tubing were made from stainless steel, owing to the corrosive nature of CTFE. Owing to the relatively high flow rate of 20 gpm, pressure line ID was kept to a minimum diameter of 0.56 in.

5.1.3 Submarine and Aircraft Component Test Stand

The system depicted in Figs. 5.8, 5.9, and 5.10 consists of five independant consoles powered by a remote hydraulic supply. The power supply contained six pumps, each powered by a 75-hp electric

Fig. 5.8. Submarine and aircraft component test stand system—front view. *(Courtesy Heco Division/Barker Rockford Co.)*

motor, which translates to a maximum system output of 450 hp. The power supply was designed to be housed in an area isolated from the consoles to reduce noise and vibration at the operator station. Depending on the flow demand at all five consoles, the system would sequentially start and stop the six pumps to reduce power draw and heat generation.

Fig. 5.9. Rear view of test stand system. *(Courtesy Heco Division/Barker Rockford Co.)*

Fig. 5.10. Single console configuration. *(Courtesy Heco Division/Barker Rockford Co.)*

The console was designed to allow ease of adjustment, even though the circuit contained many valves. This was accomplished by placing the majority of the valves just below the test area. The test area was enclosed with Lexan shields to protect the operator in the event of an oil leak. All five consoles were installed as shown, with hard plumbing between each console and the power supply.

The power supply contained the six pump/motor assemblies, each of which contained a relief valve, check valve, pressure gage, and solenoid vent valve for automatic no-load starting. A 350 gal air-pressurized reservoir provided system fluid through six suction filters at the inlet of the pumps. The reservoir included two 3 kW immersion heaters with automatic heating controller, level gage, reservoir pressure gage, high and low oil level switches, drain valve, filler-breather, temperature gage, and air control valve with regulator and gage.

The outlets of the pumps were plumbed to a common manifold containing a bank of seven high-pressure, 3 micron absolute filters.

The outlets of these filters were plumbed together and this manifold supplied the conditioned fluid out to the consoles.

Return fluid from all five consoles was through two water-to-oil heat exchangers and bank of four 10 micron return filters. The heat exchangers contained an automatic water regulator that could be set to maintain desired oil temperature. The heat exchangers were sized to remove 25% of full system capacity, while maintaining oil out temperature at 120°F.

CONSOLE SCHEMATIC

Each console contained the circuit depicted in Fig. 5.11. The inlet contained a shutoff valve for servicing, a solenoid valve for on–off cycling and emergency system isolation, and a fixed pressure compensated flow control valve set at 50 gpm. This provided maximum flow limitation at each console. A 100–5000 psi relief valve with vent valve controlled system pressure. Pressure was monitored by three pressure gages (1,000, 3,000 and 6,000 psi) with needle valves, snubbers, gage protectors, and test ports. An accumulator with shutoff valve prevented shock and dampened ripple. A bypass valve diverted all flow to the return line and a temperature gage monitored oil inlet pressure.

This inlet circuit then was divided into several subcircuits. The main outlets consisted of two $1\frac{1}{2}$ in. outlets controlled by a four-way valve, shutoff valve, and loading throttle valve. The outlet of the four-way valve tied into a leakage valve that diverted leakage flow from this point or from a separate external source into a beaker. Five additional outlets were provided; a $\frac{3}{4}$ in. outlet with shutoff and bypass valve, accumulator, pressure gage and throttling valve, a $\frac{1}{4}$ in. outlet with pressure gage and throttling valve, and three $\frac{1}{4}$ in. outlets with throttling valves. The smaller outlets were used for low flow testing and pilot supply for test units requiring pilot flow. A 0–50 psid differential pressure gage with snubbers, gage savers, and test ports allowed accurate pressure drop measurement and could be connected at any point of the test unit with external hoses.

The return circuit consisted of six $1\frac{1}{2}$ in. ports with check valves for free-return and a $\frac{3}{4}$ and 2 in. port, which connected to a dual range flowmeter with a range of 0.36–60 gpm and contained a shutoff valve to isolate the lower range rotameter during higher flow testing.

An intensifier circuit provided up to 30,000 psi for hydrostatic testing and consisted of a handpump for pressures to 10,000 psi and an intensifier for higher pressures. A three-position valve allowed selection of vent, handpump out, or handpump to intensifier inlet. A common $\frac{3}{8}$ in. outlet contained a 10,000 and 30,000 psi gage

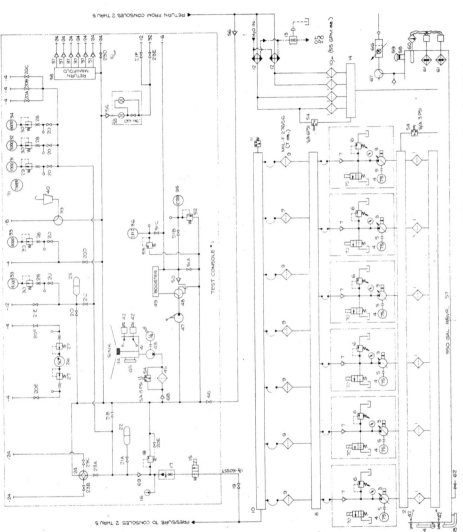

Fig. 5.11. Test stand schematic. (Courtesy Heco Division/Barker Rockford Co.)

with needle valve, test port, and relief valve. A bypass valve allowed the outlet to vent to tank.

Finally, a scavenge system collected sink oil in a separate reservoir. Once full, the reservoir high/low switches would start a small pump, which drew this oil and passed it through a 3 micron filter, returning it to the main 350 gal reservoir. This prevented any ambient contaminates from being reintroduced into the system.

5.2 Semiautomatic Testing

This control option can be added to a manual system providing increased accuracy and decreased operator error. With the use of small programmable controllers, the extra cost for this increased benefit can be surprisingly low. Automating certain test sequences or allowing the system to say "pass" or "fail" can take a great deal of burden from an operator.

The definition of which parameters encompass the category "semiautomatic" can be argued. It is easily stated that this type of system falls between the categories "manual" and "fully automatic." Any system that reduces operator tasks or allows the operator to perform other tasks with minimal test supervision can be defined as "semiautomatic." These systems are typically found on the product floor or are used for engineering-conducted cycle testing. The operator most often must be present or nearby to perform intermediate operations or to make decisions from information provided during the automated portion of the test.

Two examples of semiautomatic test stands follow for your review.

5.2.1 Solenoid Valve Test Stand

The system shown in Fig. 5.12 is a semiautomatic two-station system designed to test the function of two-way, normally closed, solenoid valves for water service. Tests performed included seat leakage, external leakage, full flow characteristics, manual operation, and reduced solenoid voltage. Each of these tests were performed at two pressures, 150 and 40 psi.

The sequence was controlled by a small programmable controller, which sensed discrete inputs from operator control switches and controlled discrete outputs to operate lights used to signify test status or to indicate the failure of any portion of the test. A pressure transducer with a range of 0–200 psi and signal conditioner with a 0–10 Vdc output fed this signal to an analog input module. Both

Fig. 5.12. Solenoid valve test stand. *(Courtesy Heco Division/Barker Rockford Co.)*

stations were to be completely independent of each other, so two separate programmable controller systems were used.

The circuit (as depicted in Fig. 5.13) was set up to draw water from an in-house water pressure source of 150 psi at 15 gpm. An incoming filter was sized at 3 microns absolute to provide the degree of filtration required for this type of test. Two pressure-reducing valves regulated test pressure at 150 and 40 psi, and each had a solenoid valve operated by the controller for on–off actuation. The outlet of each valve contained a zero leak check valve to provide a positive shutoff required for the leak tests, which used the pressure decay method.

At the inlet of the test valve, a pressure switch, the pressure transducer, and a redundant pressure gage were installed. The pressure switch acted as a loss of pressure monitor to signal the controller that a gross leak had occurred during the stabilization process of a leak test. (See Chapter 3, Leak Testing, for a more detailed description of this procedure.) The transducer provided the controller with current valve pressure at all phases of the test cycle.

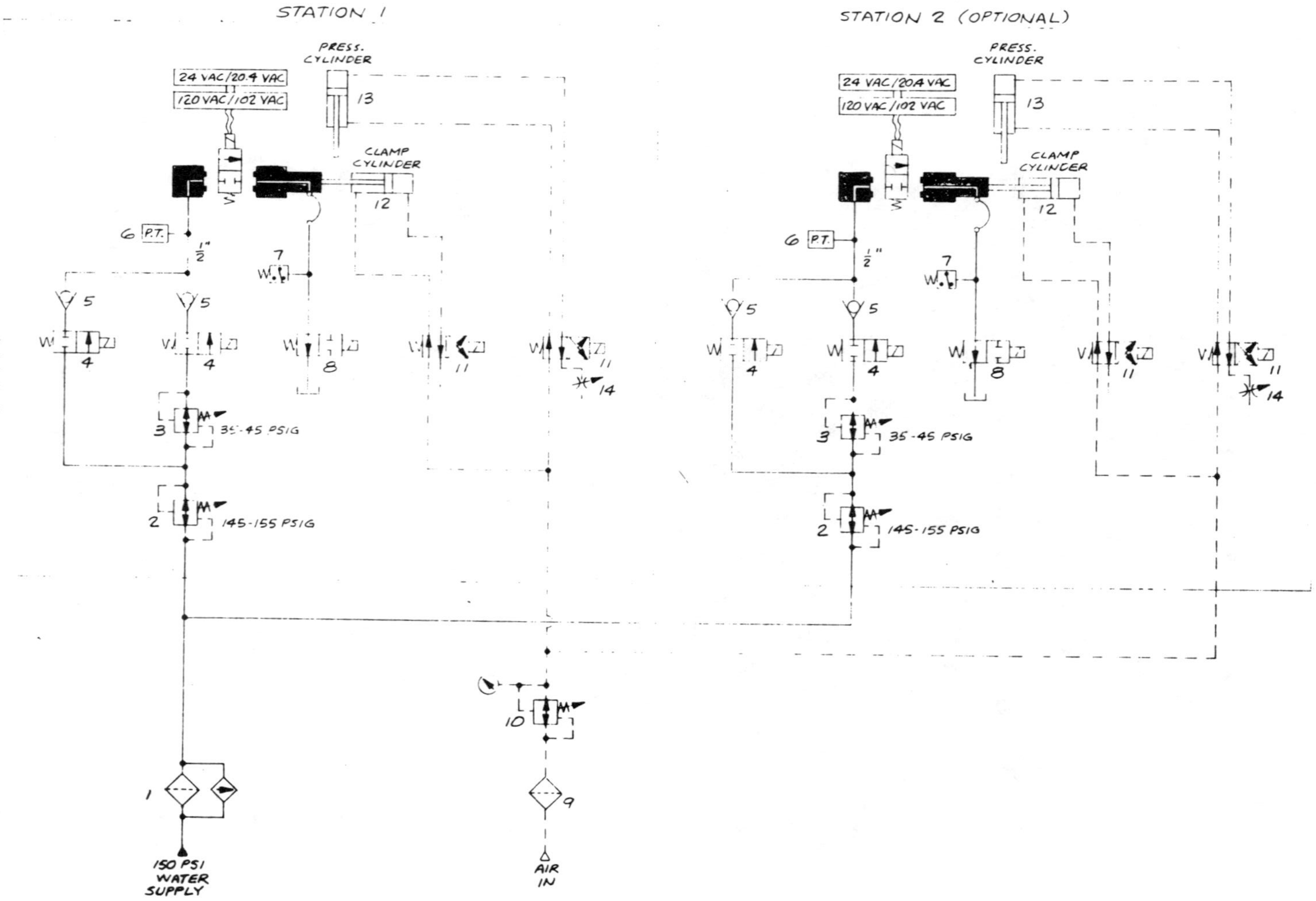

Fig. 5.13 Solenoid valve test stand. (Courtesy Hoge Division/Parker Bedford Co.)

384

The outlet of the test valve contained a zero leak solenoid valve for on–off control, and a flow switch to signal the controller that the valve was opening fully during an energized condition.

Also, included in the circuitry were two pneumatic cylinders operated by solenoid valves. One cylinder operated the port clamp fixture, which connected the test valve hydraulically as well as holding it in place. This was a manual operation during valve loading, but the controller overrode the unclamp switch during the test to prevent the operator from releasing the clamp while the test valve was still under pressure. The second cylinder actuated a lever that depressed the manual override of the valve under test.

TEST SEQUENCE

After the test valve had been loaded and electrically connected, the operator would depress a "start test" button. The test unit was subjected to 150 psi and was cycled on and off for three cycles with full operating voltage. During each of the on cycles, the controller would sense if the valve was opening completely from the flow switch output.

The second step was to repeat the three on–off cycles, checking for full flow each time, but with valve operating voltage reduced 15%. This reduced voltage test detected whether the valve would shift properly or at all. Minimum operating voltage of the valve was recommended at this nominal rating.

The third step, if the valve had passed both tests, was to check for drip-tight closure of the valve stem and seat. This was accomplished by the water pressure decay method in order to detect extremely small leaks. It was calculated that based on an internal volume of 6.23 in.^{3} during the seat leak test, a 0.05 cm^3/min leak would cause a pressure decay of 156.3 psi with a test pressure of 150 psi. This, of course, was too sensitive. In order to compensate for this, a small air-charged accumulator was added. This still allowed the system to accurately detect small leaks without unnecessary sensitivity. This required the system be calibrated to a known leak standard as fully described in Chapter 3.

The fourth step consisted of depressing the manual override, closing the outlet solenoid valve, and conducting an external leak test at 150 psi using the pressure decay method. Once 150 psi was sensed, the inlet solenoid valve was closed, trapping pressure between the inlet and outlet valves, with the test valve manually held open. Any external leak would be sensed as a pressure decay, again dampened by the accumulator. Decay rate for external leaks was deemed to be zero, except for any pressure fluctuation due to internal component elastomer expansion.

The fifth, sixth, and seventh steps repeated the first, second, and third steps, except test pressure was reduced to 40 psi. It was found that a valve might not operate properly at 40 psi even though it had passed all parameters at 150 psi, so for this reason, the extra time was used.

The eighth and final step consisted of three cycles on–off manually, checking for full flow. This detected any flaw in the mechanical linkage between the solenoid and valve stem assembly.

All the tests were automatic, but if any parameter of any test failed, the program would stall the test at that point, signaling the operator that a valve defect had been found. The operator then could push a "failure check" button to cycle the program back to the last test, so that he could observe the test and possibly fix the problem immediately. If the operator felt that the valve could not be repaired, depressing a "reset" button would clear the sequence and allow the test valve to be removed after the unit was depressurized and electric power to the valve was shut off.

Voltage selection of 120 or 24 Vac was provided, because the test valve could be supplied in either coil rating. The electric supply included panel-mounted voltmeters and individual ground fault interrupters to protect the operator from a potential shock hazard.

5.2.2 Multilevel Float Switch Test Stand

This test stand was designed to verify four switchpoint deadbands and allow operator verification of external leakage and drip-tight valve closure in a float switch. The switch assembly actually consisted of two switches, but switchpoints were based on cam actuation by a float during water level increase and decrease. These points are used to control when a burner cuts on and off and when a water supply valve energizes or deenergizes in an industrial boiler system.

The system (as shown in Fig. 5.14) was designed to test five float switches simultaneously at various flow rates and pressures. The far left station contained a reference switch assembly, which had eight switchpoint reference probes installed in the body. Each probe represented either the upper or lower deadband point for each of the four switchpoints. For example, if water were rising and the lower probe were actuated, the test switch would be good if it switched before the water actuated the upper switch. If the test switch actuated after the upper limit was reached, or before the lower switch

Fig. 5.14. Multilevel float switch test stand. *(Courtesy Heco Division/Barker Rockford Co.)*

was actuated, the test switch would be out of its acceptable dead-band specification and would fail.

A small programmable controller was used to monitor the reference deadband limits and sense if any switch at any of the five stations failed. A main test panel depicted the test status and any reason for failure of any float switch. A pass or a fail light at each of the five stations would depict which test unit had failed. The program was to allow the test to continue even though one or more units had failed midway through the complete test sequence, unless the operator hit the "reset" button. If any unit had failed, a buzzer would alert the operator, so a decision could be made if testing should continue.

Fixturing connected the units hydraulically and held them in place. Each station had its own clamp button for individual station loading. These buttons were overridden by the controller so that test units could not be removed during the test.

The circuit (Fig. 5.15) was designed to use city water as a test fluid, and a 10 micron filter was used to condition this supply. A pressure-reducing valve regulated inlet pressure to two inlet solenoid valves with needle valves used for flow control. One solenoid

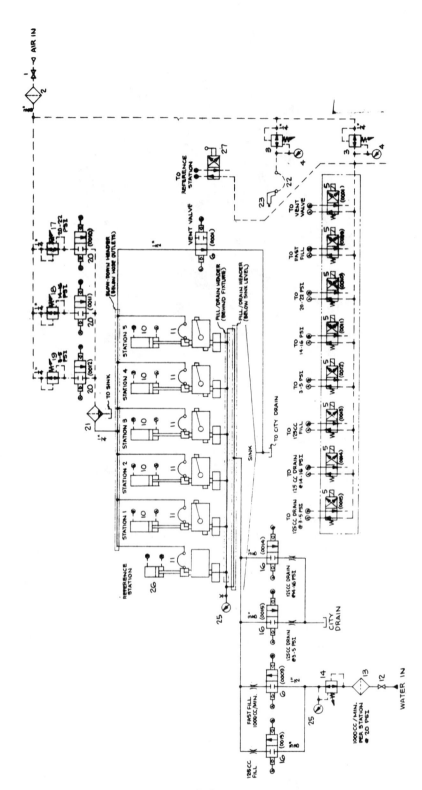

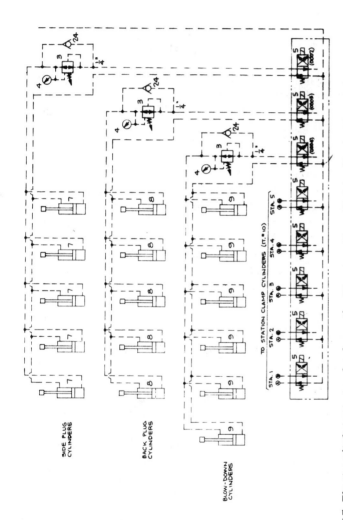

Fig. 5.15. Float switch test stand schematic. *(Courtesy Heco Division/Barker Rockford Co.)*

valve was used for a fast prefill of the reference switch and five float switches at the start of test. The second solenoid valve slowed water to a usable flow rate. All six stations were fed through a common bottom (inlet) manifold, and all were leveled for precise control. The slow water feed had to fill each float switch housing at an equal rate, otherwise the test results were useless.

Actual test pressure was selected by the controller from three air-regulator sources at 4, 15, and 21 psi. A solenoid valve on each regulator outlet provided on–off control. A fourth solenoid valve in the outlet circuit completely vented air pressure and served as a pressure vent or it allowed testing at ambient pressure.

Water level was lowered by closing the two fill solenoids and opening either of two drain solenoid valves. Both had needle valves used to control drain flow at 125 cm^3/min, one at 4 psi and the other at 15 psi.

Other system features not depicted in the simplified schematic included air valves and cylinders for clamping and plugging three different float switch designs with various port locations. A three-position selector switch was supplied to modify the program based on which unit was tested. This also included control of the various plug cylinders to seal off ports not used during the test of certain models. Change to a new test unit model required changeover of the reference unit in order to duplicate inner volumes and float movement.

Electrical connection to a test unit was accomplished by a multipin spring probe assembly which applied a 24 Vdc 750 mA supply to the switches under test. This current rating was based on the minimum current rating of the switch, so if the unit passed the test, it also passed this minimum rating.

The multipoint reference probes in the reference unit (described previously) were specially designed for this application because nothing similar was commercially available. These probes ran on 5 Vdc at low current because they were submersed in water within the reference unit. The output was minimal, so each probe had to be interfaced with a 0–5 Vdc current-sensing alarm with an output compatible with the programmable controller.

TEST SEQUENCE

The first test required after all test units were loaded was the external leak or hydrostatic test. The unit was automatically filled and pressurized at 21 psi, but the operator visually checked for any leakage caused by casting porosity or gasket failure. Once the units were completely inspected and a 60 sec timer timed out, testing would continue. A "continue test" button, depressed by the operator, signaled that the units had checked out, while the timer ensured he or she took the time to thoroughly check the units.

The system would then lower the water level to a start point below the lowest reference switch point, pressurize all units at 15 psi, and begin a controlled water level increase, monitoring switch point actuation. As the water rose, the burner switch was to turn on, and the feed valve switch was to turn off. Once water reached this high point, the fill solenoid closed and the operator would visually check for any valve seat leakage. (The valve is used for water blowdown or lowering of water level manually.) Again, if all units passed, the "continue test" button was depressed.

The next test phase consisted of lowering the water level and checking for the feed valve switch to turn on and the burner switch to go off while at 15 psi.

Once completed, the float switch was pressurized at 4 psi and the water level was raised again watching the same switch points as before. The switches had a tendency to operate at different points under different pressures, due to the diaphragm seal that sealed the float (on the water side) from the switch assembly. Pressure on its surface area caused float loading, which modified switch actuation.

Again, the unit water level was lowered at 4 psi checking for switchpoint actuation. Once completed, an automatic valve operator controlled by the programmable controller, vented the valve and testing was complete. Test results depicted by lights, remained actuated until the operator depressed the "reset" button.

Each station included an assembly fixture for another float switch. This allowed an operator to perform final assembly operations on the tested unit after the next five test units were loaded and under an automatic portion of the test sequence.

5.3 Fully Automated Testing

A fully automated type of system can provide the greatest benefit to a user because of its diverse capabilities. When fully automated fluid power test stands were first utilized, the control system was usually a custom-built microprocessor, which lead to programming and maintenance nightmares. Eventually, the machine was usually scrapped, converted to manual operation, or just not used at all. This does not mean that automated systems never worked. There were just more bad than good systems.

Now standard hardware is available with user-friendly programming and systems can be provided for long, trouble-free operation.

Unfortunately, some fluid power manufacturers remember the old days and are still reluctant to accept this new technology. However, those who choose automation are finding the many benefits derived from these complex systems. Increased speed, accuracy, and added profits can be realized from a well-planned test program employing automation.

The most important requirement in the design or purchase of a fully automated test stand is time. Time is required to document all details of test procedures, test component design, test limits, and data acquisition. The old adage "garbage in–garbage out" also applies here. If sufficient information is known up front, the program and screen can be configured to allow future additions with minimal time spent. Also, any details pertaining to manual override of an automatic system are imperative. This feature can prove to be a major benefit if a "nonstandard" component requires testing.

The following examples depict two fully automated test stands with infinite manual override capability.

5.3.1 Rotary Actuator Test Stand

This system, depicted in Figs. 5.16 and 5.17, was designed automatically to test electrically and pneumatically operated rotary actuators. The range of torque testing required was 100–100,000 in. lbf over a 90° rotation. Some test units were single acting, spring return and others were double acting. Some rotated from CW to CCW, some from CCW to CW, and others only rotated in one direction. Electric actuators could be powered 115 or 230 Vac. Pneumatic actuators could have one, two, or four pressurized chambers. With the potential of hundreds of variations possible, it was decided that total automation was the answer to solve the operator's test selection dilemma.

MICROPROCESSOR SOFTWARE

When power was turned on, the CRT would prompt the operator to select either a test mode or a system check mode. In system check, each transducer could be displayed along with its operating characteristics and its calibration data. This allowed an occasional check to determine if the transducer had drifted out of calibration. The five transducer signal conditioners next to the CRT had shunt calibration buttons, which, when depressed, would depict actual setpoint versus the calibration number stored from the last transducer test calibration. If the numbers did not match, the system allowed the quality assurance personnel to adjust the zero and span potenti-

Fig. 5.16. Rotary actuator test stand. *(Courtesy Heco Division/Barker Rockford Co.)*

ometers until the transducer was back in calibration. The calibration number also could be easily changed after each transducer was returned from occasional recalibration.

When the test mode was selected, the operator could select whether a manual test or automatic test was desired. The manual test was normally used for engineering evaluation of an actuator or for the occasional "very special" unit that did not meet the normal parameters of the automatic setup. In this manual mode, the operator made the following selections:

1. Break Torque (first rotation)?
2. Running Torque (first rotation)?
3. Finish Torque (first rotation)?
4. Single or Double Acting?
5. Break Torque (return rotation)?
6. Running Torque (return rotation)?
7. Finish Torque (return rotation)?

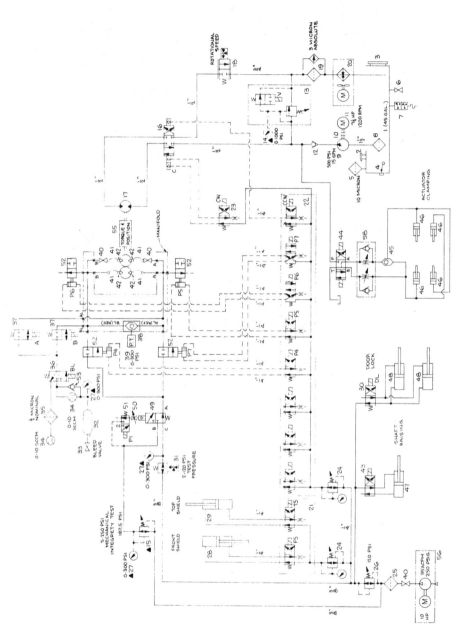

Fig. 5.17. Rotary actuator test stand schematic. (Courtesy Heco Division/Barker Rockford Co.)

8. Rotation CW to CCW? CCW to CW?
9. Rotation Speed?
10. Pneumatic Actuator?
    a. Mechanical Integrity Pressure?
       Acceptable Integrity Decay Pressure?
    b. Body Leak Pressure?
       Acceptable Body Leak Flow Rate?
    c. Piston Leak Pressure?
       Acceptable Piston Leak Flow Rate?
11. Electric Actuator?
    a. Operating Voltage 115 Vac? 230 Vac?

Once these parameters were entered, the operator could enter model number, serial number, and operator identification. The system would then depict which torque transducer and fixture adapters (if possible) to use.

If an automatic test was selected, the operator would enter the actuator model number, serial number, and operator identification. The screen would then list all test parameters as well as torque transducer and fixture adapters required for the test.

PNEUMATIC TESTING

The first test required was mechanical integrity, which was basically a proof pressure test to indicate any major leaks or defective assembly. Once the operator depressed "start test," the system would prompt the operator to close the safety shields, which sealed off the test chamber. Bulletproof Lexan and 1 in. plate steel were used to protect the operator in the event of test unit rupture. This had never occurred in the history of the company, but operator safety was of the utmost importance. Dual palmbuttons needed to be pressed and held until the top and front safety shields were in place; two air cylinders locked in the shields and remained locked until the mechanical integrity test was completed.

The test began with selector valves being positioned to ensure that all pressurized chambers within the actuator were tested at 187.5 psi. Once the pressure transducer sensed that 187.5 psi was reached, the selector valves shifted to close off the pressurized air within the actuator. Pressure decay was the method used to detect gross leaks. After the valves closed and a stabilization period was complete, the system stored the starting pressure and monitored the decay pressure versus a 1 min test time. If pressure dropped below the recommended limit, pressure was vented and the unit failed the test. If pressure dropped below a preset value during stabilization, the unit would also fail. This prevented the potential for an excessively low starting pressure.

After this test passed, the unit would vent pressure, then begin to pressurize the actuator to the body leak test pressure with a closed-loop, microprocessor-controlled air regulator. The regulator was actuated by a stepper motor with slew and step inputs. To speed up the pressure cycle, slew rate was set to bring the test pressure up to 80% of test pressure. The step function then took over and the pressure transducer monitored pressure rise until the preselected pressure was reached. The body leak test was to detect extremely small leaks, typically in the range of 5 cm^3/min. This small leak rate, which was an external leak, was a challenge to detect because actuators under test could vary in internal volume from 9.3 to 3480 in.^{3}. Using a pressure decay method was impossible owing to the ratio of leak to the largest volume. Trying to detect a 5 cm^3/min leak in a closed chamber of a larger actuator with a volume of 2160 in.^{3} at 80 psi resulted in a pressure decay of 0.0022667 psi/min.

It was decided that a mass flowmeter with a range of 0–10 cm^3/min would be used to detect the leak. This was a case where circuit design was critical to produce an accurate test. Owing to the fact that the inlet flow to the actuator had to be measured to detect external leakage, a stable air supply was required. If, for example, the unit was constantly pressurized with pressure controlled by a precision air regulator, extreme inaccuracy would occur. This would be due to the hysterisis of the regulator. The regulator would open and close to maintain a pressure setting, but this opening and closing would provide an intermittent "shot" of air that would be detected as a gross leak. In addition, the temperature effect created when first pressurizing the actuator would also provide false signals. For example, with an actuator volume of 424 in.^{3}, a 1°F temperature change of the internal pressurized air would show up as a flow rate of 8.2 cm^3/min.

To overcome these problems, the flowmeter was placed between the actuator and a volume chamber. While the actuator was being pressurized, so too was the volume chamber. Once pressure was reached, both were isolated from the rest of the system and allowed to stabilize. Any air flowing out of the actuator would have to be replaced by air from the volume chamber, passing through the flowmeter. Because there were no restrictions, temperature fluctuations were negligible and hysterisis did not exist. The volume chamber was sized at a nominal volume of 670 in.^{3}. Because the test volumes would vary, the transfer of air between the chamber and actuator would be proportional to the ratio of the volume difference. For example, the smallest unit with 9.3 in.^{3} volume would read 9.8631 cm^3/min with an actual leak rate of 10 cm^3/min, while the largest unit with 3480 in.^{3} volume would read 1.6132 cm^3/min with an actual leak rate of 10 cm^3/min. (For a more detailed explanation of this effect, see Section 3.9.) The microprocessor had these volume ratios stored and compensated the flowmeter readings for the various actuator sizes. In addition to compensating for volume differences, the microprocessor also had to convert the flowmeter reading from actual cubic feet per minute to standard cubic feet per minute, because the flowmeter was measuring air at pressure rather than free air. Pressure was sensed, and the conversion was performed, based on this reading. Once this test passed, the system would proceed to the first piston seal test.

With the actuator still in its initial position, air was applied to one rotational side of the actuator with the pressure regulator. The opposite side of the piston was vented to atmosphere through a second mass flowmeter. Once pressure was reached, the flowmeter would detect actual leak rate without volume or pressure variation conversion.

ROTATIONAL TORQUE TEST

Once a pneumatic actuator passed all the previous tests, it was ready for torque testing. Electric actuator testing started at this point. Torque versus position testing was required to determine if an actuator produced torque output equal to or in excess of its published output readings. This was accomplished by using a load actuator to stall the test actuator during its full rotation. A rotary torque transducer was used between the load and test actuator to monitor actual output torque in inch pounds. Because of the wide torque range of actuators being tested, four torque transducers were used to maintain consistent accuracy (0–500, 0–5000, 0–20,000 and 0–100,000 in.-lbf ranges were used.) The system would prompt the operator to load the proper transducer at the beginning of the test, and sense which unit was installed. In the event the wrong transducer was loaded, a light would flash, and testing would not begin.

When the test started, a rotary encoder would sense rotation position and at 0°, 45°, and 90° the microprocessor would store break torque, running torque, and finish torque values from the torque transducer, as well as storing torque output at every 1° increment for the X–Y graph printed report. Rotational speed was important (as outlined in rotary actuator testing), because if speed were too fast, torque readings would be too low and would show up as a fail. If speeds were too fast on the electric actuators, gear damage could result from excessive torque caused by the load actuator. Fast speed on the pneumatic actuators would result in too low a test pressure owing to flow rate capabilities of the pressure regulator. Lower pressure means lower torque output, which means a failed test. Each actuator had to be tested to determine the best speed range possible to prevent these problems from occurring. These speeds were entered into memory for proper set point during an automatic test.

Once this test was completed, and it was a double-acting actuator, piston leak test would again be performed on the opposite side, and torque testing would be performed in the reverse rotation. With the testing complete, the CRT would ask if a printed report or a nameplate or both were required. The report was sent with each actuator and was serialized to that particular actuator. An automatic metal nameplate stamping machine would read downloaded data from the microprocessor and stamp all necessary data on the actuator nameplate.

5.3.2 Butterfly Valve Test Stand

This system was designed to leak test butterfly valves in 3 through 8 in. sizes with deionized water at pressures up to 3000 psi. Figures 5.18, 5.19, and 5.20 depict this system, which consisted of four independent test stations with a central pumping and control station. Each station had a 100 or 300 ton hydraulic clamp, a control station with CRT and printer, and a work area to assemble a new valve while another valve was undergoing tests.

MICROPROCESSOR SOFTWARE

The system contained a menu-selection of test valve part numbers, which identified valve size, pressure rating, configuration, and component material construction. Based on the part number selection, the proper test pressures, test times, go–no-go limits, clamp pressure, and sequence were displayed on the CRT. Once the system sensed that a valve was clamped in place, a start test button would initiate automatic sequence.

Again, the system allowed for a transducer calibration mode, manual test, or automatic test. Only two transducers were required in this system: test pressure at 0–3000 psi and differential pressure at 0–150 psid. Each transducer could be displayed in calibration mode with its range, serial number, and equivalent value calibration number. A transducer, when calibrated with a dead-weight tester, produces a value that maintains accuracy when used with the

Fig. 5.18. Butterfly valve test stand system. *(Courtesy Heco Division/Barker Rockford Co.)*

shunt resistor calibration method. The calibration button on the transducer signal conditioner, when depressed, would display the actual number versus the stored calibration number. Zero and span could then be adjusted to fine tune the transducer in the event of error drift.

In a manual test selection the operator could input all the parameters required for testing a unique design or for engineering evaluation of a unit outside its normal test parameters. The system was based on a pressure-decay test method, so the following data were required for test:

1. Shell test pressure
2. Shell pressure decay limit
3. Seat test (Side A) pressure
4. Pressure decay limit
5. Seat test (Side B) pressure
6. Pressure decay limit
7. Clamp pressure

Fig. 5.19. Test stand console configuration. *(Courtesy Heco Division/Barker Rockford Co.)*

In the automatic test mode, the system would look up these data in stored tables and adjust all operating conditions based on the input model number. Once the system verified proper model number and that the valve was clamped, the microprocessor would control fill time, shell test, seat test (side A), seat test (side B), and air blowdown.

PRESSURE-DECAY TEST SELECTION

In the initial discussions of test method required to detect the leak rates desired, it was determined that water pressure decay was mandatory owing to the small leak rates. In this case, industry standards dictated that 0.06 cm^3/min leak rate was acceptable at the rated pressure of the valve. As a safety margin, 0.05 cm^3/min leak rate was used as a fail point regardless of valve size or test pressure.

Initially an air pressure-decay method was considered, but based on the fact that test volumes ranged from 10 to 60 in.^{3} during shell test and 5 to 30 in.^{3} during seat test, this method proved impractical.

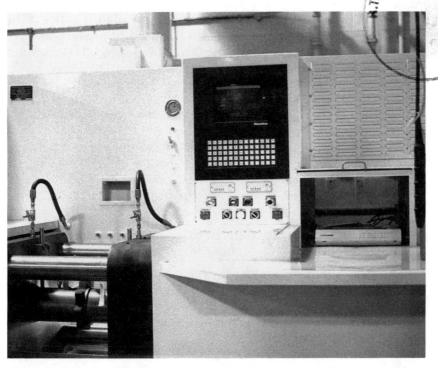

Fig. 5.20. Console controls and instrumentation. *(Courtesy Heco Division/ Barker Rockford Co.)*

To illustrate the expected results experienced with compressed air, we must use the formula $P_1 \times V_1 = P_2 \times V_2$, where pressure is expressed in absolute (= gage pressure + 14.7 psi).

If we compress a 10 in.^{3} volume to 80 psig, we find 20 in.$^3 \times 94.7$ psia = 64.4 in.$^3 \times 14.7$ psia. This means that once we have compressed the air in the 10 in.^{3} volume to 80 psig, we must vent 54.4 in.^{3} (64.4 − 10) of air to read 0 psig.

If we compress a 60 in.^{3} volume to 80 psig, we find 60 in.$^3 \times 94.7$ psia = 386.5 in.$^3 \times 14.7$ psia. We must now vent 326.5 in.^{3} to read 0 psig.

Therefore, we are working with a compressed volume range of 54.4–326.5 in.^{3} at 80 psig.

Pneumatic seat test—Again, actual volume is half of shell test volume, so we are working with volumes of 5–30 in.^{3} therefore, 5 in.$^3 \times 94.7$ psia = 32.2 in.$^3 \times 14.7$ psia and 30 in.$^3 \times 94.7$ psia = 193.3 in.$^3 \times 14.7$ psia.

Therefore, we are working with a range of 27.2 in.^{3} (32.2-5) to 163.3 in.^{3} (193.3-30) at 80 psig.

The maximum allowable leakage per ANSI standard is 0.1 ft³/hr or 2.88 in.³/min per inch of valve size. This would produce a range of 8.64 in.³/min (2.88 × 3) for the 3 in. valves and 23.04 in.³/min (2.88 × 8) for the 8 in. valves.

Using the air leak formula as defined in Chapter 3 we find that we must sense a range of 4.77 psi drop in 60 sec for an 8 in. shell test to 10.73 psi drop in 30 sec for a 3 in. seat test. Now let us evaluate the accuracy of our pressure transducer.

Pneumatic test instrumentation—Again, we are using a pressure transducer with a ± 0.1% full scale accuracy along with sensing electronics with an accuracy of ± 0.025% of reading. Because we are using a 0–100 psi transducer, this translates to ± 0.125 psi accuracy of reading:

$$4.77 \div 0.125 = 38.16 \text{ times more accuracy}$$
$$10.73 \div 0.125 = 85.84 \text{ times more accuracy}$$

This means that depending on the valve size we can expect an error in detection of leakage from 0.23 to 0.27 in.³/min (depending on valve size and the type of test) before considering the effect of temperature variations acting on the trapped volume of compressed air.

In addition, we must evaluate other causes of test inaccuracy.

Temperature effect on accuracy—This temperature effect is what makes the overall accuracy of any pneumatic test unpredictable, whether you employ pressure decay or a flowmeter to sense leakage rate.

For example, let us use the largest volume of 30 in.³ for seat test to determine the affect of 1°F temperature variation during a 30 sec test. During this test we have a compressed volume of 163.3 in.³ at 80 psig. The formula $T_1 \times V_1 = T_2 \times V_2$ is used to determine the change in volume due to temperature expressed in Rankin scale (460 + temperature in degrees Fahrenheit); (75°F) 535 × 163.3 in.³ = (74°F) 534 × 163.6 in.³. Therefore, for every degree of temperature change, we have a 0.2% change in volume, which is an additive effect on overall system accuracy. Without knowing the temperature fluctuations during the test, it is impossible to predict the overall system accuracy.

Now, we analyzed the accuracy level of the pressure decay method with water as the test media at rated valve pressures.

Hydro shell test—At 3000 psi, water volume is increased by 0.94%. Using this value: 10 in.³ × 0.94% = 0.094 in.³, and 60 in.³ × 0.94% = 0.564 in.³. Maximum allowable leakage per ANSI standard is 10 ml/hr or 0.01 in.³/min, Therefore, to find the pressure drop in 1 min: 0.01 in.^{3}/min ÷ 0.094 in.^{3} = 0.106383% × 3000 psi = 319.15 psi drop/min and 0.01 in.^{3}/min ÷ 0.564 in.^{3} = 0.01773% × 3000 psi = 53.19 psi drop/min.

Hydro seat test—At 3000 psi, and because the valve is closed, thus resulting in half the volume as during shell test, the results will be approximately twice the values of shell testing.

Therefore, a total estimated range for hydro seat testing is 106.38–638.3 psi drop/min. But, because the sensing time is 30 sec, we are back to the identical range of 53.19–319.15 psi drop/30 sec.

Hydro test instrumentation—The pressure transducer we intend to use has "worst case" error of ± 0.1% full scale. The microprocessor electronic circuitry that senses this transducer output has ± 0.025% maximum error. Together this equals a maximum of ± 3.025 psi total system accuracy (transducer range = 0–3000 psi).

Comparing this accuracy to the "worst case" leakage rates of 53.19–319.15 psi drop: 53.19 psi ÷ 3.025 psi = 17.58; 319.15 psi ÷ 3.025 psi = 105.5; 10 ml/hr ÷ 17.58 = 0.5688 ml/hr detection on 8 in. valves and 10 ml/hr ÷ 105.5 = 0.095 ml/hr detection on 3 in. valves.

This proved the point that a much greater degree of accuracy could be obtained using water over air.

SYSTEM CIRCUIT

Considering the complexity of the system architecture, the hydraulic circuit was relatively simple. As shown in Fig. 5.21, water was supplied to each station from a central fill pump. Water entered through a shutoff valve and a 1 micron nominal filter. This allowed air to be purged from the test valve and also supplied water to the inlet of an air-operated boost pump used to intensify pressure up to 3000 psi.

Two three-way valves normally blocked water flow and vented both sides of the valve under test. Two two-way valves were used to provide "bubbletight" shutoff of compressed water once the intensifier had reached test pressure. During the prefill stage, three-way valve AV was energized allowing water to flow through the test valve and back to tank. A timer then closed valve B and started the intensifier which remained on until pressure reached specified test pressure sensed by the pressure transducer.

At this point solenoid A was energized, trapping compressed fluid on both sides of the test valve, which was in an open condition. This is where the differential pressure transducer was used. A stabilization period allowed pressure to settle, and the transducer was pres-

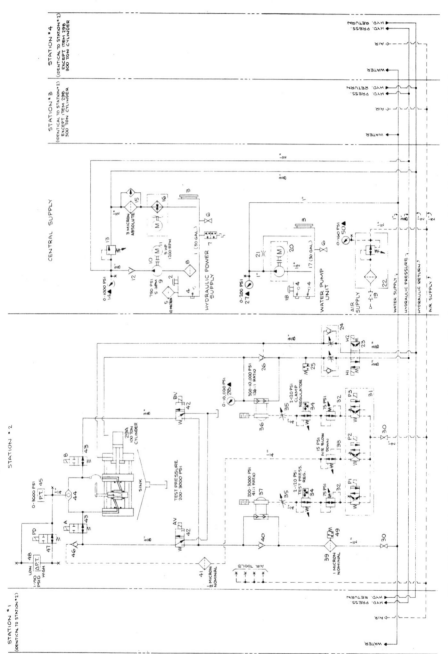

Fig. 5.21. Butterfly valve test stand schematic *(Courtesy Heco Division/Barker Rockford Co.)*

surized equally on both sides at this time, thus producing an output of 0 psi. A small accumulator was used on the high-pressure side of the transducer as a reference pressure. (See Chapter 3, Leak Testing, for more detail on this procedure.) Once stabilization was over, solenoid PD was energized, isolating the accumulator and high side of the transducer. As pressure dropped due to valve leakage, the transducer began producing an output based on the differential pressure between the accumulator reference pressure and the actual pressure in the test valve.

After a successful completion of this 60 sec shell test, leakage across the valve seat on both sides was conducted in the same manner for 30 sec with the test valve closed.

After testing was complete, the valve would be purged of water with an air blowdown circuit. The operator could request a printed report and have a metal nametag machine stamp the valve nameplate with all input data from the system. A conveyor would deliver the nameplate from the machine to the appropriate station, dropping the nameplate down a delivery chute.

Appendix 1: Seal Compatibility Chart*

LEGEND

A — RECOMMENDED
B — MINOR TO MODERATE EFFECT
C — MODERATE TO SEVERE
U — UNSATISFACTORY
BLANK — INSUFFICIENT DATA

CHEMICAL	Natural Rubber (Nitco AX-2033 / Qualitex AX-2071 / Duro AX-20750 / Amtek No. 16750 / No. 16530)	SBR, GRS (Ironflex AX-20065)	Butyl (Butene AX-90560)	EPT, EPDM (Restr-O-Norflex No. AX-60660)	Buna N, Nitrile, NBR (Oil Ace AX-1060 / Mincar No. 4045)	Hydrin Epi-Chloro-Hydrin (Compound No. GF-0860)	Neoprene (Matchless AX-1060 / Mitrlene No. 1050)	Hypalon (Chlorosol No. 19040)	Urethane, Polyurethane (Polvest No. UR-2970)	Polysulfide (Thiokol No. 23945)	Silicone (Thermoflex No. 2850R)	Fluorosilicone (Compound No. WK-8870)	Fluoro Elastomer Viton® Fluorel® (No. EV-9970)
Acetaldehyde	C	A	A	A	U		C	C	C	A	U	U	U
Acetamide	U	B	B	A	A		B	C	B	B	A	B	B
Acetic Acid 5%	B	A	A	A	B		A	A	B	A	A	A	A
Acetic Acid 30%	B	B	C	C	U		A	A	C	C	C	C	C
Acetic Acid, Hot High Press	C	U	B	B	U	B	C	C	B	U	C	C	C
Acetic Acid, Glacial	B	B	C	C	C		U	C	B	C	U	C	C
Acetic Anhydride	B	A	A	A	U		A	A	B	B	B	C	C
Acetone	U	A	A	A	U	U	U	U	C	U	U	U	U
Acetophenone	U	U	A	A	U	U	U	U	B	U	A	U	U
Acetyl Acetone	U	C	C	U	U	U	U	U		A	U	U	U
Acetyl Chloride	U	U	U	U	U	U	U	U	U	C	A	A	A
Acetylene	B	B	A	A	B	B	B	C	C	B	U	A	A
Acetylene Tetrabromide	U	U	U	U	U		C	U	U	U	A	A	A
Acrylonitrile	U	C	C	C	U		C	C	U	B	A	U	U
Adipic Acid	U	C	U	U	U		C	C	U	U	A	U	U
Aero Lubriplate	U	B	U	U	A		A	A	A	B	U	A	A
Aero Safe 2300	U	B	A	U	U		U	U	U	B	C	U	U
Aero Safe 2300W	U	B	A	U	U		U	U	U	C	C	U	U
Aero Shell IAC	U	U	U	U	A		A	A	A	B	U	A	A
Aero Shell 7A Grease	U	U	U	U	A		B	B	A	B	U	A	A
Aero Shell 17 Grease	U	U	U	U	B		U	U	U	B	B	U	U
Aero Shell 750	U	U	U	U	A		U	U	B	U	A	A	A
Aerozene 50 (50% Hydrazine 50% UDMH)	U	A	A	B	C		B	C	A	A	A	U	U
Air - Below 300°F	U	B	B	A	U		U	U	A	A	B	A	A
Air - Above 300°F	U	U	U	U	B		U	C	C	A	B	B	B
Alkazene	U	U	U	U	U		U	B	B	A	U	D	D
Alum-NH₃Cr-K	A	A	A	A	B		B	B	A	A	A	A	A
Aluminum Acetate	A	B	A	A	B	B	B	C	U	A	U	A	A
Aluminum Bromide	A	A	A	A	A		A	A	A	B	A	C	C
Aluminum Chloride	A	A	A	A	A		A	A	A	B	A	A	A
Aluminum Fluoride	B	B	B	A	A		A	C	U	U	A	A	A
Aluminum Nitrate	A	A	A	A	A		A	A	A	A	A	A	A
Aluminum Phosphate	A	A	A	A	A		A	A	B	A	A	A	A
Aluminum Salts	A	A	A	A	A		A	A	C	A	A	A	A
Aluminum Sulfate	B	A	A	A	A		A	A	C	A	C	A	A
Ambrex 33 Mobil	U	U	B	B	B		B	B	B	D	A	C	C
Amines, Mixed	B	B	B	B	U		B	U	B	B	U	D	D
Ammonia Anhydrous (Liquid)	A	U	A	A	B		U	U	C	A	A	A	A
Ammonia Gas, Cold	A	A	A	A	U		A	A	C	B	A	A	A
Ammonia Gas, Hot	U	U	B	B	B		B	B	U	U	B	D	D
Ammonia and Lithium Metalin Solution	U	U	B	B	U		U	U			B	U	U
Ammonium Carbonate	A	A	A	A	B		A	A	A	B	C	C	C
Ammonium Chloride	A	A	A	A	A		A	A	B	A	A	A	A
Ammonium Hydroxide (Concentrated)	C	U	A	A	B		B	B	C	A	B	C	C
Ammonium Nitrate	C	A	A	A	B		A	A	C	A	B	B	B

406

Ammonium Nitrite	U					U	A	A	A				
Ammonium Persulfate Solution	U	A			U	U	A	A	A				C
Ammonium Persulfate 10%		U					A	A	A				C
Ammonium Phosphate	A	B	B	B	U	A	A	A	B		A	C	A
Ammonium Phosphate, Mono-Basic	A	A	A	A	A	A	A	A	A		A	B	U
Ammonium Phosphate, Dibasic	A	A	A	A	A	A	A	A	A		A	B	U
Ammonium Phosphate, Tribasic	A	A	A	A	A	A	A	A	A		A	B	B
Ammonium Salts	A	A	A	A	A	A	A	A	A		A	B	A
Ammonium Sulfate	A	A	A	A	A	A	A	A	A		A	B	A
Ammonium Sulfide	A	B	B	U	A	A	A	A	U		U	B	U
Amyl Acetate	U	B	U	B	U	A	U	B	U		U	U	D
Amyl Alcohol	B	B	B	A	B	B	B	A	A		A	A	U
Amyl Borate	U	U	U	U	U	U	U	U	A		B	B	U
Amyl Chloride	U	U	U	U	U	U	U	U	U		U		A
Amyl Chloranaphthalene	U	U	U	U	U	U	U	U	U				A
Amyl Naphthalene	U	U	U	U	U	U	U	U	U		C	U	A
Anderol L-774 (di-ester)	U	U	U	U	U	U	B	U	U		C	B	A
Anderol L-826 (di-ester)	U	U	U	U	U	U	B	U	U		B	B	A
Anderol L-829 (di-ester)	U	B	B	U	B	U	B	U	U		B	B	A
Ang-25 (Glyceral Ester)	B	B	U	U	U	U	B	B	U		U	B	U
Ang-25 (Di-ester Base) (TG749)	U	U	U	A	J	B	B	U	B		U	U	U
Anhydrous Ammonia	U	U	C	A	A	U	A	A	A		B	B	C
Anhydrous Hydrazine	U	A	B	B	B	A	B	A	B	U	C	C	A
Anhydrous Hydrogen Fluoride	A	U	A	A	A	A	A	A	A		U	A	A
Aniline	U	U	B	B	U	U	B	J	U		U	B	A
Aniline Dyes	B	U	B	B	U	U	U	B	J		A	B	A
Aniline Hydrochloride	B	C	U	B	U	U	D	U	B	U	C	C	U
Aniline Oils	U	U	U	U	U	U	U	U	U		U	C	C
Animal Fats	U	U	U	U	U	U	U	C	U		A	A	A
Animal Oils (Lard Oil)	U	U	U	U	U	U	U	C	J	A	A	A	A
AN-O-3 Grade M	U	U	U	U	U	U	A	A	A		A	A	A
AN-O-6	U	U	U	U	U	U	A	U	A		A	A	A
AN-O-366	U	U	U	C	U	U	C	U	A		A	C	A
AN-VV-O-366b Hydr. Fluid	U	U	C	U	U	C	O	U	A		A	C	U
Ansul Ether	U	U	U	B	B	U	C	B	C	A	B	B	B
Aqua Regia	U	U	U	J	B	U	U	J	B		C	B	A
Argon	U	U	U	A	U	A	A	A	A		A	A	A
Aroclor 1248	U	U	U	U	J	U	D	U	U		D	J	A
Aroclor 1254	U	U	U	U	J	U	B	B	C		A	B	A
Aroclor 1260	U	U	U	U	U	U	U	U	J		C	B	U
Aromatic Fuel 50%	A	U	A	A	A	A	A	A	A		A	A	A
Arsenic Acid	U	U	U	U	B	U	U	U	B	A	U	B	A
Arsenic Trichloride	B	A	A	A	A	U	B	B	A		B	A	C
Askarel	U	U	U	U	U	U	A	A	A		B	B	A
ASTM Oil #1	U	U	U	U	J	B	B	A	C		B	A	A
ASTM Oil #2	U	U	U	U	U	U	U	A	C		A	C	A
ASTM Oil #3	U	U	U	U	U	U	U	B	C		C	B	A
ASTM Oil #4	U	U	U	U	U	U	U	B	J		C	B	A
ASTM Reference Fuel A	U	U	U	U	U	B	A	J	U		A	A	A
ASTM Reference Fuel B	U	U	U	U	U	U	A	U	B		B	B	A
ASTM Reference Fuel C	U	U	U	U	U	U	C	U	B		A	A	A
ATL-857	U	U	U	J	U	J	B	B	B		B	B	A
Atlantic Dominion F	B	B	B	U	U	U	B	C	B		B	A	A
Aurex 903R Mobil	J	J	J	J	C	A	B	B	U		A	U	A
Automatic Transmission Fluid		U	B	B	U	B	B	C	B		U	U	A
Automotive Brake Fluid		B	U	J	J	U	U	B	B		J	U	U
B													
Bardol B	U	U	U	U	U	U	U	U	U		B	B	A

CHEMICAL	Fluoro Elastomers Vitone® V/chem. No. EV-9970	Fluorosilicone Compound No. WK-8870	Silicone Thermoflex No. 2850R	Polysulfide Thiokol® No. 3345	Urethane, Polyurethane Polyest. No. UR-2970	Hypalon® Chloroel. No. 19040	Neoprene® Matchless Miprene No. AX-1060	Hydrin® Epi. Chloro-Hydrin Compound No. GF-0960	Buna N, Nitrile, NBR Oil Ace. AX-4060 Mincar No. 4045	EPR EPT, EPDM Resist. (N/cide) No. AX-60660	Butyl Butane - AX-90560	SBR, GRS Ironide No. AX-20065	Natural Rubber Mirce No. 18050 Quietex AX-2001 Durax AX-16750 Amtex No. 15030
Barium Chloride	A	A	A	A	A	A	A	A	A	A	A	A	A
Barium Hydroxide	A	A	A	A	A	A	A	A	A	A	A	A	A
Barium Salts	A	A	A	A	A	A	A	A	A	A	A	A	A
Barium Sulfate	A	A	A	A	A	A	A	A	A	A	A	A	A
Barium Sulfide	A	A	A	U	B	U	U	U	A	U	U	U	U
Bayol D	A	A	A	A	A	A	A	A	A	A	A	A	
Beer	A	A	A	A	A	A	A	A	A	A	A	A	A
Beet Sugar Liquors	A	U	U	U	U	C	C	C	U	U	U	C	U
Benzaldehyde	U	C	C	C	C	B	U	B	U	U	U	U	U
Benzene	A	B	B	C	C	A	U	U	C	U	U	U	U
Benzenesulfonic Acid	A	A	B	U	A	A	A	B	C	A	U	U	U
Benzine	A	A	A	U	B	U	B	U	A	C	U	U	U
Benzochloride	A	C	C	U	C	U	U	U	B	U	B	U	U
Benzoic Acid	A	B	C	B	B	U	U	U	B	B	B	B	U
Benzophenone													
Benzyl Alcohol	A	A	B	U	B	B	B	C	U	C	A	A	U
Benzyl Benzoate	A	B	C	C	U	U	U	U	B	B	A	U	U
Benzyl Chloride	A	A	B	B	U	U	U	U	B	U	U	U	U
Black Point 77	A	C	B	A	C	C	C	C	C	A	A	A	C
Black Sulphate Liquors	A	U	U	B	U	U	B	B	B	U	A	U	B
Blast Furnace Gas	A	U	B	A	B	A	U	A	B	A	A	A	U
Bleach Solutions	A	U	C	B	A	A	A	C	B	A	A	A	B
Bordeaux Mixture	A	U	A	A	A	A	A	A	A	A	A	A	B
Boric Acid	A	B	U	U	A	A	A	U	A	C	A	A	U
Boron Fluids (HEF)	A	B	B	U	U	U	U	A	U	B	U	A	U
Brake Fluid (Non-Petroleum)	A	C	U	A	A	B	B	A	C	U	B	U	U
Bray GG-130	U	U	U	B	B	U	B	U	U	A	U	U	U
Brayco 719-R (VV-H-910)	B	B	B	B	U	B	B	B	B	U	A	B	B
Brayco 885 (MIL-L-6085A)	U	B	B	C	U	U	U	U	B	U	U	U	U
Brayco 910	A	A	A	C	A	A	A	B	B	A	A	A	A
Bret 710	A	B	B	B	B	A	A	A	B	U	A	A	A
Brine													
Brom - 113	B	C	C	U	A	U	B	A	C	U	U	U	U
Brom - 114	A	C	C	B	B	B	C	U	B	B	U	U	U
Bromine													
Bromine Anhydrous	C	U	U	C	B	C	C	C	U	U	U	U	U
Bromine Pentafluoride	U	U	U	C	U	U	C	C	U	U	U	C	C
Bromine Trifluoride	A	U	U	U	C	U	U	C	U	U	U	C	C
Bromine Water	A	U	B	U	U	C	C	U	A	A	B	U	U
Bromobenzene	A	U	M	U	C	U	B	B	U	U	A	U	U
Bromochloro Trifluoroethane	U	A	B	B	A	B	B	B	B	B	A	A	U
Bunker Oil	U	B	U	C	U	U	B	B	B	C	U	A	U
Butadiene	A	A	A	U	A	B	B	B	U	A	C	A	U
Butane	A	A	A	U	A	B	B	B	U	A	U	U	U
Butane 2,2-Dimethyl	A	A	A	U	A	A	A	A	A	A	U	C	U
Butane 2,3-Dimethyl	B	B	B	B	B	B	B	A	A	A	U	U	U
Butanol (Butyl Alcohol)	A	A	A	U	B	B	A	A	A	B	B	A	A

Chemical	1	2	3	4	5	6	7	8	9	10	11	12	13
1-Butene, 2-Ethyl	U	U	U	U	A		U	U	U	A	U	C	A
Butter	U	U	B	A	A	A	B	B	A	U	B	C	A
Butyl Acetate	U	U	B	B	U	U	U	U	U	U	U	A	U
Butyl Acetyl Ricinoleate	U	U	A	A	B		B	B	U			B	A
Butyl Acrylate	U	U	U	U	U		U	U		B		U	U
Butyl Alcohol	A	A	B	B	A		A	A	U	B	B	A	A
Butyl Amine	U	U	U	U	C		U	U	U		B	U	U
Butyl Benzoate	U	U	B	A	U		U	U		U		A	A
Butyl Butyrate	U	U	A	A	U		U	U				A	A
Butyl Carbitol	U	U	A	A	B		B	B		U	U	U	C
Butyl Cellosolve	U	U	A	A	C		C	B	U			U	U
Butyl Cellosolve Adipate	U	U	B	B	U		U	U	U	B	B	B	B
Butyl Ether	U	U	C	C	C		U	U	B	A	U	C	U
Butyl Oleate	U	U	B	B	U		U	U				B	A
Butyl Stearate	U	U	B	B	B		U	U		A		B	A
Butylene	U	U	U	U	B		C	B	U	B	U	B	A
Butyraldehyde	U	U	B	B	U		U	U	U	C	U	U	A
Butyric Acid		U	B	B	U		U	U					B
C													
Calcine Liquors			A	A	A				U			A	A
Calcium Acetate	A	U	A	A	B		B	U	U	U	U	U	U
Calcium Bisulfite	U	U	U	U	A		A	A	A	U	A	A	A
Calcium Carbonate	A	A	A	A	A		A	A	C	U	A	A	A
Calcium Chloride	A	A	A	A	A	A	A	A	A	A	A	A	A
Calcium Cyanide	A	A	A	A	A		A	A		A	A		
Calcium Hydroxide	A	A	A	A	A	A	A	A	A	U	A	A	A
Calcium Hypochloride	U	U	A	A	U	B	U	A	U	U		A	A
Calcium Hypochlorite	U	U	A	A	C	B	C	A	U	U	B	B	A
Calcium Nitrate	A	A	A	A	A	A	A	A	A	A	B	A	A
Calcium Phosphate	A	A	A	A	A		B	A	A		A		A
Calcium Salts	A	A	A	A	A		A	A	A	A	B	A	A
Calcium Silicate	A	A	A	A	A		A	A					A
Calcium Sulfide	B	B	A	A	B	B	A	A	A	U	B	A	A
Calcium Sulfite	B	B	A	A	A		A	A	A	B	A	A	A
Calcium Thiosulfate	B	B	A	A	B		A	A	A	B	A	A	A
Caliche Liquors	A	A	A	A	A		A	A	A	A	B	A	A
Cane Sugar Liquors	A	A	A	A	A	A	A	A	U	U	A	A	A
Caproic Aldehyde	B		B	B					U	B	B	U	U
Carbamate	U	U	B	B	C		B	B	U	B		A	A
Carbitol	B	B	B	B	B		B	B	U	B	B	B	B
Carbolic Acid	U	U	B	B	U		C	C		U	U	A	A
Carbon Bisulfide			U	U	C	U	U	U		C		A	A
Carbon Dioxide, Dry	B	B	B	B	A	A	B	B	A	B	B	B	B
Carbon Dioxide, Wet	B	B	B	B	A		B	B		B	B	B	B
Carbon Disulfide	U	U	U	U	U		U	U				A	A
Carbon Monoxide	B	B	A	A	A	A	A	A	A	U	A	B	A
Carbon Tetrachloride	U	U	U	U	C	B	U	U	U	C	U	A	A
Carbonic Acid	A	B	A	A	B		A	A	A	A	A	A	A
Castor Oil	A	A	B	B	A	A	A	A	A	C	A	A	A
Cellosolve	U	U	B	B	U		U	U	U	B	U	U	U
Cellosolve Acetate	U	U	B	B	U		U	U	U	B	U	U	U
Cellosolve Butyl	U	U	B	B	U		U	U	U	B	U	U	U
Celluguard	A	A	A	A	A		A	A	U	A	A	A	A
Cellulube A60 Now Fyrquel)	U	U	B	A	U		U	U	U	B		C	B
Cellulube 90, 100, 150, 220, 300, 500	U	U	A	A	U		U	U	U	U	A	B	A
Cellutherm 2505A	U	U	U	U	B		U	U	U	U		B	A
Cetane (Hexadecane)	U	U	U	U	A		B	B	U	A	U	C	A
China Wood Oil (Tung Oil)	U	U	C	U	A		B	C	C	C	U	B	A
Chloracetic Acid	U	U	B	B	U		U	U	U	U		U	U

Column legend (left → right): Fluoro Elastomer® Viton® Fluorel® V/Chem.® No. EV-9970 · Fluorosilicone Compound No. WK-8870 · Silicone Thermoflex - No. 2850R · Polysulfide Thiokol® - No. 2394S · Urethane, Polyurethane Polycast - No. UR-2970 · Hypalon® Chlorosol - No. 19040 · Neoprene® Mirprene - AX-1060 · Epi. Chloro-Hydrin Hydrin Compound No. GF-0960 · Buna N, Nitrile, NBR Oil Acet. - AX-1060 Mincer - No. 4045 · EPR, EPT, EPDM Restr. O (Nordel®) No. AX-6060© · Butyl Butene - AX-9060 · SBR, GRS Irondex - AX-2006S · Natural Rubber Dalatex - AX-1520 Buro - AX-3201 Amtex - No. 15030

CHEMICAL	Fluoro Elast.	Fluorosilicone	Silicone	Polysulfide	Urethane	Hypalon	Neoprene	Hydrin	Buna N	EPDM	Butyl	SBR	Nat. Rubber
Chlorodane	A	B	B	U	U	C	C		B	U	U	U	U
Chlorextol	A	B	A	U	U	B	B		B	U	U	U	B
Chlorinated Salt Brine	A	A	A	C	C	C	C		C	U	U	U	U
Chlorinated Solvents, Dry	A	A	A	C	C	C	C		C	U	U	U	U
Chlorinated Solvents, Wet	A	A	A	C	C	C	C		C	A	U	U	U
Chlorine, Dry	A	B	A	C	B	U	U	B	C	C	C	C	U
Chlorine, Wet	A	B	B	C	C	C	C	B	C	C	C	C	U
Chlorine Dioxide	U	B	B		U	U	U		C	U	U	C	C
Chlorine Dioxide (8% Cl as NaClO2 in solution)	U	U	U			C	C		C	U	U	C	C
Chlorine Trifluoride		U						U					
Chloroacetone	A	B	B	A	U	U	U		U	U	U	U	U
Chloroacetic Acid	A	B	B	B	C	A	B		U	A	B	B	B
Chlorobenzene	B	B	B	C	C	U	U		C	B	U	U	U
Chlorobenzene, (Mono)	A	B	B			U	U		C	B	U	B	U
Chlorobromo Methane	A	A	A	C	C	U	B		C	U	U	B	U
Chlorobutadiene	A	A	A	C	U	U	U		U	U	U	U	U
Chlorododecane	A	B	B	C	U	U	U		U	U	U	U	U
Chloroform	C	A	A	C	U	U	U		U	U	U	U	U
O-Chloronapthalene	C	B	B	C	U	B	B		U	U	U	U	U
1-Chloro 1-Nitro Ethane	A	U	A	D	C	U	U		U	U	U	U	U
Chlorosulfonic Acid	A	U	B	A	A	B	B		B	B	B	B	U
Chlorotoluene	A	B	B	C	U	U	U		U	U	U	U	U
O-Chlorphenol	A	A	B	D	U	U	B	A	B	A	B	A	A
Chrome Alum	A	B	A	A	C	A	A		A	A	C	A	A
Chrome Plating Solutions	A	C	B	B	U	B	B		U	U	C	U	U
Chromic Acid	A	C	C	C	U	U	U		U	C	C	U	A
Chromic Oxide .88 Wt. % Aqueous Solution	A	A	A		A	A	A	A	A	A	B	A	A
Circo Light Process Oil	A	A	A	A	B	B	B		A	B	B	B	A
Citric Acid	A	A	A	U	U	U	U		A	U	U	A	A
City Service Koolmotor-AP Gear Oil 140-E.P. lube	A	A	B		A	B	B		A	U	U	U	U
City Service Pacemaker #2	A	A	A		B	B	B		A	U	U	U	U
City Service #65, #120, #250	A	A	A		B	B	B		A	U	U	U	U
Cobalt Chloride	A	A	A	B	U	A	A		A	A	A	A	A
Cobalt Chloride, 2N	A	A	A	A	U	A	C		A	A	A	B	U
Cocoanut Oil		C	C		C	U	U		A	C	B	U	U
Cod Liver Oil	A	B	A	A	C	B	B		A	U	A	A	A
Coke Oven Gas	A	A	A	A	U	A	A		A	A	A	U	U
Coffee	C	C	B	A	C	U	U		U	C	U	U	A
Coliche Liquors	A	C	B	B	D	A	A		B	B	B	B	A
Convelex 10	U	U	U		U	B	B		B	B	U	U	U
Coolanol (Monsanto)	A	A	A		C	C	C		A	U	U	U	U
Coolanol 45 (Monsanto)	C	C	U		U	U	U		B	A	A	A	A
Copper Acetate	A	B	A	B	C	A	A		A	A	A	A	A
Copper Chloride	A	C	C	U	U	B	B		A	B	A	A	B
Copper Cyanide	A	D	A		C	U	U		A	A	A	A	A
Copper Salts	A	A	A	A	U	A	A		A	A	A	A	A
Copper Sulfate	A	A	A	U	U	A	A		A	A	A	A	B

Fluid																
Copper Sulfate 10%	A	A	A	U	B	A	A	B			B	B	B	A		
Copper Sulfate 50%	A	A	A	A	C	A	A	A			B	B	A	U		
Corn Oil	A	A	A	U	A	C	U	U			C	U	U	U		
Cottonseed Oil	A	B	U	U	C	U	U	U			C	U	U	U		
Creosols	A	A	U	C	U	U	U	U			U	U	U	U		
Creosote	A	A	U	C	U	U	U	U			U	U	U	U		
Creosote, Coal Tar	A	B	U	C	U	U	U	U			U	U	U	U		
Creosote, Wood	A	A	C	C	U	U	U	U			U	U	U	U		
Creosylic Acid	B	B	U	U	B	B	B	B			A	U	U	U		
Crude Oil	B	A	U	U	U	U	U	U			B	B	B	U		
Cumene	B	B	U	A	B	U	C	C			U	U	U	U		
Cutting Oil	A	C	U	B	A	A	U	U			A	U	U	U		
Cyclohexane	A	A	U	B	U	U	U	U			B	C	C	U		
Cyclohexanol	A	A	U	C	U	U	U	U			U	U	U	U		
Cyclohexanone	A	U	U	B	B	B	B	B			B	U	U	U		
P-Cymene	A	B	U	D	U	U	U	U			U	U	U	U		

D																
Decalin	U	U	U	D	U	U	U	U			U	U	U	U		
Decane	A	C	U	B	U	C	A	A	A		B	U	U	U		
Delco Brake Fluid	A	A	A	A	B	A	A	B	A		A	A	B	C		
Denatured Alcohol	A	B	A	A	A	A	A	B			C	A	A	A		
Detergent Solutions	A	B	A	B	A	A	A	A			U	B	U	U		
Developing Fluids (Photo)	A	U	B	B	A	A	A	A			U	U	U	U		
Dextron	A	U	U	A	B	B	U	U			U	U	U	U		
Diacetone	A	U	U	U	A	U	U	U			C	A	A	A		
Diacetone Alcohol	B	U	U	U	B	U	U	U			U	U	U	U		
Diazinon	U	U	B	B	B	C	C	C			U	U	U	U		
Dibenzyl Ether	A	U	U	U	U	U	U	U			U	U	U	U		
Dibenzyl Sebacate	B	U	U	U	U	U	U	U			U	U	U	U		
Dibromoethyl Benzene	A	U	U	U	U	U	U	U	B		U	U	U	U		
Dibutylamine	A	U	U	U	U	U	U	U			U	U	U	U		
Dibutyl Ether	A	U	U	U	C	U	C	U			C	U	U	C		
Dibutyl Phthalate	C	U	A	B	A	A	A	A			U	U	U	U		
Dibutyl Sebacate	B	B	U	U	B	C	U	U			A	A	A	B		
O-Dichlorobenzene	A	U	U	U	A	A	B	B			U	U	U	U		
P-Dichlorobenzene	A	U	U	U	A	C	U	U			U	U	U	U		
Dichloro-Butane	U	U	U	U	B	U	U	U			C	U	U	U		
Dichloro-Isopropyl Ether	U	U	U	U	U	U	U	U	A		U	U	U	U		
Dicyclohexylamine	U	U	U	U	U	U	U	U			C	U	U	B		
Diesel Oil	A	B	U	B	B	B	B	B			A	B	B	A		
Di-ester Lubricant MIL-L-7808	U	U	U	U	C	U	U	B			B	C	C	U		
Di-ester Synthetic Lubricants	U	U	U	U	C	U	U	U			C	U	U	C		
Diethylamine	U	B	B	B	B	U	U	U			U	B	B	A		
Diethyl Benzene	A	U	U	U	U	U	U	U	A		C	A	C	C		
Diethyl Ether	U	U	U	U	B	U	U	U			D	B	B	A		
Diethyl Sebacate	U	B	B	A	U	U	U	U			U	U	U	U		
Diethylene Glycol	A	A	A	A	A	A	A	A			A	B	B	A		
Difluorodibromomethane	U	U	U	U	C	C	C	C			A	U	U	U		
Diisobutylene	U	U	U	B	B	A	A	B			U	U	U	B		
Diisooctyl Sebacate	U	U	U	U	U	U	U	U			C	C	C	B		
Diisopropyl Benzene	U	U	U	U	U	U	U	U			B	B	B	U		
Diisopropyl Ketone	A	B	B	B	U	U	U	U			U	U	U	U		
Dimethyl Aniline	U	B	B	B	C	B	B	B			B	B	B	B		
Dimethyl Formamide	U	U	B	B	B	U	U	U			U	U	U	U		
Dimethyl Phthalate	U	U	U	U	U	U	U	U			B	B	B	B		
Dimethyl Phthalate	U	U	U	U	U	U	U	U			U	U	U	U		
Dinitro Toluene	U	U	U	U	U	U	U	U			U	B	B	U		

CHEMICAL	Natural Rubber Pure... AX-1093, Duo... AX-2201, Amtex No. 15030	SBR, GRS Ironside AX-20065	Butyl Butene AX-90560	EPR EPT EPDM Restr'O (Norde..) No. AX-50660	Buna N Nitrile NBR Oil Ace AX-4.1060, Mincar No. 4045	Epi-Chloro-Hydrin Compound No. GF-0960	Neoprene Matchless AX 1060, Miprene No. 1050	Hypalon Chlorosol No. 19040	Urethane, Polyurethane Polycast No. UR-2970	Polysulfide Thiokol® No. 3394S	Silicone Thermoflex No. 2850R	Fluorosilicone Compound No. WK-8870	Fluoro Elastomers Viton® Fluorel® Vi-Chem No. EV-9970
Dioctyl Phthalate	U	B	B	B	U		U	U	B	C	B	B	B
Dioctyl Sebacate	U	B	B	B	C		B	B	C	C	C	C	B
Dioxane	U	C	B	B	C		U	U	C	U	C	C	U
Dioxolane	U	C	C	D	C		U	U	C	U	C	C	C
Dipentene	U	D	C	U	B		U	U	A	U	C	C	A
Diphenyl	U	D	D	D	A		U	U	B	A	B	B	U
Diphenyl Oxides	U	B	B	A	U		U	B	A	U	B	B	A
Dow Chemical 50-4	U	D	U	A	C		B	U	U	B	U	U	U
Dow Chemical ET378		A	A	A	A		B	B	A	A	A	A	A
Dow Chemical ET588	A	B	A	A	C		B	B	A	U	B	A	A
Dow Corning - 3	A	A	A	A	A		A	A	A	C	A	A	A
Dow Corning - 4	A	A	A	A	A		A	A	A	C	A	A	A
Dow Corning - 5	A	A	A	A	A		A	A	A	C	A	A	A
Dow Corning - 11	A	A	A	A	A		A	A	A	C	A	A	A
Dow Corning - 33	A	A	A	A	A		A	A	A	C	A	A	A
Dow Corning - 44	A	A	A	A	A		A	A	A	C	A	A	A
Dow Corning - 55	A	A	A	A	A		A	A	A	C	A	A	A
Dow Corning - 200	A	A	A	A	A		A	A	A	C	A	A	A
Dow Corning - 220	A	A	A	A	A		A	A	A	C	A	A	A
Dow Corning - 510	A	A	A	A	A		A	A	A	C	A	A	A
Dow Corning - 550	A	A	A	A	A		A	A	A	C	A	A	A
Dow Corning - 704	A	A	A	A	B		A	A	A	C	A	A	A
Dow Corning - 705	A	A	A	A	B		A	A	A	C	A	A	A
Dow Corning - 710	A	A	A	A	A		A	A	A	C	A	A	A
Dow Corning - 1208	A	A	A	A	A		A	A	A	C	A	A	A
Dow Corning - 4050	A	A	A	A	A		A	A	A	C	A	A	A
Dow Corning - 6620	A	A	A	A	A		A	A	A	U	A	A	A
Dow Corning - F60	A	A	A	A	A		A	A	A	B	A	A	A
Dow Corning - F61	A	A	A	A	A		A	A	A	C	A	A	A
Dow Corning - XF60	U	A	A	A	U		A	A	A	A	A	A	A
Dow Guard	U	U	U	U	U	U	U	C	U	B	A	U	A
Dowtherm Oil	U						U	B	U	D	B		U
Dowtherm A or E	A	A	A	A	A	B	B	U	U	C	U	A	A
Dowtherm 209, 50% Solution	U	U	U	D	C		B	B	U	A	A	B	B
Drinking Water	U	U	U	A	C		B	U	U	B	B		A
Dry Cleaning Fluids				B			U	U	U	U	U		A
DTE Light Oil	U	U	U	U	A		B	B	B	U	U		B
E													
Elco 28-EP Lubricant		U	U	U	A		C	A	U	U	B	A	A
Epichlorohydrin	U	B	B	U	C		C	C	C	D	C	U	U
Epoxy Resins		A	A	A			A	C	C	A			C
Esam-6 Fluid	U	B	B	D	U		B	B	B	B	B	D	D
Esso Fuel 208	U	D	A	D	A		B	C	A	U	B	A	A
Esso Golden Gasoline	U	D	U	U	B		U	U	U	U	U	A	B
Esso Motor Oil	U	D	U	U	A		C	C	B	C	U	A	A
Esso Transmission fluid (type A)	U	U	U	U	B		B	U	A	U	A	A	B
Esso WS3812 (MIL-L-7808A)	U	U	U	U	A		U	C	B	U	A	A	U

Esso XP90-EP Lubricant	U	U	U	U	A		B	B	A	A	U	A	A
Esstic 42, 43	U	U	U	U	A		B	U	B		U	A	A
Ethane	U	U	U	U	A		B	B	C	A	U	B	A
Ethanol	A	A	A	A	C		A	A	U	A	B	A	C
Ethanol Amine	B	B	B	B	B	B	B	C	C	B	B	U	U
Ethers	U	U	U	C	U		U	U	B	A	U	C	C
Ethyl Acetate-Organic Ester	U	U	B	B	U	U	U	U	U	B	B	U	U
Ethyl Acetoacetate	C	C	B	B	U		U	U	U	B	B	U	U
Ethyl Acrylate	U	U	B	B	U	U	U	U	U	B	B	U	U
Ethyl Acrylic Acid	U	U	B	B	U		B	U	U		U	U	
Ethyl Alcohol	A	A	A	A	C	A	A	A	U	A	B	A	A
Ethyl Benzene	U	U	U	U	U	U	U	U	U	U	U	A	A
Ethyl Benzoate	U	U	U	U	U		U	U	U	B	U	A	A
Ethyl Bromide	U		U	U	B		U	U				A	A
Ethyl Cellosolve	U	U	B	B	U		U	U	U	B	U	U	U
Ethyl Cellulose	B	B	B	B	B		B	B	B	U	C	U	U
Ethyl Chloride	B	B	A	A	A	B	B	U	B	U	U	A	A
Ethyl Chlorocarbonate	U	U	U	U	U		U	U	U	U	U	B	A
Ethyl Chloroformate	U	U	U	U	U		U	U	U	U	U	B	A
Ethyl Cyclopentane	U	U	U	U	A		C	U	A	B	U	A	A
Ethyl Ether	U	U	C	C	C	B	U	U	B	A	U	C	U
Ethyl Formate	U	U	B	B	U	U	B	B		U		A	A
Ethyl Hexanol	A	A	A	A	A		A	A	U	B	B	A	A
Ethyl Mercaptan	U	U	U	U	U	U	C	C		U	C		B
Ethyl Oxalate	U	U	U	U	U	U	U	U	A	A	U	B	A
Ethyl Pentachlorobenzene	U	U	U	U	U	C	U	U	U	U	U	B	A
Ethyl Silicate	B	B	A	A	A	A	A	B		B		A	A
Ethylene					A							A	A
Ethylene Chloride	U	U	U	U	U		U	U	U	B	U	C	B
Ethylene Chlorohydrin	B	B	B	B	U		B	B	U	B	C	B	A
Ethylene Diamine	B	B	A	A	A	A	A	B	U	U	A	U	U
Ethylene Dibromide	U	U	C	C	U		U	U	U	U	U	C	A
Ethylene Dichloride	U	U	C	C	U	U	U	U	U	U	U	C	A
Ethylene Glycol	A	A	A	A	A	A	A	A	B	C	A	A	A
Ethylene Oxide	U	U	C	C	U	U	U	U	U		U	U	U
Ethylene Trichloride	U	U	C	C	U		U	U	U	U	U	C	A
Ethylmorpholene Stannous Octoate (50/50 mixture)		U	B	B	U								U
F													
F-60 Fluid (Dow Corning)	A	A	A	A	A		A	A	A	A	U	A	A
F-61 Fluid (Dow Corning)	A	A	A	A	A		A	A	A	A	U	A	A
Fatty Acids	C	U	U	U	B		B	U		U	C		A
FC-43 Heptacosofluorotri-butylamine		U	A	A	A		A	A		A	A	A	A
FC75 Fluorocarbon		U	A	A	A		A	A		A	A	B	B
Ferric Chloride	A	A	A	A	A	A	B	B	A	A	B	A	A
Ferric Nitrate	A	A	A	A	A	A	A	A	A	A	C	A	A
Ferric Sulfate	A	A	A	A	A	A	A	A		A	B	A	A
Fish Oil					A						A	A	A
Fluoroboric Acid	A	A	A	A	A		A	A					
Fluorine (Liquid)			C	C						U	U		B
Fluorobenzene	U	U	U	U	U		U	U			U	B	A
Fluorocarbon Oils			A	A									
Fluorolube		U	A	A	A		A	A		A	A	B	B
Fluorinated Cyclic Ethers			A	A									
Fluosilicic Acid	A				A		A	A					
Formaldehyde	B	C	A	A	C	B	C	C	U	B	B	U	U
Formic Acid	A	A	A	A	B	B	A	A	U		B	C	C
Freon, 11	U	U	U	U	B		U	A	U	A	U	B	A
Freon, 12	B	A	B	B	A	A	A	A	A	A	U	U	B
Freon, 12 and ASTM Oil #2 (50/50 mixture)	U	U	U	U	A		B	B			U	B	A

CHEMICAL	Natural Rubber	SBR, GRS	Butyl	EPR EPT, EPDM	Buna N, Nitrile, NBR	Hydrin	Neoprene	Hypalon	Urethane, Polyurethane	Polysulfide	Silicone	Fluorosilicone	Fluoro Elastomer
Freon, 12 and Suniso 4G (50/50 mixture)	D	D	D	D	A		B	B	A	D	B	A	A
Freon, 13	A	A	A	A	A		B	B	A	U	C	C	A
Freon, 13B1	C	B	C	C	C		A	A	C	U	B		U
Freon, 14	A	A	A	A	A		A	A	C	U			U
Freon, 21	U	U	C	C	C	B	C	U	C	U			C
Freon, 22	A	A	A	A	A	A	A	A	B	U			B
Freon, 22 and ASTM Oil #2 (50/50 mixture)	B	B	A	A	A		B	A	C	U	U		B
Freon, 31	A	A	A	A	A		A	B	A	U		B	U
Freon, 32	U	U	C	C	C		A	A	A	C			U
Freon, 112	U	U	C	C	C	B	A	A	A	C	B		B
Freon, 113	A	A	A	A	A	A	A	A	B	C	C		B
Freon, 114	A	A	A	A	A		A	A	A	C			B
Freon, 114B2	A	A	A	A	A		A	A	B	A			B
Freon, 115	A	A	A	A	A	A	A	A	A	C		A	A
Freon, 142b	A	A	A	A	A		A	A	A	C			U
Freon, 152a	A	A	A	A	A		A	A	A	A		A	A
Freon, 218	A	A	A	A	A	A	C	C		A		A	
Freon, C316	A	A	A	A	A		A	A	A	D	A		A
Freon, C318	C	C	C	C	B		A	U	A	C	A		B
Freon, 502	U	D	U	U	B		B	U	U	U	A		B
Freon, BF	U	U	C	D	B	U	U	B	U	C	A		B
Freon, MF	U	U	C	C	B		D	D	B	U	U		U
Freon, TF	D	B	U	A	B		A	A	A	C		A	A
Freon, TA	A	A	A	B	A		B	B	A	U		A	A
Freon, TC	B	B	B	B	B	A	B	B	A	A	A	A	A
Freon, TMC	C	U	U	U	U		U	U	A	U	U		B
Freon, T-P35	A	A	A	A	A	U	B	B	A	C		B	A
Freon, T-WD602	C	B	B	B	B		B	B	A	C	A		A
Freon, PCA	U	U	U	U	U		U	U	U	C	A		A
Fuel Oil	U	U	U	U	U		U	U	B	C		U	
Fuel Oil, Acidic	U	U	U	C	U		B	B	U	C		D	A
Fuel Oil, #6	U	B	B	B	B		U	B	B	A			B
Fumaric Acid	U	B	B	B	A	U	U	B	U	D	A	B	
Fuming Sulphuric Acid (20/25% Oleum)	U	U	U	C	C		D	D	C	C			U
Furan (Furfuran)	U	U	U	U	U		C	B	B	C	C	U	A
Furfural	D	D	D	U	D		U	U	B	C	C	B	
Furfuraldehyde	U	B	B	B	B		U	B	B	C	C		D
Furfuryl Alcohol	U	B	B	B	C	A	U	B	B	C	C		A
Furyl Carbinol		B	A	A	B		U	B	B				
Fyrquel A60	A	A	A	B	B	A	B	A	A	A	A	D	U
Fyrquel 90, 100, 150, 220, 300, 500													
G													
Gallic Acid	A	B	B	B	B		B	B		D	A		U
Gasoline	U	U	U	D	U		B	B	C	A	A	U	A
Gelatin	A	A	A	A	A	A	A	A	A	B	C	A	A
Girling Brake Fluid		B	B	B	B		A	A	C	U	C	C	U
Glacial Acetic Acid	B	A	A	B	C		B	D	B	B		A	U

Glauber's Salt	B				B	B			B	U	B	B		B	A	B
Glucose	A	U	U		A	A	A	D	A	A	A	A	A	A	A	A
Glue (Depending on type)	A	A	A		A	A	A	A	A	A	A	A	A	A	A	A
Glycerine-Glycerol	A	A	A		A	A	A	A	A	A	A	A	A	A	A	A
Glycols	A	B	A		A	A	A	A	A	A	A	A	A	A	A	A
Green Sulphate Liquor	B					B	D	D	A	B	B	A	A	B	A	A
Gulfcrown Grease	U	U	D		D	U	U	A	U	D	U	U		D	A	A
Gulf Endurance Oils	U	U	D		D	U	U	A	U	U	U	U		D	B	A
Gulf FR Fluids (Emulsion)		U	B			A	A		A	A	A	A	A	A	B	A
Gulf FRG-Fluids	A	U	A		A	B	U		A	B	B	A	B	A	A	A
Gulf FRP-Fluids	U	U	B		B	U	U	A	U	U	U	U		U	A	A
Gulf Harmony Oils	U	U	U		U	U	U	U	U	U	U	U		U	A	A
Gulf High Temperature Grease	U	U	U		U	U	U	A	U	U	U	U		U	A	A
Gulf Legion Oils	U	U	U		U	U	U	A	U	U	U	U		U	A	A
Gulf Paramount Oils	U	U	U		U	U	U	A	U	U	U	B		B	A	A
Gulf Security Oils	U	U	U		U	U	U	A	U	U	B	B		B	A	A
H																
Halothane	U	U	U		U	D	D		U	D	D	U		D	B	A
Halowax Oil	U	U	U		D	D	D		D	A	A	A		B	A	A
Hannifin Lube A	U	B	U		A	A	A		A	U	B	A		A	A	A
Heavy Water	A	U	A		A	A	A	A	A	A	A	A	A	A	A	A
HEF-2 (High Energy Fuel)	A	U	A		B	A	A		A	D	U	A		A	U	A
Helium	A	D	D		U	A	A		A	A	A	A	A	A	A	A
N-Heptane	U	U	U		A	A	B		B	A	B	B		A	A	A
N-Hexaldehyde	U	U	U		U	U	C		C	B	C	U		C	U	U
Hexane	A	A	C		C	C	C		C	B	B	B	A	B	A	A
N-Hexane-1	A	A	U		A	A	U		U	U	B	D		D	A	A
Hexyl Alcohol	A	A	A		A	A	D		D	A	A	A		A	A	A
High Viscosity Lubricant, U4		A	A		A	A	A		A	A	A	A		A	A	A
High Viscosity Lubricant, H2	U	A	A		B	B	B	A	B	B	C	B		U	A	U
Hilo MS#1		B	B		B	B	B		B	C	B	B		C	U	B
Houghto-Safe 271 (Water and Glycol base)	U	A	A		A	A	A		D	D	A	A	D	B	A	B
Houghto-Safe 620 (Water/Glycol)		A	A		A	A	A		C	U	D	B		B	A	B
Houghto-Safe 1010, Phosphate Ester	D	U	A		A	A	A		C	C	U	U	U	B	A	B
Houghto-Safe 1055, Phosphate Ester	D	U	A		A	A	B		C	C	U	U	U	B	B	A
Houghto-Safe 1120, Phosphate Ester	D	U	A		A	A	A		D	A	U	U	U	B	A	A
Houghto-Safe 5040 (Water/Oil emulsion)	D	U	A		A	A	A	A	A		U	U	U	B	A	A
Hydraulic Oil (Petroleum Base)	U	U	U		U	U	A		U		U	U	U	A	A	A
Hydrazine	A	B	A		A	A	B		B	A	B	A		C	C	U
Hydrobromic Acid	A	U	A		A	U	U		U	U	A	C		C	A	D
Hydrobromic Acid 40%	U	U	C		C	U	U		U	U	A	C		D	U	A
Hydrocarbons (Saturated)	D	B	A		U	U	B		B	B	U	A		D	B	A
Hydrochloric Acid Hot 37%	B	U	C		C	U	U		B	B	D	U		B	C	A
Hydrochloric Acid Cold 37%	C	C	A		A	A	C		A	A	C	B		C	A	A
Hydrochloric Acid 3 Molar	U	U	A		A	A	U		U	C	B	U		C	C	A
Hydrochloric Acid Concentrated	C	B	C		C	A	U		A	A	D	A		B	B	A
Hydrocyanic Acid	B	B	A		A	A	U		B	A	B	C		A	A	A
Hydro-Drive, MIH-50 (Petroleum Base)	U	U	U		U	U	A		A	A	A	A		A	A	A
Hydro-Drive, MIH-10 (Petroleum Base)	U	B	C		C	U	B		B	U	C	C		C	B	B
Hydrofluoric Acid, 65% Max. Cold	B	U	U		U	U	U		U	U	B	U		B	U	D
Hydrofluoric Acid, 65% Min. Cold	U	U	U		U	U	U		D	U	D	U		A	C	U
Hydrofluoric Acid, 65% Max. Hot	U	U	U		U	U	U		U	U	B	U		A	C	C
Hydrofluoric Acid, 65% Min. Hot	U	U	C		C	U	B		C	U	U	U		A	B	C
Hydrofluosilicic Acid	A	B	A		A	A	A		B	A	A	A		A	A	A
Hydrogen Gas, Cold	B	B	A		A	A	A		U	A	U	C		B	A	A
Hydrogen Gas, Hot	B	U	C		C	U	A		B	B	C	U		C	A	B
Hydrogen Peroxide (1)	U	A	A		A	A	A		A	A	A	B		A	B	U
Hydrogen Peroxide 90% (1)	A	A	A		A	A	A		A	A	A	C			C	U
Hydrogen Sulfide Dry, Cold	A		A		A	A	A		A	U	A	A		U	A	U

CHEMICAL	Natural Rubber (Micros No. 2031, Qualatex AX-2031)	SBR, GRS (Ironside AX-20065, Duraflex No. 18030, Amtex No. 18030)	Butyl (Butene AX-90560)	EPR EPT EPDM (Restr.O(Nordel®) No. AX-60660)	Buna N, Nitrile NBR (Oil Resist. AX-40060, Mincer. No. 4060)	Epi-Chloro-Hydrin, Hydrin (Compound No. GF-0960)	Neoprene® (Matchless AX-1060, Miprene No. 1060)	Hypalon® (Chlorosol No. 19040)	Urethane, Polyurethane (Polyest No. UR-2970)	Polysulfide (Thiokol® No. 2394S)	Silicone (Thermoflex No. 2850R)	Fluorosilicone (Compound No. WK-8870)	Fluoro Elastomers (Viton® Fluorel® ViChem No. EV-9970)
Hydrogen Sulfide Dry, Hot	U	U	A	A	U		B	C	A	A	C	C	U
Hydrogen Sulfide Wet, Cold	U	C	A	A	U	B	A	B	A	A	C	C	U
Hydrogen Sulfide Wet, Hot	U	C	B	A	U		B	C	C		C	C	C
Hydrolube-Water/Ethylene Glycol		A	B	U	C		B	B	C	B	B	A	A
Hydroquinone	B	U	B	U	B		U	U	C		U	U	U
Hydyne	B	B	B	A	U		B	U	C		B	U	U
Hyjet	U	U	B	A	U		U	D	C		C	C	U
Hyjet III	C	U	B	A	U		U	C	C		U	C	U
Hyjet S	C	U	B	A	U		U	C	C		U	C	U
Hyjet W	U	U	B	A	U		U	U	C		U	B	U
Hypochlorous Acid	B	C	B	B	C	B	U	B	C		A	C	A
I													
Industron FF44		U	U	C	A		B	B	B		C	A	A
Industron FF48		U	U	C	A		B	B	B		C	A	A
Industron FF53		U	U	C	A		B	B	B		C	A	A
Industron FF80		B	B	C	B		C	B	A		U	A	U
Iodine	U	U	B	D	U		U	C	U		C	U	C
Iodine Pentafluoride													
Iodoform	A	B	A	A	B	U	B	U	B	U	U	B	A
Iso-Butyl Alcohol	U	U	U	C	U		A	B	C	A	A	A	A
Iso-Butyl N-Butyrate	U	U	U	C	U		U	B	B	U	A	A	A
Isododecane	U	U	U	B	A		B	U	A	B	D	D	D
Iso-Octane	U	D	A	A	A	A	U	A	C	U	A	A	A
Isophorone (Ketone)	A	B	U	B	B		B	A	U	A	B	B	U
Isopropanol	U	U	U	A	U		D	U	B	U	U	U	C
Isopropyl Acerate	A	B	A	B	U		B	A	U	A	J	A	A
Isopropyl Alcohol	U	U	U	A	U		J	A	C	A	B	A	A
Isopropyl Chloride	U	U	U	U	B		C	U	A	U	C	C	U
Isopropyl Ether	U	U	U	U	B		C	U	A	U	C	C	U
J													
JP 3 (MIL-J-5624)	U	U	U	U	A		U	U	B	U	U	U	A
JP 4 (MIL-J-5624)	U	U	U	U	A		U	U	B	U	U	C	A
JP 5 (MIL-J-5624)	U	U	U	U	A		U	U	B	U	U	C	A
JP 6 (MIL-J-25656)	U	U	U	U	A		B	U	A	U	U	U	D
JP X (MIL-F-25604)	U	U	U	U									
K													
Kel F Liquids	U	C	B	A	A	A	C	B	B	A	B	B	A
Kerosene	U	C	C	C	A		C	A	B	C	B	A	A
Keystone #87HX-Grease					A		B		A		B	A	A
L													
Lactams-Amino Acids	U	U	U	B	U		B	U	U	U	U	U	U
Lactic Acid, Cold	A	A	A	A	C		A	A	B	A	A	A	A
Lactic Acid, Hot	U	C	C	C	U		C	U	B	C	B	A	A
Lacquers	U	U	U	U	U		C		A	U	U	C	C

Lacquer Solvents	U	U	U	U	U	U	U	U	U	A	U	U	U
Lactic Acids	A	A	A	A	A		A	A		U	A	A	A
Lard, Animal Fats	U	U	U	U	A	A	C	U	A	U	B	A	A
Lavender Oil	U	U	U	U	B		U	U	U	B	U	B	A
Lead Acetate	A	U	A	A	B	B	B	U	U	U	U	U	U
Lead Nitrate	A	A	A	A	A		A	U		U	B	A	
Lead Sulphamate	B	B	A	A	B		A	A		U	B	A	A
Lehigh X1169	U	U	U	U	A		B	B	A	A	U	A	A
Lehigh X1170	U	U	U	U	A		B	B	A	A	U	A	A
Light Grease	U	U	U	U	A		U	U	A	A	U	A	A
Ligroin (Petroleum Ether or Benzine)	U	U	U	U	A		B	C	B	A	U	A	A
Lime Bleach	A	A	A	A	A		B	B		U	B	A	A
Lime Sulphur	U	U	A	A	U		A	A		U	A	A	A
Lindol, Hydraulic Fluid (Phosphate ester type)	U	U	A	A	U		U	U	U	U	C	C	B
Linoleic Acid	U	U	U	U	B		U	U		U	B		B
Linseed Oil	U	U	C	C	A		C	C	B	B	A	A	A
Liquid Oxygen	U	U	U	U	U		U	U	U	U	U	U	U
Liquid Petroleum Gas (LPG)	U	U	U	U	A	A	B	U	A	A	C	C	A
Liquimoly	U	U	U	U	A		B	U	B		U	A	A
Lubricating Oils, Di-ester	U	U	U	U	B		C			C	U	B	A
Lubricating Oils, Petroleum Base	U	U	U	U	A	A	B	U	B	C	U	A	A
Lubricating Oils, SAE 10, 20, 30, 40, 50	U	U	U	U	A		B	U	B	C	U	A	A
Lye Solutions	B	B	A	A	B		B	A	U	B	B	B	B
M													
Magnesium Chloride	A	A	A	A	A	A	A	A	A	C	A	A	A
Magnesium Hydroxide	B	B	A	A	B	A	B	A	B	C			A
Magnesium Sulphate	B	B	A	A	A	A	A	A		B	A	A	A
Magnesium Sulphite	B	B	A	A	A		A	A		B	A	A	A
Magnesium Salts	A	A	A	A	A		A	A	A	A	A	A	A
Malathion	U	U	U	U	B						U	B	A
Maleic Acid	U	U	U	U	U		U	U		B			A
Maleic Anhydride	U	U	U	U	U		U	U					A
Malic Acid	A	B	U	U	A		B	B			B	A	A
MCS 312	U	U	U	U	U		U				A	A	A
MCS 352	U	U	B	A	U		U	U	U	U	C	C	U
MCS 463	U	U	B	A	U		U	U	U	U	C	C	U
Mercuric Chloride	A	A	A	A	A	A	A	A					A
Mercury	A	A	A	A	A	A	A	A	A	A			A
Mercury Vapors	A	A	A	A	A		A	A					A
Mesityl Oxide (Ketone)	U	U	B	B	U		U	U	U	B	U	U	U
Methane	U	U	U	U	A	A	B	B	C	A	U	B	A
Methanol	A	A	A	A	A		A	A	U	B	A	A	A
Methyl Acetate	U	U	B	B	U	U	B	U	U	B	U	U	U
Methyl Acetoacetate			B	B	U		U	U	U	B	B	U	U
Methyl Acrylate	U	U	B	B	U		B	U	U	B	U	U	U
Methylacrylic Acid	U	U	B	B	U		B	U	U		U	U	C
Methyl Alcohol	A	A	A	A	A	B	A	A	U	B	A	A	U
Methyl Benzoate	U	U	U	U	U		U	U	U	B	U	A	A
Methyl Bromide	U		U	U	B		U	U				A	A
Methyl Butyl Ketone	U	U	A	A	U		U	U	U	A	U	U	U
Methyl Carbonate	U	U	U	U	U		U	U	U	U	U	B	A
Methyl Cellosolve	U	U	B	B	C		B	U	U		U	U	U
Methyl Cellulose	B	B	B	B	B		B	B	B	U	B	U	U
Methyl Chloride	U	U	C	C	U		U	U	U	C	U	B	A
Methyl Chloroformate	U	U	U	U	U		U	U	U	U	U	B	A
Methyl D-Bromide	U	U			U		U	U	U	U	U	B	A

CHEMICAL	Natural Rubber (Milco No. 7033 / Dotex No. X-7201 / Duro No. 7203)	SBR, GRS (Ironsides AX-20065)	Buna N Nitrile (Amrex No. 15050)	Butyl Butene AX-90560	EPR, EPT, EPDM Resist-C (Nordel®) No. AX-60660	Buna N Nitrile, NBR (Oil Ace AX-41060 / Mincer No. 4045)	Hydrin Epi-Chloro-Hydrin Compound No. GF-0960	Neoprene (Merchias AX-1060 / No. 1060)	Hypalon® Chlorosol No. 19040	Urethane, Polyurethane Polycast No. UR-2970	Polysulfide Thiokol® No. 2394S	Silicone Thermolax No. 2850R	Fluorosilicone Compound No. WK-8970	Fluoro Elastomers Viton® Fluorel® Vi-Chem No. EV-9970
Methyl Cyclopentane	U	U	U	U	U			U	U	U	U	B	B	A
Methylene Chloride	U	U	U	U	U			U	U	U	U	B	B	B
Methylene Dichloride	A	A	A	A	A	U		C	U	C	A	C	C	C
Methyl Ether	U	U	U	U	U	C		U	U	U	U	U	U	U
Methyl Ethyl Ketone (MEK)	U	B	B	A	B	C		U	U	U	U	U	U	U
Methyl Ethyl Ketone Peroxide	U	C	C	U	C	C		U	U	U	U	U	U	U
Methyl Formate	U	B	B	B	B	U		B	B	B	U	U	U	U
Methyl Isobutyl Ketone (MIBK)	U	B	B	B	B	U		U	U	U	U	U	U	U
Methyl Isopropyl Ketone	U	U	U	U	U			U	U	U	U	U	U	U
Methyl Methacrylate	U	U	U	U	U	U		U	U	U	C	C	C	A
Methyl Oleate	U	C	C	C	A	C		C	C	C	B	B	B	A
Methyl Salicylate		C	C	C	C			B	A					
MIL-L-644B	C	C	C	U	A	A	A	C	C	C	C	C	A	A
MIL-L-2104B	U	U	U	U	U	A	A	A	A	A	U	A	A	A
MIL-L-2105B	U	U	U	U	U	A	A	B	B	A	A	A	A	A
MIL-G-2108	U	U	U	U	U	A	A	U	A	A	U	B	B	A
MIL-S-3136B, Type I Fuel	U	U	U	U	B	A	A	U	A	A	U	A	A	A
MIL-S-3136B, Type II Fuel	U	U	U	U	U	A	A	U	B	A	U	A	A	A
MIL-S-3136B, Type III Fuel	U	U	U	U	U	A	A	U	C	A	U	A	A	A
MIL-S-3136B, Type IV	U	U	U	U	U	A	A	U	A	A	U	A	A	A
MIL-S-3136B, Type V	U	U	U	U	U	A	A	U	B	A	U	A	A	A
MIL-S-3136B, Type VI	U	U	U	U	U	A	A	C	U	A	U	A	B	A
MIL-S-3136B, Type VII	U	U	U	U	U	A	A	U	C	B	U	B	A	A
MIL-L-3150A	U	U	U	U	U	A	A	U	C	A	B	A	A	A
MIL-G-3278	U	U	U	U	U	A	A	U	U	A	C	A	A	A
MIL-L-3503	C	U	U	C	A	A	A	C	C	A	C	B	B	B
MIL-L-3545B	U	U	U	U	B	B	B	U	C	C	C	B	A	A
MIL-C-4339C	U	U	U	U	A	A	A	B	A	A	U	A	A	A
MIL-G-4343B	U	U	U	U	A	A	A	B	A	A	U	A	A	A
MIL-L-5020A	U	U	U	U	A	A	A	C	C	C	U	A	B	A
MIL-J-5161F	U	U	U	U	A	A	A	U	B	A	B	A	A	A
MIL-C-5545A	B	A	A	A	A	A	B	C	C	A	C	B	A	A
MIL-H-5559A	U	U	U	U	A	A	B	C	A	A	U	A	A	A
MIL-F-5566	A	A	A	A	A	A	A	B	A	A	B	B	B	A
MIL-G-5572	U	U	U	U	A	A	A	C	B	A	U	A	B	A
MIL-F-5602	U	U	U	U	A	A	A	U	C	A	C	A	B	B
MIL-H-5606B	U	U	U	U	A	A	A	C	C	C	U	A	A	A
MIL-J-5624G, JP-3	U	U	U	U	A	A	A	U	U	B	U	A	A	A
MIL-J-5624G, JP-4	U	U	U	U	A	A	A	U	B	B	U	A	B	A
MIL-J-5624G, JP-5	U	U	U	U	A	A	A	U	C	B	C	A	B	A
MIL-L-6081C	U	U	U	U	A	A	A	U	C	A	C	A	A	A
MIL-L-6082C	U	U	U	U	A	A	A	B	B	A	D	A	A	A
MIL-H-6083C	C	U	U	U	B	B	B	U	U	A	U	B	B	B
MIL-L-6085A	U	U	U	U	B	B	B	U	C	B	C	B	A	B
MIL-L-6086B	U	U	U	U	A	B	A	U	C	C	C	A	A	A
MIL-A-6091	U	A	A	A	A	B	B	A	A	A	A	A	A	A
MIL-L-6387A	A	U	U	U	U	B	U	U	D	U	U	A	A	A
MIL-C-6529C	U	U	U	U	B	A	A	C	C	C	A	B	A	A

Specification													
MIL-F-7024A	A	A	D	A	B	D	D	B	A	D			
MIL-H-7083	B	B	B	C	D	B	B	B	A	A	A	B	B
MIL-G-7118A	A	A	D	B	C	C	C	B	B	D	C	D	D
MIL-G-7187	A	B	D	A	A	D	C	A	B	D	D	D	D
MIL-G-7421A	A	B	D	B	B	D	B	B	B	D	B	A	B
MIL-H-7644	A	A	D	C	C	B	D	B	B	A	D	D	D
MIL-L-7645	A	A	D	C	C	D	D	B	B	D	D	D	D
MIL-G-7711A	A	B	C	A	D	D	D	A	A	D	D	D	D
MIL-L-7808F	B	B	D	B	D	D	D	B	B	D	D	D	D
MIL-L-7870A	B	B	B	C	D	D	D	A	B	A	D	A	B
MIL-C-8188C	A	A	D	A	K	A	B	A	A	A	D	D	D
MIL-A-8243B	A	A	D	B	U	C	B	C	A	D	A	D	A
MIL-L-8383B	A	C	D	C	A	U	U	A	B	C	D	A	C
MIL-H-8446B	A	B	C	A	L	U	U	U	A	U	D	A	D
MIL-L-8660B	A	A	C	B	C	U	U	B	A	D	A	U	U
MIL-L-9000F	A	A	D	U	U	A	A	A	A	A	D	A	A
MIL-T-9188B	A	A	C	A	C	A	A	A	A	A	D	A	A
MIL-L-9236B	A	A	U	A	B	B	B	B	A	A	A	C	A
MIL-E-9500	A	A	C	A	B	B	B	B	A	A	A	A	A
MIL-L-102395A	A	A	D	A	C	B	B	A	A	D	A	D	A
MIL-L-10324A	A	B	D	D	U	B	D	B	A	A	A	A	A
MIL-G-10924B	A	A	C	A	B	C	C	C	A	A	D	A	A
MIL-L-11734B	A	A	C	A	B	C	C	C	A	A	A	A	A
MIL-O-11773	A	A	C	U	C	U	C	A	A	A	C	A	A
MIL-P-12098	A	A	D	U	B	B	B	B	B	C	B	B	B
MIL-H-13862	A	A	C	C	B	B	B	U	U	A	U	U	U
MIL-H-13866A	A	C	C	U	B	U	B	A	B	U	A	A	B
MIL-H-13910B	A	A	D	A	A	B	B	U	A	B	B	B	B
MIL-H-13919A	A	A	D	C	A	A	A	A	C	U	U	U	U
MIL-L-14107B	A	A	C	U	C	B	B	B	A	U	U	U	U
MIL-L-15016	A	A	C	A	C	A	B	A	A	U	U	U	U
MIL-L-15017	A	A	U	D	U	A	B	A	A	U	U	U	U
MIL-L-15018B	A	B	D	B	C	C	C	A	A	B	U	B	B
MIL-L-15019C	A	A	C	A	U	U	U	C	A	U	U	C	U
MIL-L-15719A	A	A	D	B	C	C	C	B	A	U	C	A	U
MIL-G-15293	A	A	C	A	C	B	B	B	A	U	U	A	U
MIL-F-16884	A	A	C	C	A	A	C	A	A	U	U	U	U
MIL-F-16929A	A	A	D	D	B	U	C	D	U	U	C	A	U
MIL-L-16958A	A	B	U	B	C	B	A	U	B	U	U	U	U
MIL-F-17111	A	A	C	A	A	A	A	U	A	U	U	U	U
MIL-L-17331D	A	A	U	A	A	A	A	U	A	U	A	U	U
MIL-L-17353A	A	A	U	D	A	D	D	A	U	U	U	U	U
MIL-L-17672B	A	A	C	A	C	U	U	A	A	U	U	U	U
MIL-L-18486A	A	A	C	B	A	C	C	A	U	U	U	U	U
MIL-G-18709A	U	D	C	A	C	B	C	D	D	A	A	A	A
MIL-H-19457B	A	A	C	A	A	A	A	B	A	U	U	U	U
MIL-F-19605	A	B	D	B	A	U	A	A	B	U	U	U	U
MIL-L-19701	A	A	C	A	U	U	B	U	A	A	A	A	A
MIL-L-21260	A	A	D	A	A	A	A	A	B	U	U	U	U
MIL-G-21568A	U	D	C	A	C	C	B	B	B	A	A	A	A
MIL-H-22072	A	B	D	B	A	A	A	A	A	A	A	A	A
MIL-H-22251	A	B	B	M	C	B	B	B	B	A	A	A	A
MIL-L-22396	A	A	C	B	C	B	B	A	A	D	C	D	D
MIL-L-23699A	A	A	C	B	C	C	C	A	A	D	C	D	D
MIL-G-23827A	A	A	C	A	C	B	C	B	A	D	U	D	D
MIL-G-25013D	A	A	U	B	C	C	C	A	A	D	U	D	D
MIL-F-25172	A	A	C	A	C	C	C	A	A	D	U	D	D
MIL-L-25336B	A	A	D	A	C	U	U	A	A	D	C	D	D
MIL-F-25524A	A	A	U	A	C	U	U	B	A	D	U	D	D
MIL-G-25537A	A	A	D	A	B	B	B	B	A	D	U	D	U

CHEMICAL	Fluoro Elastomers Viton®/Fluorel® EV-9970	Fluorosilicone Compound WK-8870	Fluorosilicone Thermoflex 2850R	Silicone / Polysulfide Thiokol® 23945	Urethane Polycast UR-2970	Urethane Polyurethane Chloronol® 19040	Hypalon® AX-1060 / Matchliss 1060	Neoprene® Epi-Chloro-hydrin GF-0960	Hydrin (Oil Ace AX-41060 / Mincar 4045)	Buna N Nitrile NBR Resist-O AX-60660	EPR, EPT, EPDM Butane AX-90560	Butyl Ironsides AX-20065	SBR, GRS Qualitex AX-2011 / Dunlex AX-16650 / Amtex 15030	Natural Rubber
MIL-F-25558B	A	A	A	D	A	B	B	B	A	A	C	D	D	C
MIL-F-25576C	A	A	A	D	A	C	C	C	A	A	C	D	D	C
MIL-H-25598	A	A	A	C	A	B	B	B	A	A	C	D	D	C
MIL-F-25656B	A	A	B	D	B	C	C	C	A	B	A	A	B	D
MIL-L-25681C	A	B	B	D	B	C	B	B	B	B	C	D	D	D
MIL-G-25760A	A	B	B	D	A	B	C	C	A	A	C	D	C	D
MIL-L-25968	A	A	A	C	A	A	A	A	B	A	C	A	A	D
MIL-L-26087A	A	A	A	C	U	A	B	B	A	A	C	D	D	D
MIL-G-27343	A	A	A	C	C	C	C	B	A	A	A	D	A	A
MIL-P-27402			B	B					B	B	A	B	B	U
MIL-H-27601A	A	B	A	D	C	C	B	B	B	A	A	D	U	B
MIL-G-27617	B	B	B	D	A	C	C	C	A	A	A	A	A	B
MIL-I-27686D	A	A	A	B	A	A	B	B	A	A	A	C	C	A
MIL-L-27694A	A	A	A	C	B	C	A	C	B	A	D	D	D	D
MIL-L-46000A	A	A	A	C	B	A	C	C	A	A	A	D	D	D
MIL-H-46001A	A	A	A	C	A	A	A	A	A	A	A	A	A	A
MIL-L-46002	A	A	A	C	A	B	B	B	A	A	C	B	A	B
MIL-H-46004	A	A	B	C	C	C	B	B	A	A	A	D	U	U
MIL-P-46046A	A	A	A	A	A	B	A	A	A	A	A	D	D	A
MIL-H-81019B	A	A	A	B	B	C	B	B	A	A	D	D	D	D
MIL-S-81087	A	A	A	D	B	U	C	B	A	A	U	D	D	D
MIL-H-83282	A	A	A	C	B	A	D	D		A	D	D	D	A
Milk	A	A	A	A	B	B	B	B	A	A	A	A	A	A
Mineral Oils	A	A	A	D						A	D	D	D	D
Mobil 24 DTE	A	B	B	C			B	A	A	A	D	A	A	D
Mobil HF	A	B	B	C	B	B	B	B		A	D	D	D	D
Mobil Delvac 1100, 1110, 1120, 1130	A	A	A	C	B	B	B	B		A	D	A	B	A
Mobil Nyvac 20 and 30				A	B	A	A	A		A	D	D	D	U
Mobil Velocite C				B	A	B	B	B		A	D	D	D	U
Mobilgas WA200, Type A, Automatic trans. fluid					B	B	B	B	U	A	B	B	B	U
Mobil Oil SAE 20	A	B	B	A	B	B	B	B		A	D	D	D	D
Mobiltherm 600	A	A	A	B	C	B	B	B		A	D	D	D	D
Mobilux	A	B	B	D	C	B	B	B		A	D	D	D	D
Mono Bromobenzene	A	C	C	D	C	C	C	C		D	D	D	D	D
Mono Chlorobenzene	A	C	C	D	C	C	C	C		D	D	D	D	D
Mono Ethanolamine	A	B	B	C	B	C	A	A		B	B	D	B	B
Monomethyl Aniline	A	A	A	A	A	D	B	A	A	A	A	A	A	D
Monomethylether	A	A	A	B	C	U	B	B	B	B	U	A	U	B
Monomethyl Hydrazine	A	A	A	A		D	U	U	B	C	A	B	A	B
Mononitrotoluene & Dinitrotoluene (40/60 Mix.)	A	A	B	B			D	D						U
Monovinyl Acetylene	C	C	C	U	B	B	B	B						B
Mopar Brake Fluid	C	C	B	B	U	U	B	B						
Mustard Gas	A	U	U	C	C	C	B	B						A
N				A										

Naptha	U	U	U	B	U	U		C	U	U	U	B	B	A
Napthalene	U	U	U	A	U	U		B	U	U	U	B	A	A
Napthenic Acid	C	C	U	U	U	U		A	U	U	U	A	C	A
Natural Gas	A	A	B	B	A	B	A	A	U	B	A	B	A	A
Neatsfoot Oil	U	U	B	A	A	U		A	A	A	A	A	B	A
Neon	A	A	A	B	B	B	A	B	B	B	B	A	B	U
Neville Acid	A	A	B	A	A	A		A	A	A	A	A	A	A
Nickel Acetate	A	A	A	B	B	C		B	B	B	C	A	U	A
Nickel Chloride	A	A	A	A	A	B		B	B	B	B	A	A	A
Nickel Salts	A	A	A	A	A	A		A	A	A	A	C	A	A
Nickel Sulfate	B	B	A	A	A	B		A	A	A	A	C	A	A
Niter Cake	A	U	A	C	A	A		A	A	A	A	A	C	C
Nitric Acid (1) 3 Molar	U	C	C	U	U	U	U	U	U	U	U	U	U	B
Nitric Acid (1) Concentrated	U	U	U	U	U	U	U	U	U	U	U	B	U	B
Nitric Acid Dilute	U	U	U	U	U	U	U	U	U	U	U	U	U	A
Nitric Acid (1) Red Fuming (RFNA)	U	U	U	U	U	U	U	U	U	U	U	U	U	A
Nitric Acid (1) Inhibited Red Fuming (IRFNA)	U	U	U	U	U	U	U	U	U	U	U	U	U	A
Nitrobenzene								U	C	C	C	C	K	A
Nitrobenzine	B	B	C	B	C	B		A	A	D	A	A	A	A
Nitroethane	A	A	B	B	A	A	A	U	U	U	U	D	C	C
Nitrogen	U	C	U	C	C	U		U	U	U	U	D	U	A
Nitrogen (Textroxide (N2O4) (1)	B	U	U	U	C	C		C	C	C	C	A	U	U
Nitromethane	B	C	B	B	B	B	B	U	C	U	U	D	U	U
Nitropropane	U	U	U	U	U	U		U	U	U	U	U	U	U
O														
O-A-548A	B	A	B	A	B	B	B	A	B	B	B	B	B	B
O-T-634b	U	U	U	U	U	U	C	C	U	U	U	U	U	A
Octachloro Toluene	U	U	U	U	U	U		A	U	U	U	C	U	A
Octadecane	U	U	U	U	B	U		B	U	B	B	A	B	A
N-Octane	B	B	B	A	B	U		B	B	A	B	B	B	A
Octyl Alcohol	U	U	U	B	B	D		C	U	C	B	U	B	B
Oleic Acid	U	U	U	U	U	U		U	C	U	C	U	D	U
Oleum (Fuming Sulfuric Acid)	U	U	U	U	U	B		B	B	B	C	A	C	C
Oleum Spirits	U	U	U	U	U	U		A	U	B	B	U	U	U
Olive Oil	U	B	B	B	A	U		A	A	A	A	B	A	A
Oronite 8200	U	U	U	U	U	U		B	U	U	U	A	A	A
Oronite 8515	U	U	U	U	U	U		U	U	U	U	A	A	A
Orthochloro Ethyl Benzene	U	U	U	U	U	U		U	U	U	U	U	B	A
Ortho-Dichlorobenzene	U	U	U	U	U	U		B	U	B	B	U	B	A
OS 45 Type III (IOS45)	U	U	U	U	U	U		B	B	B	B	A	A	A
OS 45 Type IV (IOS45-1)	U	U	U	U	U	U		B	B	B	B	A	A	A
OS 70	B	U	B	A	A	D	C	B	B	B	C	B	B	A
Oxalic Acid	B	B	C	A	A	A	B	B	A	A	A	A	A	U
Oxygen, Cold	U	U	U	A	A	U	U	U	U	D	U	U	C	C
Oxygen, Cold 200-400°F	U	U	U	B	A	U	A	U	A	U	C	A	D	U
Ozone								D	A	C	A	A	B	U
P														
P-S-661b	U	U	U	U	U	U	A	A	C	C	C	U	A	A
P-D-680	U	U	U	U	U	U	A	A	C	C	C	U	A	A
Paint Thinner, Duco	B	B	B	U	U	B	B	U	U	U	U	B	B	B
Palmitic Acid	U	U	U	B	C	C		A	D	B	B	B	A	A
Para-dichlorobenzene	U	U	U	U	U	U		A	U	U	U	U	D	A
Par-al-Ketone	U	U	B	C	U	B		A	C	A	A	B	A	A
Parker O Lube	B	B	U	U	B	B		B	B	B	B	A	A	A
Peanut Oil	U	U	U	C	C	B		B	A	B	B	A	C	A
Pentane, 2 Methyl	U	U	U	U	U	D		U	U	U	U	U	C	A
Pentane, 2,4 dimethyl	U	U	U	U	U	D		U	U	U	U	U	C	A
Pentane, 3 Methyl	U	U	U	U	U	U		U	U	U	U	U	C	A

CHEMICAL	Viton® Fluorel® Fluoro Elastomer (Vi-chem EV-9970)	Fluorosilicone (Compound WK-8870)	Silicone (Thermoflex 2650R)	Polysulfide (Thiokol® 2945)	Urethane, Polyurethane (Polycast UR-2970)	Hypalon® (Chlorosol 19040)	Neoprene® (Mirprene AX-1060)	Hydrin (Epi-Chloro-Hydrin GF-0960)	Buna N Nitrile NBR (Oil Ace AX-4160 / Miner 4045)	Buna N Nitrile NBR (AX-60660 Resistoflex)	EPR, EPT, EPDM (Butrane AX-90560)	Butyl (Buryl AX-90560)	SBR GRS (Ironsides AX-20065)	Natural Rubber (Amtex 16500 etc.)
N-Pentane	A	A	C	D	A	D	B		A	D	D	D	D	D
Perchloric Acid	A		A	D	B		B	C	C	B	B	B	C	D
Perchloroethylene	A		B	D	B	C	D	B	U	D	D	D	D	D
Petroleum Oil, Crude	A	A	A	D	D	A	B		A	A	D	D	B	D
Petroleum Oil, Below 250°F	A	B	D	D	D	B	D	B	C	D	D	D	D	U
Petroleum Oil, Above 250°F	A		D	D	D	D	D		C	D	D	D	D	D
Phenol	A	A	B	C	D	D	D		C	D	D	D	D	D
Phenol, 70%/30% H2O	A	A	B	C	D	D	D		C	D	D	D	D	D
Phenol, 85%/15% H2O	A	A	C	C	D	D	D		C	D	D	D	D	D
Phenylbenzene	C			C	C	C	C		D	D	D	D	D	A
Phenyl Ethyl Ether	D	A	D	D	B	D	B		D	D	C	C	U	B
Phenyl Hydrazine	A		B	B	B	B	B	U	B	B	A	B	B	A
Phorone	A	A	B	C	B	B	B		B	B	A	B	B	B
Phosphoric Acid 20%	A		B	C	C	C	B		B	B	B	B	B	C
Phosphoric Acid 45%	A	A	D	C	A	C	A		C	C	B	B	C	D
Phosphoric Acid, 3 Molar	A	A	B	C	B				U		B	B	C	B
Phosphoric Acid, Concentrated	B	C	D		C	B	C		B	B	A	U	U	D
Phosphorous Trichloride	A		A			U	U		B	B	C	C	C	U
Pickling Solution	A	C	D	D		C	C	U	B	C	C	C	C	B
Picric Acid, H2O Solution	A	D	B	D	B	B	B		B	B	A	B	B	B
Picric Acid, Molten	A		B	C	C	B	B		B	B	A	A	B	B
Pinene	A	A	D	C		C	C		B	D	D	D	U	D
Pine Oil	A	A	D	D	C	D	C		B	C	D	D	D	U
Piperidine	C	C	C	D	D	D	D		B	D	U	C	C	U
Plating Solutions, Chrome							A		A	A	A	A	A	
Plating Solutions, Others	A					A	A	A	A	A	A	A	A	A
Pneumatic Service	A	A	A		A	A	A		A	B	A	A	A	U
Polyvinyl Acetate Emulsion	D		D	B	B	B	B		B	A	A	A	A	A
Potassium Acetate	A	A	A	C	A	D	A	A	D	A	A	A	C	U
Potassium Chloride	A	A	A	A	A	A	A	A	A	A	A	A	A	A
Potassium Cupro Cyanide	A	A	A	A	B	A	A		A	A	A	A	A	A
Potassium Cyanide	A	A	A	C	A	B	A	A	A	A	A	A	A	B
Potassium Dichromate	A	A	C	A	B	A	A		A	A	A	A	B	B
Potassium Hydroxide	C	C	A	A	D	A	A	A	B	B	A	A	A	B
Potassium Nitrate	A	A	A	A	A	A	A	A	A	A	A	A	A	A
Potassium Salts	A	A	A	A	A	A	A		A	A	A	A	A	A
Potassium Sulphate	A	A	A	B	B	B	B		A	A	A	A	B	B
Potassium Sulphite	A	A	A	B	B	B	B		A	A	A	A	B	B
Prestone Antifreeze	A	A	D	D	C	C	A	A	A	C	A	A	D	D
PRL-High Temp. Hydr. Oil	A	A	B	B	B	A	U		B	B	U	U	U	U
Producer Gas	A	A	B	D	D	C	B		B	B	B	B	B	U
Propane	A	A	C	D	D	C	A	A	A	A	U	C	B	U
Propane Propionitrile	A		C	D	C	C	U		A	C	U	U	U	U
Propyl Acetate	A	A	C	D	D	C	U	U	D	C	C	U	U	U
N-Propyl Acetone	A	A	A	D	A	C	D		C	C	C	B	U	U
Propyl Alcohol	A	A	A	A	A	C	A	A	A	A	A	A	A	A
Propyl Nitrate	A		D	D	D	C	D		U	D	B	C	D	D
Propylene	A	D	B	D	B	C	D		D	D	U	U	D	U

Fluid														
Propylene Oxide	U	U	C	B	C	U	U	U		U	U	U	U	U
Pyranol, Transformer Oil	U	U	U	U	U	A	B	B		U	A	A	A	A
Pyranol	U	U	U	B	B	U	U	U		U	U	U	U	U
Pydraul, 10E, 29ELT	U	U	B	B	U	U	U	B		U	U	U	A	A
Pydraul, 30E, 50E, 65E, 90E	U	U	B	B	B	U	B	U		U	U	A	C	A
Pydraul, 115E	U	U	U	U	U	U	U	U		U	U	U	A	A
Pydraul, 230E, 312C, 540C	U	U	U	B	U	U	U	U		U	U	U	U	U
Pyridine Oil	U	U	B	B	A	U	U	U	U	B	B	B	U	U
Pyrogard 42, 43, 53, 55 (Phosphate Ester)	U	U	B	U	A	A	A	A		U	U	U	B	A
Pyrogard, C, D	U	U	U	U	B	U	U	U		B	B	B	U	A
Pyroligneous Acid	U	U	U	B	B	U	U	U		U	U	B	U	U
Pyrolube										B	B	B	A	A
Pyrrole	C	C	C	U	D	U	U	U		C		B	U	U
R														
Radiation	C	C	U	U	C	C	C	C		C	C	B	U	A
Rapeseed Oil	U	U	A	A	B	B	B	B	A	B	A	U	A	A
Red Oil (MIL-H-5606)	U	U	U	U	A	B	B	B		A	A	A	A	A
Red Line 100 Oil	U	U	U	U	A	B	B	B	A	A	B	B	A	A
RJ-1 (MIL-F-25558)	U	U	U	U	A	B	B	B	A	A	A	U	A	A
RP-1 (MIL-R-25576)														
S														
Sal Ammoniac	A	A	A	A	A	A	A	A		A	A	B	A	A
Salicylic Acid	A	B	A	A	B	B	A	A		A	A	A	A	A
Salt Water	A	A	A	A	U	U	A	A	A	U	U	A	A	A
Santo Safe 300	U	C	B	C	U	U				A	U	A	A	A
Sewage	B	B	B	B	C	A	B	B	A	B	D	U	A	A
Shell Alvania Grease #2	U	U	U	U	A	U	U	B	A	A	U	A	A	A
Shell Carnea 19 and 29	U	U	U	U	A	U	B	U		A	B	B	A	A
Shell Diala	U	U	U	U	A	U	B	U		B	U	B	A	A
Shell Iris 905	U	U	U	U	A	B	B	B		A	U	D	A	A
Shell Iris 3XF Mine Fluid (Fire resist. hydr.)	U	U	U	U	A	A	U	U	A	U	A		A	A
Shell Iris Tellus #27, Pet. Base	U	U	U	U	A	B	B	B		A	B	U	A	A
Shell Iris Tellus #33	U	U	U	U	A	U	U	A		A	U	U	A	A
Shell Iris UMF (5% Aromatic)	U	U	U	U	A	A	A	A		A	U	U	A	A
Shell Lo Hydrax 27 and 29	U	U	U	U	A	B	B	B		A	U	U	A	A
Shell Macoma 72	U	U	U	U	B	B	A	A		A	B	U	A	A
Silicate Esters	A	A	A	A	B	U	A	A		C	A	U	A	A
Silicone Greases	A	A	A	A	B	A	A	A	A	C	A	C	A	A
Silicone Oils	U	U	U	U	B	U	A	A	A	A	B	C	A	A
Silver Nitrate	B	B	A	A	B	A	B	A	A	A	B	A	A	A
Sinclair Opaline CX-EPLube	U	U	U	U	A	U	B	B		B	U		U	C
Skelly, Solvent B, C, E	U	U	U	U	U	U	U	U	U	C	C	C	C	B
Skydrol 500	B	B	A	A	A	U	A	A	D	A	A	A	A	A
Skydrol 7000	U	U	U	U	A	A	A	A	U	A	U	U	U	U
Soap Solutions	B	B	A	A	U	B	B	B	A	U	D	A	B	B
Socony Mobile Type A	U	U	U	U	A	A	U	B		A	U	A	B	B
Socony Vacuum AMV AC781 (Grease)	U	U	U	U	A	U	B	B	A	A	U	A	A	A
Socony Vacuum PD959B	U	U	U	U	B	B	A	A		U	B	A	A	A
Soda Ash	A	A	A	A	B	U	A	A		C	C	U	U	U
Sodium Acetate	A	U	B	A	U	B	U	A	A	A	C	A	U	A
Sodium Bicarbonate (Baking Soda)	A	A	A	A	A	A	A	A	A	A	U	A	A	A
Sodium Bisulfite	B	A	A	A	A	A	A	A	A	C	C	C	C	C
Sodium Borate	A	A	A	A	A	A	A	A	A	A	A	A	A	A
Sodium Carbonate (Soda Ash)	A	A	A	A	A	A	A	A		A	A	U	A	A
Sodium Chloride	A	A	A	A	B	B	B	B	A	C	C	U	A	A
Sodium Cyanide	A	A	A	A	A	A	A	A		U	A	A	A	A
Sodium Hydroxide	A	A	A	A	B	U	B	B	B	A	U	A	B	B
Sodium Hypochlorite	C	C	B	B	A	A	B	B	A	U	C	A	A	A

CHEMICAL	Natural Rubber Melbor No. 2037, Qualitex AX-2011, Durco AX-10300, Amtex No. 15000	SBR, GRS Ironsides AX-20065	Butyl Butene AX-90560	EPR, EPT, EPDM Resist-O (Noresist) No. AX-60660	Buna N Nitrile NBR Mincer No. 4045, Oil Ace AX-41060	Hydrin Epi-Chloro-hydrin Compound No. GF-0960	Neoprene Matchless AX-1060, Mitprene No. 1060	Hypalon Chlorosol No. 19040	Urethane Polyurethane Polycast No. UR-2970	Poly sulfide Thiokol No. 23945	Silicone Thermolax No. 2850R	Fluorosilicone Compound No. WK-8870	Fluoro Elastomers Viton, Fluorel, Vit-Chem No. EV-9970
Sodium Metaphosphate	A	A	A	A	A		B	B		U	A	A	A
Sodium Nitrate	B	B	A	A	B		B	B		B	A	A	A
Sodium Perborate	B	B	A	A	B	A	B	A		U	A	A	A
Sodium Peroxide	A	A	A	A	A		A	A	U	U		A	A
Sodium Phosphate (Mono)	A	A	A	A	A		B	A	A	U	A	A	A
Sodium Phosphate (Dibasic)	A	A	A	A	A		B	A	A	A	A	A	A
Sodium Phosphate (Tribasic)	A	A	A	A	A	A	A	A	A	A	A	A	A
Sodium Salts	A	A	A	A	A		A	A	A	U	A	A	A
Sodium Silicate	B	B	A	A	B		A	A	A	U	A	A	A
Sodium Sulphate	B	B	A	U	A		B	B	B	U	A	A	A
Sodium Sulphide	B	U	U	U	B		B	B	B	B	A	A	A
Sodium Sulfite	U	C	C	C	A	A	C	C	B	B	A	A	A
Sodium Thiosulfate	U	B	B	B	B		B	U	A	U	A	A	A
Sovasol #1, 2, and 3	U	U	U	U	U		U	U	C	U	U	A	A
Sovasol #73 and 74	A	A	B	U	A	A	A	A	A	B	B	A	A
Soybean Oil	U	U	U	C	B		A	U	B	B	A	A	A
Spry	U	U	U	U	A		C	C	C	U	U	C	C
SR-6 Fuel	U	U	U	U	B		B	U	A	C	C	U	U
SR-10 Fuel	U	U	U	U	U	U	U	U	C	U	C	U	U
Standard Oil Mobilube GX90-EP Lube	A	A	A	A	A	B	U	U	A	U	A	A	A
Stannic Chloride	U	B	U	U	A	A	U	U	A	U	A	B	B
Stannic Chloride 50%	U	U	U	U	C		U	B	B	U	A	B	B
Stannous Chloride	U	U	U	U	A		U	C	C	B	A	A	A
Stauffer 7700	U	U	U	J	B		U	A	A	B	U	A	A
Steam, Below 350ºF	U	B	U	U	U		B	U		U	A	C	C
Steam, Above 350ºF	U	B	U	U	U		C	C		U	C	C	C
Stearic Acid	B	B	B	C	B	A	C	B		U	A	A	A
Stoddard Solvent	U	U	U	U	C		U	U		B	A	A	A
Styrene (Monomer)	U	U	U	U	C		U	U		U	C	B	B
Sucrose Solutions	U	U	U	U	C		U	U		U	C	B	B
Sulfite Liquors	A	A	A	A	B		A	A	U	A	A	A	A
Sulfur	B	B	A	A	U		U	B	U	A	U	A	A
Sulfur Chloride	C	C	U	U	U	A	U	A		U	A	U	C
Sulfur Dioxide, Wet	B	B	U	U	U		U	C		C	B	U	U
Sulfur Dioxide, Dry	C	U	U	B	B	B	B	U	U	B	B	B	C
Sulfur Dioxide, Liquidified under pressure	U	U	U	A	U	C	U	U		B	B	B	C
Sulfur Hexafluoride	B	B	A	B	B	C	A	B		U	B	U	A
Sulfur Liquors	B	B	A	B	B		B	B		A	B	B	B
Sulfur Molten	U	U	U	C	U		C	U	U	C	C	C	C
Sulfur Trioxide	C	B	B	U	U		U	U	C	B	B	B	B
Sulfuric Acid Dilute	C	C	B	B	C		B	U	C	B	C	C	C
Sulfuric Acid Concentrated	U	U	U	U	U		U	C	U	U	U	U	U
Sulfuric Acid 20% Oleum	C	C	B	B	U		U	A	U	U	U	U	C
Sulfuric Acid 3 Molar	C	C	B	B	U		C	C	U	U	C	C	C
Sulfurous Acid	B	B	B	B	B		B	A		B	C	A	A
Sunoco SAE 10	U	U	U	U	A		U	B		U	B	A	A
Sunoco #3561	U	U	U	U	A		B	B	A	U	C	A	A

Sunoco All purpose grease
Sunsafe (Fire resist. hydr. fluid)
Super Shell Gas
Swan Finch EP Lube
Swan Finch Hypoid 90

T
TT-N-95a
TT-N-97B
TT-I-735b
TT-S-735, Type I
TT-S-735, Type II
TT-S-735, Type III
TT-S-735, Type IV
TT-S-735, Type V
TT-S-735, Type VI
TT-T-656b

Tannic Acid
Tannic Acid, 10%
Tar Bituminous
Tartaric Acid
Terpineol
Tertiary Butyl Alcohol
Tertiary Butyl Catechol
Tertiary Butyl Mercaptan
Tetrabromomethane
Tetrabutyl Titanate
Tetrachloroethylene
Tetraethyl Lead
Tetraethyl Lead "Blend"
Tetrahydrofuran
Tetralin
Texaco 3450 Gear Oil
Texaco Capella A and AA
Texaco Meropa #3
Texaco Regal B
Texaco Uni-Temp. Grease
Texamatic "A" Transmission Oil
Texamatic 1581 Fluid
Texamatic 3401 Fluid
Texamatic 3525 Fluid
Texamatic 3528 Fluid
Texas 1500 Oil
Thiokol TP 90B
Thiokol TP-95
Thionyl Chloride
Tidewater Oil-Beedol
Tidewater Oil-Multigear 140, EP Lube
Titanium Tetrachloride
Toluene
Toluene Diisocyanide
Transformer Oil
Transmission Fluid Type A
Triacetin
Triaryl Phosphate
Tributoxyethyl Phosphate
Tributyl Mercaptan

CHEMICAL	Natural Rubber (Meloro No. 1033, Dulatex No. AX-5011, Dura No. 16750, Amtex No. 15050)	SBR, GRS (Ironsides AX-20065)	Butyl AX-90560	EPR, EPT, EPDM (Resist-O Norprel No. AX-60660)	Buna N Nitrile NBR (Oil Ace AX-41060, Mincer No. 4045)	Epi-Chloro-Hydrin Compound No. GF-0960 (Hydrin)	Neoprene (Mercholess AX-1060, Mirprene No. 1050)	Hypalon® Chlorosol No. 19040	Urethane, Polyurethane Polycast No. UR-2970	Polysulfide Thiokol® No. 23945	Silicone Thermolex No. 2850R	Fluorosilicone Compound No. WK-8870	Fluoro Elastomers Viton® Fluorel® Vi-Chem No. EV-9970
Tributyl Phosphate	B	B	A	A	U		U	U	A	A	U	C	U
Trichloroacetic Acid	C	B	B	B	B		B	B	B	U	B	B	C
Trichloroethane	U	U	U	U	U		U	U	U	U	U		A
Trichloroethylene	U	U	A	A	U		U	U	B	U	B	A	B
Tricresyl Phosphate	B	B	B	B	C		U	U	C	U	U	U	U
Triethanol Amine													
Triethyl Aluminum	U	U	U	U	U		U	U	U	U	B	B	A
Triethyl Borane	U	U	U	A	U		B	B	U	U	B	B	B
Trifluoroethane	U	U	U	U	U		U	U	U	C	B	B	B
Trinitrotoluene	U	U	U	U	U		C	C	A	C	B	B	B
Trioctyl Phosphate	U	A	A	A	U	A	B	U	B	U	B	B	A
Tripoly Phosphate	U	C	U	U	B	U	B	U	U	B	B	A	A
Tung Oil (China Wood Oil)	U	C	U	U	B	U	U	C	A	U	A	B	B
Turbine Oil	U	C	U	U	A	A	B	U	B	U	U	B	B
Turbine Oil #15 (MIL-L-7808A)	U	C	U	U	A	A	B	U	U	A	U	B	B
Turbo Oil #35	U	U	U	U	B	A	U	U	B	U	B	A	A
Turpentine	U	U	U	U	A	A	C	B	B	U	B	U	A
Type I Fuel (MIL-S-3136)													
Type II Fuel (MIL-S-3136)													
Type III Fuel (MIL-S-3136)													
U													
Ucon Hydrolube J-4	B	A	A	A	A	B	B			A	B	A	A
Ucon Lubricant LB-65	A	B	A	A	A	A	A			A	A	A	A
Ucon Lubricant LB-135	A	A	A	A	A	A	A			A	A	A	A
Ucon Lubricant LB-285	A	A	A	A	A	A	A			A	A	A	A
Ucon Lubricant LB-300	A	A	A	A	A	A	A			A	A	A	A
Ucon Lubricant LB-625	A	A	A	A	A	A	A			A	A	A	A
Ucon Lubricant LB-1145	A	A	A	A	A	A	A			A	A	A	A
Ucon Lubricant 50-HB55	A	A	A	A	A	A	A			A	A	A	A
Ucon Lubricant 50-HB100	A	A	A	A	A	A	A			A	A	A	A
Ucon Lubricant 50-HB260	A	A	A	A	A	A	A			A	A	A	A
Ucon Lubricant 50-HB660	A	A	A	A	A	A	A			A	A	A	A
Ucon Lubricant 50-HB5100	A	A	A	A	A	A	A			A	A	A	A
Ucon Oil LB-385	A	A	A	A	A	A	A			A	A	A	A
Ucon Oil LB-400X	A	A	A	A	A	A	A			A	A	A	A
Ucon Oil 50-HB-280X (Polyacrylon Glycol Deriv.)	A	A	A	A	A	A	A			A	A	B	A
Univis 40 (Hydr. Fluid)	U	U	U	U	B	B	B	C	A	U	U	U	U
Univolt #35 (Mineral Oil)	B	B	A	A	B	B	B	C	U	U	A	U	A
Unsymmetrical Dimethyl Hydrazine (UDMH)	A	A	A		A	A	A	A	A	C	C	C	C

V

VV-B-680	B		B		B		B	B	C	C	B	A
VV-G-632	U	A	U	A	A	A	A	A	A	A	A	A
VV-G-671c	U	U	U	U	A	C	A	A	A	A	B	A
VV-H-910	B	U	B	U	A	A	A	A	D	B	A	B
VV-I-530a	U	U	U	A	C	A	C	B	B	B	A	A
VV-K-211d	U	U	U	U	A	A	C	U	C	C	C	A
VV-K-220a	U	U	U	U	A	B	B	B	U	C	C	A
VV-L-751b	U	U	U	U	A	B	B	B	B	B	C	A
VV-L-800	U	U	U	U	A	A	A	A	A	A	C	A
VV-L-820b	U	U	U	U	A	A	A	A	B	A	C	A
VV-L-825a, Type I	U	U	U	U	A	B	B	C	A	C	U	A
VV-L-825a, Type II	U	U	U	U	A	B	B	C	C	A	C	A
VV-L-825a, Type III	U	U	U	U	A	A	A	A	A	C	A	A
VV-O-526	U	U	U	U	A	B	A	B	B	B	C	A
VV-P-216a	U	A	U	C	C	B	B	U	C	C	U	A
VV-P-236	U	A	A	A	A	D	C	U	C	A	D	A
Varnish	U	A	B	A	A	A	U	A	B	C	A	A
Vegetable Oil	U	A	C	C	A	C	C	A	A	U	A	A
Versilube	A	A	A	A	A	A	A	B	A	D	B	C
Vinegar	B	B	A	A	B	B	B	B	U	B	B	
Vinyl Chloride			B		U	D	D	A				A

W

Wagner 21B Brake Fluid	A	A	B	A	C	B	B	B	A	C	U	U
Water	U	A	A	A	A	A	A	A	U	A	A	A
Wemco C	A	U	U	U	A	A	B	U	U	A	A	A
Whiskey and Wines	U	A	A	A	A	A	A	A	A	B	D	A
White Pine Oil	U	A	U	U	B	A	U	B	C	A	B	A
White Oil	A	A	A	A	A	A	B	A	A	A	A	A
Wolmar Salt	A	A	A	A	A	A	A	U	A	B	A	A
Wood Alcohol	U	U	C	U	A	B	A	C	U	B	C	U

X

Xylene	U	U	U	U	C	U	U	U	U	B	A	U
Xylidenes-Mixed-Aromatic Amines	U	U	U	U	U	U	U	U	U	U	U	U
Xylol	A	A	A	A	A	A	A	A	A	A	A	A
Xenon												

Z

Zeolites	A	A	A	A	B	A	A	A	U		D	D
Zinc Acetate	A	U	A	B	A	B	U	U	C	A	U	A
Zinc Chloride	A	A	A	A	A	A	A	A	C	C	A	A
Zinc Salts	A	A	A	A	A	A	A	A	U	U	U	A
Zinc Sulfate	B	B	A	A	A	A	A	A			A	A

*Courtesy of the Minor Rubber Co., Bloomfield, NJ 07003. Note that the rating for Neoprene with respect to ozone should be an "A."

Appendix 2
Professional Organizations

NFPA National Fluid Power Association
3333 North Mayfair Road
Milwaukee, Wisconsin 53222
(414) 778-3344

ANSI American National Standards Institute
1430 Broadway
New York, New York 10018
(212) 868-1220

ISO International Organization for Standardization
1 Rue de Varembe, Case Postale 56
1211 Geneve 20, Switzerland
34 12 40

ASME American Society of Mechanical Engineers
245 East 47th Street
New York, New York 10017
(212) 354-3300

SAE Society of Automotive Engineers
400 Commonwealth Drive
Warrendale, Pennsylvania 15096
(412) 776-4970

ISA Instrument Society of America
P.O. Box 12277
Research Triangle Park, North Carolina 27709
(919) 549-8411

JIC Joint Industrial Conference
7901 Westpark Drive
McLean, Virginia 22102
(703) 893-2900

SME Society of Manufacturing Engineers
One SME Drive
Dearborn, Michigan 48121
(313) 271-1500

Index